Atomic Diffusion in Disordered Materials:

Theory and Applications

Atomic Diffusion in Disordered Materials:

Theory and Applications

Edited by

M Balkanski
Université de Paris VI

R Elliott
University of Oxford

World Scientific
Singapore • New Jersey • London • Hong Kong

Published by

World Scientific Publishing Co. Pte. Ltd.

P O Box 128, Farrer Road, Singapore 912805

USA office: Suite 1B, 1060 Main Street, River Edge, NJ 07661

UK office: 57 Shelton Street, Covent Garden, London WC2H 9HE

British Library Cataloguing-in-Publication Data
A catalogue record for this book is available from the British Library.

ATOMIC DIFFUSION IN DISORDERED MATERIALS: THEORY AND APPLICATIONS

ISBN 981-02-2735-3

Printed in Singapore.

Preface

The diffusion of atoms within solids is responsible for a wide variety of important physical phenomena. In particular, if the moving entities are ions they give rise to an ionic (electrolytic) conductivity which can be exploited in a number of useful ways. Some solids contain a high concentration of mobile ions and achieve conductivities comparable to those found in an ionic melt. Such systems are called superionic, or fast ion conductors. Because of the high density, interaction between the ions is important and gives rise to correlation effects. Some of the best known systems involve relatively small ions like silver (for example in AgI) or copper (for example in CuI and CuBr). Other materials, notably fluorides and oxides have mobile anions but they tend to achieve high conductivities only at relatively higher temperatures. All these lattices have relatively open structures with extra atomic sites into which the mobile ions can move. There are other classes of materials where components of the system form a fixed and stable framework with insterstices through which mobile ions can move. The most successful practical crystal ionic conductor is of this type – sodium beta aluminate where the sodiums move comparatively freely in sheets between the alumina skeleton. There are several other systems which use lithium as the mobile ion.

Although crystalline materials have advantages in terms of quality control and allow a detailed description it is often inconvenient to use them as real electrolytes. For this reason a great deal of attention has concentrated on glassy materials with a disordered but rigid framework structure. The most studied, and potentially the most useful system, is lithium borate glass where the change of valence of the boron occurring with the introduction of lithium modifies the coordination number and opens up the structure, in a way that allows some lithium atoms to move comparatively freely. Aspects of this system will be reviewed in this volume.

The motivation for much of the work described here derived from an ultimate goal to optimize the design of solid state batteries on a micro scale which could be used, for example, in integrated circuitry. Such systems clearly require a solid rather than a liquid electrolyte, while problems of manufacture suggested that a glass would be easier to use than a crystal. The work ranged widely from fundamental descriptions of the atomic diffusion process, and of the structural properties of the glass through experimental and phenomenological studies of the optical, vibrational and electrical properties, to a study on the barrier layers which would be present in a circuit, and the actual fabrication of batteries themselves. The work was largely supported by the European Community through two grants – one an extensive European network devoted to the study of microbatteries, and the subsequent grant to allow collaboration with colleagues in Mexico. The work was centred in Paris where Professor Balkanski's group carried out a wide range of experiments, while the theoretical programme was coordinated from Oxford. At the end of these projects, we had available a comprehensive series of studies which we believed it would be valuable to make more widely available.

The pattern of this volume is six chapters which review different aspects of the work. Each review is undertaken by a single author, although it reviews the work which was originally performed by a wider group and sets it in a broader context. Chapter 1 deals with the fundamental theory of atomic diffusion in the model where the mobile atoms jump between vacant sites. In particular it takes account of the strong correlation arising from the restrictions against multiple occupancy of such sites and also the effects of disorder of the type which appear in glasses. Chapter 2 develops a fundamental theory of the structure and coordination of glassy materials derived by building the systems according to specific prescriptions. In chapter 3 the properties of these structures are examined in more detail with particular attention to the signatures of small rings of atoms which occur in silica and borate glasses. These signatures are related to vibrations and are observed optically. It also considers a more realistic phenomenological model of diffusion in borate glasses which allows for double occupancy of some vacant sites. Chapter 4 reviews the experimental situation with respect to properties of the most important material, lithium borate glass. Optical, vibrational and electrical properties are described. In chapter 5 we move closer to the problems involved in the actual construction of devices. This work describes the effect of barrier layers on controlling the entry and exit of ions from the electrodes into the electrolyte. Finally in chapter 6 progress with application to microbatteries sensors and other devices is reviewed.

Many people other than the authors were involved in this work and appropriate references to the individual contributions are given in each chapter. However, we would particularly like to thank Richard Wallis and Christian

Julien for their interest and help at various stages of the work.

Finally we would all wish to acknowledge our debt to Dr. Ivo S. Nachev who died tragically in August 1995 after a short illness. It was originally envisaged that he would write chapter 5 on the area where he made such significant contributions.

Authors

1. Minko Balkanski – Professor of Physics, University of Paris VI, Laboratoire de Physique des Solides, 4 Place Jussieu, 75252 Paris, France

2. Rafael A. Barrio – Professor of Physics, Instituto de Fisica, UNAM, Apartado Postal 20–364, CD Universitaria, 01000 Mexico DF, Mexico

3. J. Deppe – Abacus Technology Corporation, 5454 Wisconsin Avenue, Ste. 1100, Chevy Chase, MD, USA

4. Roger J. Elliott – Professor of Physics, University of Oxford, Theoretical Physics, 1 Keble Road, Oxford, OX1 3NP, England

5. Richard Kerner – Professor of Physics, Laboratoire de Gravitation et Cosmologie Relativistes University of Paris VI, 4 Place Jussieu, 75005 Paris, France

6. M. Massot – University of Paris VI, Laboratoire de Physique des Solides, 4 Place Jussieu, 75252 Paris, France

Contents

Chapter 1
Models of Atomic Diffusion

R.J. Elliott

1 Introduction

1.1 Background

The basic concept of a solid as a system in which the atoms oscillate about fixed positions is only approximate. Atoms can migrate from site to site and this gives rise to a wide variety of phenomena, which can be loosely classed under the heading of diffusion. If the mobile atoms are charged ions then this diffusing motion gives rise to an ionic conductivity which is the basis of the useful properties of solid electrolytes. The diffusion (and conductivity) is enhanced in the presence of defects such as vacancies as well as more complicated structural defects such as dislocations.

In many systems, and in particular in those which show high conductivities, there are usually two types of atom. Some of these move rarely and remain largely fixed in their lattice sites. The others move relatively easily between sites made available by this underlying lattice. If the concentration of mobile ions is high, a high conductivity results. Good examples of this are seen in the so called superionic conductors like AgI. Here the iodines forms a fixed lattice and the Ag ions sit in interstitial positions in that lattice. Since there proves to be more interstitial sites energetically available than are needed for charge neutrality the Ag ions have a choice of vacant sites between which they move relatively freely. A similar situation exists in the β-aluminates where the alumina skeleton forms a layer structure through which the relatively mobile ions like Na can move. Again there are more sites energetically available than there are ions to fill them so that diffusion along the planes is relatively easy.

In all of these systems there is a concentration of mobile ions which occupy a finite fraction of the available sites. Since interatomic repulsion due to both to Coulomb and hardcore forces prevents the multiple occupation of sites it does inhibit motion to an important extent. One of the purposes of this article is to review theoretical work which studies diffusion in the presence of a finite concentration of moving interacting atoms. Such theories are approximate at intermediate concentrations but it is desirable that they reduce to the exact results which are known for the limits of high and low concentrations. In the latter case, usually called interstitial diffusion, atoms move freely through a lattice with only rare interactions. In the former an atom can only diffuse if it finds an adjacent vacancy into which to move. Since the number of such vacancies is small in this limit the diffusing motion is highly correlated with that of the vacancy and these correlation effects will be the focus of attention in later sections.

This picture of atomic diffusion and associated ionic conductivity also holds in glassy materials. Here the underlying network of fixed atoms is irregular so that any mobile atoms moving through this network see a variety of different sites which are not arranged on a perfect lattice. It is the detailed treatment of this problem which will be the main concern of this article. There has been relatively little work in this field which includes both the effect of the disorder and the interactions arising from a high concentration of mobile ions. The material of most practical interest to date is lithium doped borate glass. This is a complex system which has been described in more detail in other chapters. For theoretical purposes we shall introduce more idealised models of disorder in order to ascertain the factors which have the greatest influence on diffusion in this type of material.

1.2 Diffusion Effects

There are many types of measurement which reflect the diffusive motion of atoms. For a comprehensive review see the book by Allnatt and Lidiard [1]. We shall concentrate our attention on the tracer diffusion or self diffusion coefficient, since this contains full information about the system and other properties can be derived from it.

Self diffusion applies in a situation where the motion of an individual atom is followed in time and space. We define $G_s(r,t)$ as the self correlation function which gives the probability that the particle would be at r at time t if it began at the origin at time $t = 0$. For a system which obeys the classical diffusion equation this takes the form.

$$G_s(r,t) = (4\pi Dt)^{-3/2} exp(-r^2/4Dt) \tag{1.1}$$

Where D is the self diffusion co-efficient. This result is perhaps more familiar in its Fourier transform where

$$G_s(k,t) = exp(-Dk^2 t) \tag{1.2}$$

Such a function determines the motion of a radioactive tracer, and it is also readily related to the incoherent scattering of neutrons [2]. Many experiments have been performed on hydrogen which has a large incoherent neutron scattering cross section and which often diffuses readily in metals. In some systems such as Pd high concentrations of diffusing H can be obtained [3].

In the case of bulk diffusion there is a relationship between a current, such as the number of atoms crossing a unit area and a "driving force" such as a concentration gradient. This is given by Fick's Law in the form.

$$J_i = -D_i \nabla n_i \tag{1.3}$$

In the general case D_i as defined here is different to that in the self diffusion as discussed above, although the two are related. More complicated coefficients occur if there are several fluxes J_i such as heat flow, and electrical current, as well as particle flow of different atomic types, each driven by appropriate forces X_i; thermal gradients, electric fields or gradients of composition. In this case there is always a set of equations of the form.

$$J_i = \sum_j L_{ij} X_j \tag{1.4}$$

with a series of coefficients L_{ij} [4, 5]. In the case of electrical conductivity for ions with charge q there is a relationship, first given by Nernst and Einstein between conductivity and the diffusion constant which takes the form.

$$\sigma = \sum_i n_i q_i^2 D_i / kT \tag{1.5}$$

1.3 Hopping Models

All the properties associated with atomic diffusion are found experimentally to vary with temperature according to an Arrhenius Law.

$$A exp(-U/kT) \tag{1.6}$$

where, in its simplest form, the pre-exponential factor A and the activation energy U are independent of the temperature. In practice some variation of these factors may be observed but overall the exponential factor is dominant. This behaviour is consistent with a model where the atom oscillates in a potential well formed by the underlying static lattice, and makes occasional hops to an adjacent site by surmounting a potential barrier of height U. On this picture the residence time in the well is much longer than the transit time between wells which can be regarded as effectively instantaneous. Thus the hopping rate $W(i,j)$ is expected to be of the form.

$$W(i,j) = \nu(i) exp - (U(i,j)/kT) \tag{1.7}$$

where ν can be regarded as the attempt frequency with which the particle, oscillating in its well, approaches the barrier. Much work has been done with microscopic theories [5] to refine this treatment and obtain realistic values for W. In our work we shall assume that W is known and we shall concentrate our efforts on the details of the mechanics relating the diffusion coefficient D to W. In a perfect lattice the W will be regular and depend only on the distance between the two sites but in disordered glassy materials W will vary in a complicated way from place to place in the material. In studying the effects of disorder we shall be compelled to make relatively simple approximations for the distribution of the hopping rates.

2 Basic Theory

2.1 Equations of Motion

The theory developed to provide a working approximation to tracer motion in a system containing a finite concentration of moving background atoms can be conveniently approached from two points of view, which are in fact closely related. One is to consider the motion of the tracer as a random walk which involves correlated movements with the vacancies which allow hops to take place. This can be formulated in terms of a master equation relating to the probability of the occupancy of sites. We can proceed directly to this type of formula.

Let $p(l,t)$ be the probability that the tracer particle is at site l at time t. This is a stochastic variable with values 1 or 0 depending on whether the site is occupied or not. We similarly define $n(l,t)$ to be the probability that a background particle is at l at t and

$$v(l,t) = 1 - p(l,t) - n(l,t) \tag{2.1}$$

as the probability that there is a vacancy there. The state of the system at any time is defined by an array [p, n] of $p(l)$ and $n(l)$. Transitions to a state $[p^1, n^1]$ take place at rate $W(pn \leftarrow p^1 n^1)$ say which can be related to the hopping rates $W(i,j)$ introduced in section (1.3). To obtain observable functions describing diffusive processes we need the correlation functions $< p(r,t)p(0,0) >$ the average probability that the tracer has reached site r at time t when it began at the origin at time 0, which is equivalent to $G_s(r,t)$of (1.1). For convenience we write the equation for $< p(r,t) >$. This takes the form

$$\frac{\delta < p(r) >}{\delta t} = \sum_j [W(j,r) < p(j)v(r) > - W(r,j) < p(r)v(j) >] \tag{2.2}$$

where the explicit dependence of the variables on t has been dropped. This is not an equation of motion in the normal sense in that it does not hold at very short times, but only on scales long compared with the time between hops.

In a perfect lattice with all sites equivalent the hopping rates must be symmetric $W(i.j) = W(j,i)$. In this case we see using (2.1) that

$$\frac{\delta < p(r) >}{\delta t} = \sum^{j} W_0(j,r) < p(j)(1 - n(r)) - p(r)(1 - n(j)) > \tag{2.3}$$

since the terms on the right in $p(j)p(r)$ cancel. Similarly

$$\frac{\delta < n(r) >}{\delta t} = \sum_j W(j,r) < n(j) - n(r) >$$

and

$$\frac{\delta < v(r) >}{\delta t} = \sum_j W(j,r) < v(j) - v(r) > \tag{2.4}$$

if we neglect the terms in p because the system is assumed to have only one marked tracer atom while there is finite density of background atoms n and vacancies, v. In these equations we have allowed the tracer to have a hopping rate W_0 different from that of the background particle W.

The solution of the linear equations (2.4) is straightforward. Taking the Fourier transform

$$< n(k) >= \sum_r < n(r) > exp(ikr) \tag{2.5}$$

and the Laplace transform $exp(-st)$ we find

$$< n(k,t) >= exp - (W(k)t) \tag{2.6}$$

where

$$W(k) = \sum_j W(j,i)(1 - e^{i\underline{k}(\underline{i}-\underline{j})}) \tag{2.7}$$

$$= Dk^2 \text{ as k} \to 0 \tag{2.8}$$

and

$$D = \frac{1}{6} \sum_j W(j,i)(\underline{i} - \underline{j})^2 \tag{2.9}$$

Thus for bulk diffusion in a single component system the diffusion constant is simply related to the hopping rate and the hopping distance. However, if there are several different components n_α say, equation (2.4) retains non-linear terms of the form $n_\alpha(1-\sum_\beta n_\beta)$ similar to those which appear in the tracer diffusion equation (2.3). These terms give rise to the so called correlation factor in the relevant diffusion coefficient L_{ij}.

To proceed further we specify the system by including only z nearest neighbour hops over a distance a of rate J for background atoms and hence for the vacancies and J_0 for the tracer. The simplest approximation to the non-linear equation (2.2) is to replace $v(r)$ by its average value v in a form of mean field approximation. The tracer diffusion constant is then

$$D = D_0 v$$

where

$$D_0 = \frac{1}{6} z J_0 a^2 \tag{2.10}$$

A more correct treatment will be written in the form

$$D = D_0 v f(v) \tag{2.11}$$

where f is the correlation factor.

Tahir-Kheli and Elliott [6] showed that a greatly improved result could be obtained by considering the properties of $< p(r)v(j) >$ and truncating the equations of motion at that stage. In particular it is convenient to expand the result in terms of the fluctuation around the mean field result, writing

$$< u(r) > = < v(r) > - v$$

and neglecting all terms involving two or more factors of u. Then we find after some manipulation that the basic equation (2.3) becomes

$$\frac{\delta < p(r) >}{\delta t} = J_0 \sum_\delta v < p(r+\delta) - p(r) > + < p(r+\delta)u(r) - p(r)u(r+\delta) > \tag{2.12}$$

and the two particle correlation is determined from

$$\frac{\delta}{\delta t} < p(r)u(j) > =$$

$$= \sum_\delta [vJ_0 < p(r+\delta)u(j) - p(r)u(j) > + J < p(r)u(j+\delta) - p(r)u(j) >][1-\delta(r,j)]$$

$$- \sum_\delta xJ_0 < p(r)u(j+\delta) - p(r+\delta)u(j) > + v < p(r+\delta) - p(r) > \delta(r,j)$$

$$+J_0\left[(v-x)<p(r)u(j)>+x<p(j)u(r)>+vx<p(r)>\right]\delta<r,j>$$
$$+J\left[<p(r)u(j)+x<p(r)>\right]\delta<r,j> \qquad (2.13)$$

where $\delta<r,j>=1$ if both r and j are near neighbours of the same site. The first two terms represent the independent motion of the tracer p and the vacancy u and the other terms are corrections due to the hardcore interaction of the particles. They only occur when r and j are within a near neighbour cluster. x is the particle density $x=1-v$.

Performing a double Fourier space transform on the two particle correlation functions, together with a Laplace time transform we define

$$F_k(\lambda)=\int\int\int<p(r,t)u(j,t)p(0,0)>e^{-st}dte^{i(\underline{k}-\lambda)\underline{r}}e^{i\lambda\underline{j}}drdj \qquad (2.14)$$

The first two terms in (2.13) give the result for free tracer and vacancy motion as

$$F_k^0(\lambda)=\frac{1}{s-vW_0(k-\lambda)-W(\lambda)} \qquad (2.15)$$

with $W(k)$ defined as (2.7). For the present case

$$W(\lambda)=J\sum_{\delta}(1-e^{i\underline{\lambda\delta}}) \qquad (2.16)$$

The motion of a tracer vacancy pair with total momentum k and relative distance b is

$$F_k(b)=\int F_k(\lambda)e^{i\underline{\lambda b}}d\lambda \qquad (2.17)$$

Going back to equation (2.13) gives

$$(s-vW_o(k))G(k)=J_o\sum^{\xi}(F_k(\delta)-e^{i\underline{k.\delta}}F_k(-\delta)) \qquad (2.18)$$

while $F_k(\delta)$ can be obtained from (2.14). The inhomogeneous terms on the right give a result proportional to $vxG(k)$. After some manipulation one obtains

$$F_k(\delta)=vxJ_0G(k)\sum^{\delta^1}(1-e^{i\underline{k\delta}})\Pi(\delta,\delta^1) \qquad (2.19)$$

where Π is the matrix

$$\Pi=(1-F^o(k)V)^{-1}F^o(k) \qquad (2.20)$$

ie it describes the pair motion with an interaction term derived from the terms in $\delta(r,j)$ and $\delta < r,j >$ in (2.13). In the limit $k \to 0$ $s \to 0$

$$G(k) = (s + J_0 k^2 a^2 v f)^{-1} \tag{2.21}$$

where

$$f = \frac{1}{6a^2} \sum_{\delta} (\delta^2 + 2xJ_0 \sum_{\delta} (\underline{\delta}.\underline{\delta}^1)\Pi(\delta, \delta^1)) \tag{2.22}$$

We note that the factor $(\underline{\delta}.\delta^1)$ gives the cosine of the angle between two near neighbour vectors and allows the widely used definition

$$< cos\theta >= \frac{1}{\delta^2} \sum (\underline{\delta}.\underline{\delta}^1) F^o(\underline{\delta} - \underline{\delta}^1) \tag{2.23}$$

as the average angle between successive jumps in the motion. The weighting of Π in (2.22) shows that only p-wave scattering is significant and in the simple case

$$\Pi(\delta, \delta) - \Pi(\delta, -\delta) = F^o(p)/1 - (J + J_0 v - 2J_0 x)F^o(p) \tag{2.24}$$

and

$$F^o(p) =< cos\theta > /(J + J_0 v) \tag{2.25}$$

Substituting back into (2.22) gives

$$f(v) = \left[1 - \frac{2xJ_0 cos\theta}{(J + J_0 v)(1 + cos\theta)}\right]^{-1} \tag{2.26}$$

This gives the well known results which are correct in the limit $x \to 0$ [7].

$$f(x) = 1 + \frac{2xJ_0 M}{(J + J_0)} \tag{2.27}$$

while in the limit $v \to 0$ [8]

$$f(v) = \left[1 - \frac{2J_0}{J} M\right]^{-1} \tag{2.28}$$

where $M = cos\theta/1 + cos\theta$

2.2 Random Walk Theory

These results can also be considered from the point of view of a restricted random walk [9, 10, 11]. The self correlation function gives the probability that the particle reaches r at time t, starting at the origin, and the average distance covered

$$< r^2 >= 2Dt \tag{2.29}$$

Regarding this as a walk of n steps as $n \to \infty$

$$R_n = \sum_i^n \underline{\delta}_i \tag{2.30}$$

so

$$D = Lt\frac{1}{2n}(\sum_i \delta_i^2 + 2\sum_{i>j} \underline{\delta}_i.\underline{\delta}_j) \tag{2.31}$$

The first term gives the mean field result $\frac{1}{6}J_0a^2v$ since the average probability of making any step is v. The second term represents the correlation between successive steps. In the limit $v \to 0$ this term will be of order v, but as $v \to 1$ we expect uncorrelated motion except for collisions between the tracer and the other particle of density $x = (1 - v)$. This suggests that the term has the form $xv\Pi(p)$ where $\Pi(p)$ is the combination of pair propagators given in (2.22). From (2.24) we see that the general formula allows for multiple interactions between the particles after successive propagators and hence includes contributions for all $\underline{\delta}_i.\underline{\delta}_{i+j}$. A more detailed analysis of the random walk in the limits $v \to 1$ and $v \to 0$ gives the results (2.27) and (2.28) and the formula (2.26) can be obtained by interpolation [11].

2.3 Generalisations

The interpolation formula (2.26) gives good agreement over the whole range of concentration v except in certain extreme cases. One of the most interesting of these is the limit $J = 0$ when the background particles are fixed. It is well known that in this situation percolation effects occur and D falls to zero at some finite concentration x_p. However in formula (2.26) this occurs only at $x = 1$.

The theory can be improved by making it self consistent. We noted in the study of (2.12) that the effective rate of the relative tracer-vacancy motion was $(J + J_0v)$ as in (2.25). But the full rate for the tracer is $J_0vf(v)$ as (2.11) and (2.21). Tahir-Kheli [12] therefore proposed that this result should be substituted in (2.26) to give the self consistent result

$$f = \left[1 - \frac{2zJ_0M}{(J + J_0vf)}\right]^{-1} \tag{2.32}$$

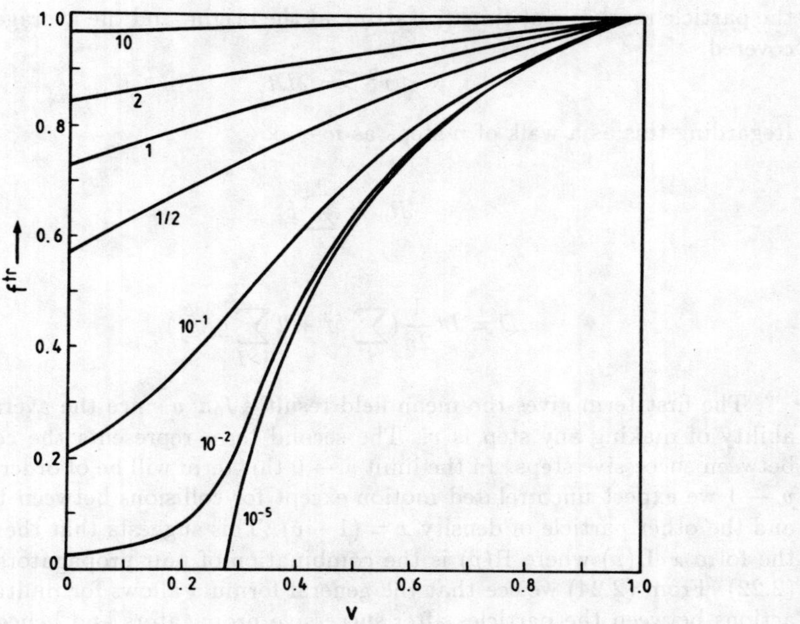

Figure 1: Variation of the correlation factor f against vacancy concentration v for various ratios (J/J_o) (of the hopping rates in a s.c. lattice using the self-consistent formula (2.32). (after ref. 12)

When J=0 this gives the percolation limit $f = 0$ at

$$x_p = (1 - 2M)^{-1} = \frac{1 - cos\theta}{1 - cos\theta} \tag{2.33}$$

This result agrees well with exact values of x_p [13]. Results for $f(v)$ at various ratios of (J/J_0) are shown in figure 1.

The fully generalised formula has been used in a number of other situations, for example in one-dimension [14] where the diffusion is known to be anomalous. Also Holdsworth [15] generalised the result to a multicomponent alloy to find the correlation factor f for a tracer moving with J_0 to be

$$f = \left[1 - \frac{2(1-v)J_0 M}{(J_{eff} + J_0 v f)}\right]^{-1} \tag{2.34}$$

when $v = 1 - \sum_\alpha n_\alpha$ and J_{eff} is the effective hopping rate of the vacancies

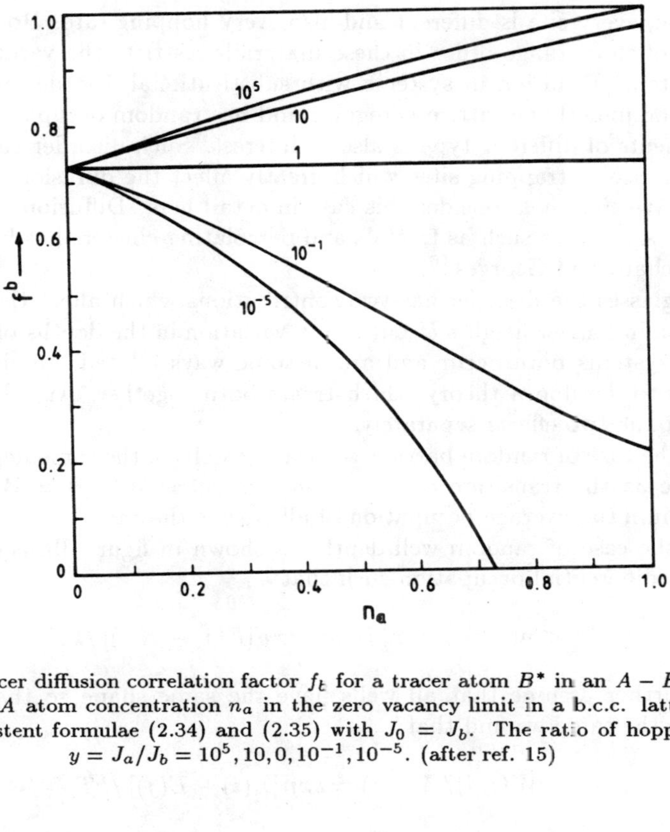

Figure 2: Tracer diffusion correlation factor f_t for a tracer atom B^* in an $A-B$ alloy as a function of A atom concentration n_a in the zero vacancy limit in a b.c.c. lattice using the self consistent formulae (2.34) and (2.35) with $J_0 = J_b$. The ratio of hopping rates $y = J_a/J_b = 10^5, 10, 0, 10^{-1}, 10^{-5}$. (after ref. 15)

given by

$$J_{eff} = \left[\sum_\alpha n_\alpha J_\alpha/(J_0 f + J_\alpha)\right] \left[\sum n_\alpha/(J_0 f + J_\alpha)\right] \tag{2.35}$$

Results for a binary alloy are shown in figure 2.

3 Disordered Systems

3.1 Types of Disorder

In the glassy systems of direct concern to us here the disorder is of the structural type in which the underlying static lattice displays no symmetry. In principle, every site is different and also every hopping rate. However the existence of short range order in these materials restricts the variation of these parameters. Diffusion in systems with substitutional disorder such as alloys where the underlying lattice is regular and has random occupation by atomic components of different type is also of interest. Such disorder can, for example, give rise to trapping sites which greatly affect the diffusion of the mobile atoms. We shall not consider this case in detail here. Diffusion in special non periodic structures such as fractals and percolating clusters has been reviewed by Bouchard and Georges [16].

In glasses the disorder has two contributions which affect the diffusion, a variation in barrier heights U but also a variation in the depths of the wells [11]. In real systems both occur and are in some ways related, but it has proved difficult to develop a theory which treats both together. We shall therefore consider the two effects separately.

In the case of random barriers separating wells of the same depth as shown in figure 3a the transition rates remain symmetric $W(i,j) = W(j,i)$ and in equilibrium the average occupation of all sites is the same.

In the case of random well depths as shown in figure 3b, equilibrium requires a differential occupation such that

$$< n(i) > / < n(j) >= exp\left[E(j) - E(i)\right]/kT \tag{3.1}$$

If we further assume that all wells have the same shape so that ν in (1.7) remains the same we find that

$$W(i,j)/W(j,i) = exp\left[E(i) - E(j)\right]/kT \tag{3.2}$$

and the transition rates depend only on the initial site. In other words W depends on the height of the barrier facing the oscillating particle. In both cases we shall assume that the underlying lattice is regular so that the disorder is manifested only through a variation in W.

3.2 Random Barriers

For particles moving in an empty lattice a solution of linear equations of the type (2.4) is required with a distribution of W. The simplest approximation is to take an independent average over W. If the range is still fixed at a this gives for a distribution $P(J)$ of rates J

$$\bar{J} = \int JP(J)dJ \Big/ \int P(J)dJ \tag{3.3}$$

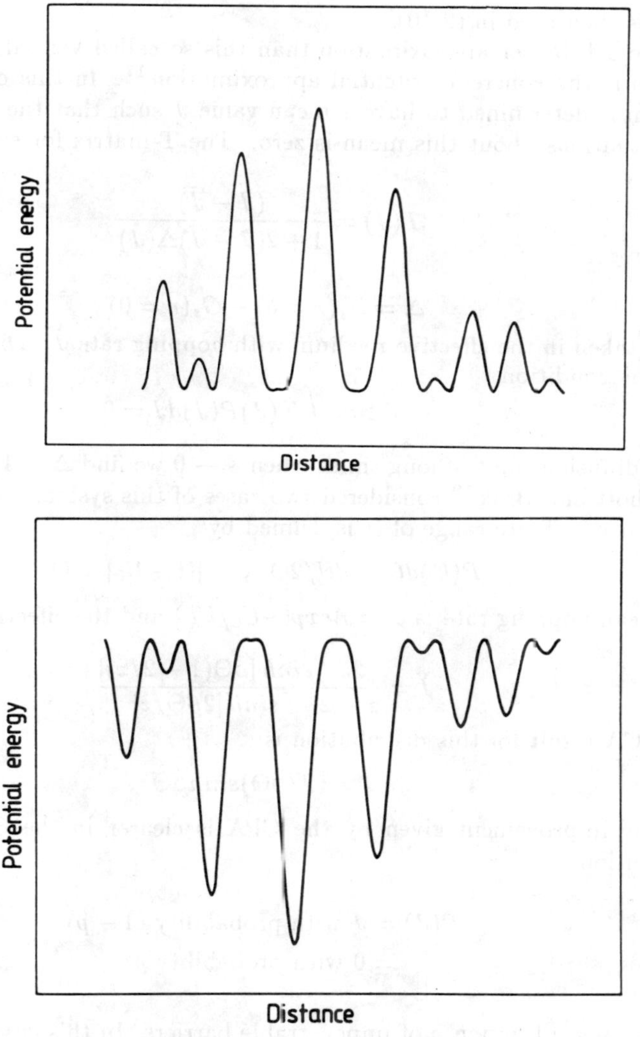

Figure 3: Schematic representation of the potential in a linear system (a) with random barrier heights for wells of the same depth and (b) with random well depths.

which is then used in (2.10).

A much better approximation than this so called virtual crystal approximation is the coherent potential approximation[17]. In this case an effective medium is determined to have a mean value $\bar{J}$ such that the total scattering by fluctuations about this mean is zero. The T-matrix for such scattering is given by

$$T(J) = \frac{(J - \bar{J})}{1 + 2(J - \bar{J})\Delta(\bar{J})} \tag{3.4}$$

where

$$\Delta = G_s(r = \delta) - G_s(r = 0)$$

for G_s taken in the effective medium with hopping rate $\bar{J}$. This is determined from the condition

$$< T >= \int T(J)P(J)dJ = 0 \tag{3.5}$$

In the diffusion limit of long times when $s \longrightarrow 0$ we find $\Delta = 1/z\bar{J}$[18].

Elliott and Brak[19] considered two cases of this system. A uniform distribution over a finite range of U is defined by

$$P(U)dU = dU/2\Theta \quad ; \qquad |U - U_0| < \Theta \tag{3.6}$$

The mean hopping rate is $J = Aexp(-U_0/kT)$ and the effective medium rate is

$$\bar{J} = \frac{2J}{z-2} \frac{sinh\,[\beta\Theta(1 - 2/z)]}{sinh\,[2\beta\Theta/z]} \tag{3.7}$$

The VCA result for this distribution is

$$\bar{J} = (J/\beta\Theta) sinh\beta\Theta \tag{3.8}$$

The improvement given by the CPA is clearer in the case of a binary distribution.

$$\begin{aligned} P(J) &= J \text{ with probability } (1 - p) \\ &= 0 \text{ with probability } p \end{aligned}$$

representing a fraction p of impenetrable barriers. In this case (2.5) gives

$$\bar{J} = J(1 - zp/(z - 2)) \tag{3.9}$$

and $\bar{J} = 0$ at $p_c = (1 - 2/z)$. This is another form of the percolation problem appropriate to bond disorder (as opposed to site disorder referred to in section (2.2)). The value of p_c obtained is in reasonable agreement with the known exact results[13]. In this case the VCA result $\bar{J} = J(1 - p)$ does not predict any percolation effects.

3.3 Random Traps

The situation in a lattice of sites with random well depths is somewhat more complex, since at equilibrium the relative occupation of the sites, by both tracer and background particles varies with the depth as given by (3.1). If we specify a site of a given type by E say, then the equilibrium values of $< p(r) >_E$ and $< n(r) >_E$ will vary although it is clear from the form of (3.1) that

$$< p(r) >_E / < n(r) >_E = C \tag{3.10}$$

In our model $W(j,r)$ depends only on the site j at which the particle begins its hop. For equilibrium we see from (2.2) and (2.3) that if site j is type E and site r is type E^1 that

$$W_E < p >_E (1- < n >_{E^1}) = W_{E^1} < p >_{E^1} (1- < n >_E) \tag{3.11}$$

or that $W_E < p >_E /(1- < n >_E)$ is a constant independent of E. Similarly from (3.10)

$$W_E < n >_E /(1- < n >_E) = K \tag{3.12}$$

At low concentrations when $< n >$ is negligible a particle moving in an empty lattice has $< p >_E \sim W_E^{-1}$. Normalising to the total probability

$$< p > = \int P(E) < p >_E dE \tag{3.13}$$

we find that the effective rate $\bar{W}$ defined by

$$\int P(E) W_E < p >_E dE = \bar{W} < p >$$

is given by [20]

$$\bar{W}^{-1} = \int P(E) W_E^{-1} dE \tag{3.14}$$

For nearest neighbour hops only and a concentration c_N of sites of type N

$$\bar{J}^{-1} = \sum_N c_N J_N^{-1} \tag{3.15}$$

and $D = \frac{1}{6} a^2 z \bar{J}$

The same result may be obtained by determining an effective medium as in section (2.2). Since the $W(i,j)$ depends only on the initial site we find that the equation equivalent to (3.4) takes the form

$$T(J_E) = (J_E - \bar{J})/[1 + z(J_E - \bar{J})\Delta]$$

$$= \bar{J}(1 - \bar{J}/J_E) \tag{3.16}$$

and the condition $< T >= 0$ reproduces (3.14) or (3.15).

In the presence of a finite concentration x of particles the resulting equations have no simple solution [21]. We still require $\bar{W}$ as defined in (3.13) but (3.14) no longer holds and the functions in (3.11) are required. After some manipulation we find

$$\bar{W} = \frac{1-x}{x} \int \frac{< n >_E}{1- < n >_E} P(E) dE \Big/ \int P(E) W_E^{-1} dE \tag{3.17}$$

For a binary distribution with concentration $(1-c)$ of A sites and c of B sites this gives

$$\bar{J} = \frac{1-x}{x} \left(\frac{(1-c)n^A}{1-n^A} + \frac{cn^B}{1-n^B} \right) / \left(\frac{1-c}{J_A} + \frac{c}{J_B} \right) \tag{3.18}$$

4 Correlations on Disordered Systems

Determination of the correlation factor f in the presence of disorder presents a difficult problem for which no completely satisfactory theory currently exists. As was shown in section (2.1) the effects arise from correlated two particle motion. The individual particle motion is affected by disorder as discussed in section 3, but the disorder also introduces correlations between pairs of particles which is additional to that arising from their direct interaction.

The simplest approximation ignores these extra correlations. In that case the self consistent best value for f is given by (2.32) with effective hopping rates $\bar{J}_0$ and $\bar{J}$ substituted for the perfect crystal values. If the disorder distribution of both J_0 and J is the same, as for example with missing bonds in (3.9), or if $J_0 = J$ for self diffusion, the effect cancels and the factor f is unchanged.

Further improvements have been made by two approaches. The first, by Brak and Elliott [19] considers the two particle functions F in the presence of disorder including the vertex correction which arises when both particles scatter from the disorder potential at the same site. The function F in (2.14) which leads to the value for $cos\theta$ in (2.23) is replaced by

$$P = F [1 - \Lambda F]^{-1} \tag{4.1}$$

and Λ is related to $< T^2 >$ [22] defined in analogy with (3.5) and by the Ward Indentity $\Lambda = \delta\bar{J}/\delta\Delta$. However the formula requires involved calculation since the factors ΛF require a convolution over two Laplace variables s, s^1. Nevertheless the result of the formula is an improved value for $cos\theta$. The formula

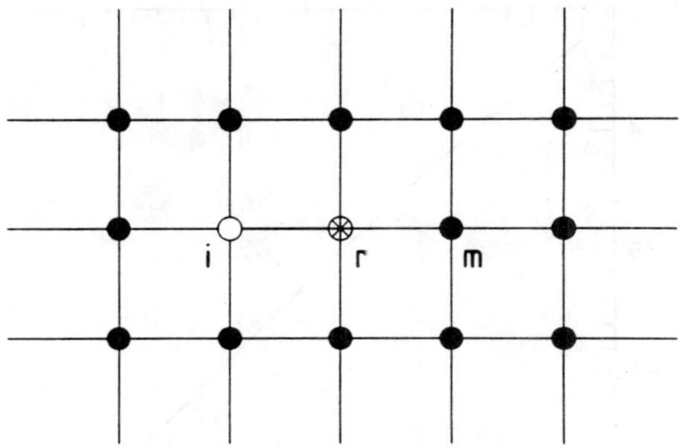

Figure 4: Situation just after the exchange of the tagged particle with a vacancy. The next hop depends on the rates for adjacent bonds in the environment.

suggests, as a crude approximation

$$< cos\theta >= cos\theta/(1 - \Lambda_0 cos\theta) \tag{4.2}$$

where Λ_0 is some average of Λ.

Even substitution of this value of $< cos\theta >$ omits certain correlations since the interparticle interactions given by V in (2.20) which result in the use of Π in (2.22) and (2.24) are not randomly distributed with respect to the disorder terms $< T^2 >$ which contribute to Λ. Recently Schwabe and Elliott [23] have shown how to include such effects by defining weighted two particle Green functions. To date this method has not been applied to this problem.

An alternative approach is to focus on the calculation of $< cos\theta >$ which is interpreted in section (2.2) as the angle between adjacent jumps in a random walk. By examining all environments of a particle in a disordered systems immediately after a jump, it is possible to obtain an improved average. The situation following a particle vacancy exchange is shown in figure 4. The probability of a second forward jump depends on the hopping rate between the sites concerned.

Applying this method to the bond disordered square lattice Perondi [11, 24] finds, that if the bond for a second forward jump is defective (as it is with

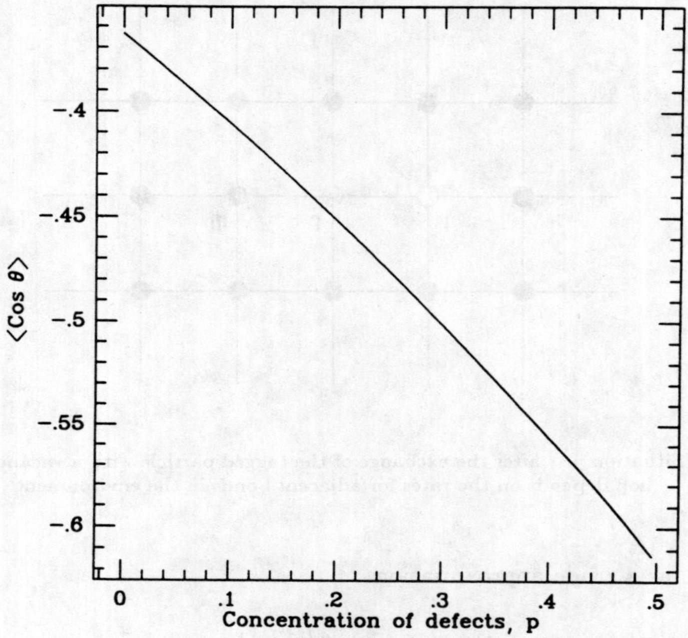

Figure 5: Variation of $< cos\theta_1 >$ as given by (4.5) in a bond disordered square lattice. (after ref. 11)

probability, p) the chance of a return jump is enhanced. In this situation

$$< cos\theta_1 >_I = 1 - \frac{2(1-p)cos\theta(1+cos\theta)}{1+2pcos\theta(1+cos\theta)} \tag{4.3}$$

If the forward bond is present

$$< cos\theta_1 >_{II} = \frac{cos\theta}{(1-p)-pcos\theta} \tag{4.4}$$

and

$$< cos\theta_1 > = p < cos\theta_1 >_I + (1-p) < cos\theta_1 >_{II} \tag{4.5}$$

The variation of $< cos\theta_1 >$ with p is shown in figure 5 for a square lattice (where $p_c = \frac{1}{2}$). As expected it becomes more negative as p increases indicating an increased probability of backward hops since after the initial hop that bond must be present, although the forward bond may be missing.

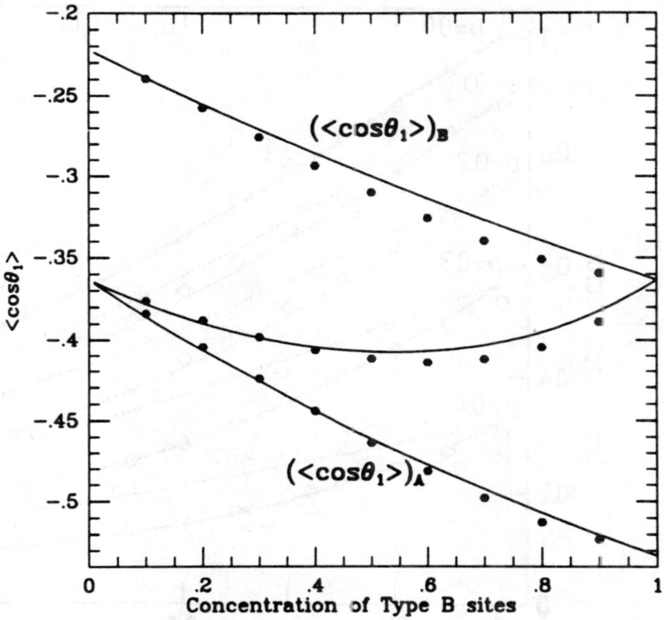

Figure 6: Variation of $< cos\theta_1 >$ in a binary alloy model on a square lattice with $J_A = 2J_B$ as a function of B site concentration in the small v limit. The curves are derived from (4.6) and (4.8) while the points are the results of a simulation. (after ref. 11)

In the case of the random trap model it is straightforward to see that, for a jump from a site of type E

$$< cos\theta_1 >_E = \frac{J_E cos\theta}{\bar{J} - (J_E - \bar{J} cos\theta)} = \frac{J_E}{\bar{J}} \frac{M}{1 - J_E M} \tag{4.6}$$

and since a vacancy is required to allow the jump

$$< cos\theta_1 > = \int P(E) < v >_E < cos\theta_1 >_E dE \tag{4.7}$$

and $< v_E > = 1 - < n_E >$ is given by (3.12). In the small vacancy limit $V_E \sim J_E$. For the binary distribution considered previously

$$< cos\theta_1 > = \left[\frac{(1-c)J_A^2}{\bar{J} - J_A M} + \frac{cJ_B^2}{\bar{J} - J_B M} \right] \frac{M}{[(1-c)J_A + cJ_B]} \tag{4.8}$$

Results for this case are shown in figure 6.

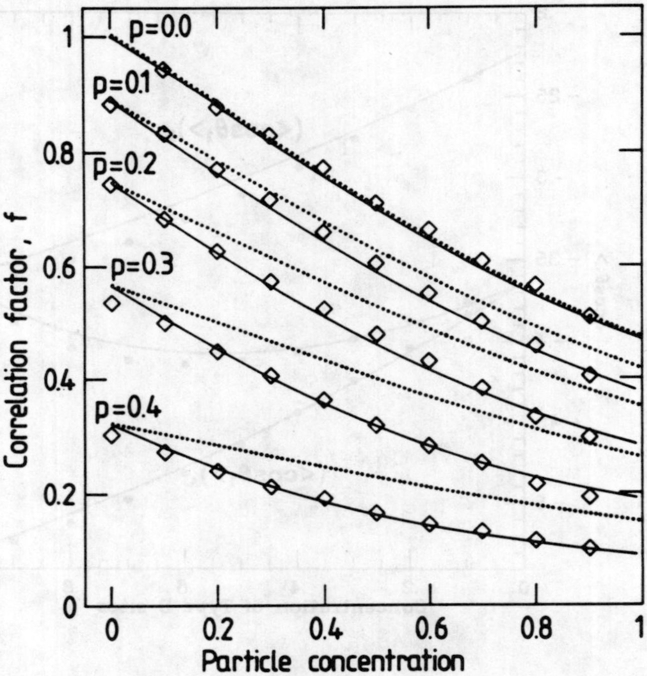

Figure 7: The correlation factor f^1 as defined in (5.2) as a function of particle concentration x and bond concentration $(1-p)$ for a bond diluted square lattice compared with the results of the effective medium theory and the improved approximation for $< cos\theta >$. (after ref. 21)

5 Results

The predictions of these theoretical results have been tested on idealised systems where the "experimental" results have been obtained by computer simulation. These have been generated by following a random walk of N steps and obtaining $< R^2(N) >$. After a sufficiently large number of steps this is linear with N. Since the average number of steps for a system with N_0 particles hopping at rate J to z nearest neighbour sites is

$$\bar{N} = N_0 z J t \tag{5.1}$$

after a time t, D can be determined using (2.29). Several simulations of this type have been carried out on perfect lattices [25] and disordered lattices [26, 27].

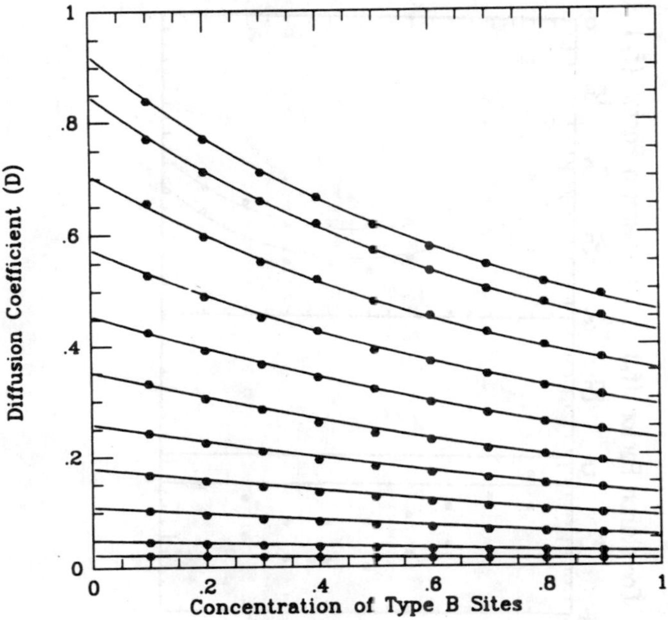

Figure 8: Comparison between theoretical (curves) and simulation (points) results for a square lattice with $J_A = 2J_B$ as a function of the concentration c of B sites at various particle concentrations using the effectve medium theory for the diffusion constant D. (after ref. 24)

In a disordered system the calculation must be made using all lattice points as the initial site in order to average over all possible environments. The systems studied by us [21, 24] were a square net of $N_s \times N_s$ sites with $N_s = 100$ although larger values were tested to ensure that size effects had been eliminated. Periodic boundary conditions are used to create a regular array of $N_s \times N_s$ blocks of the path leads outside the initial block. A series of $N_s \times N_s$ blocks were generated for a random distribution of bonds or sites. Paths with $N \sim 50,000$ were followed.

Results for the bond disordered lattice are shown in figure 7. The factor f^1 relating the true D to that expected from mean field theory, defined by,

$$D = D_0 v(1-p)f^1 \tag{5.2}$$

is plotted. The dashed curves are the results predicted by the effective medium

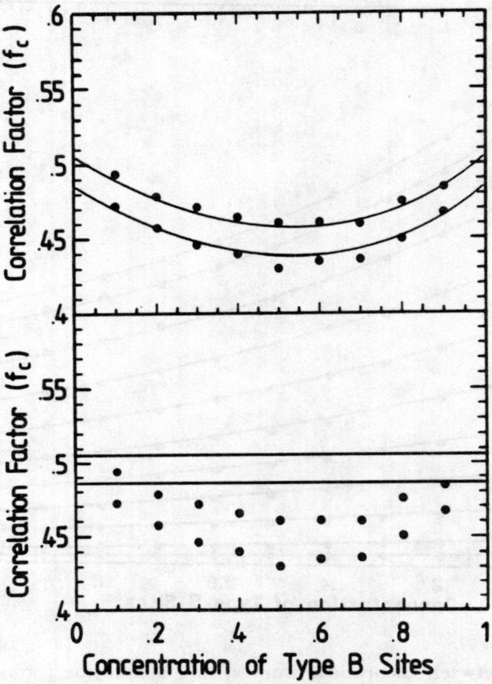

Figure 9: Comparison between simulation and theoretical results for the correlation factor in the systems shown in fig. 7 for x=0.9 and 0.95. The curve is obtained using $< cos\theta_1 >$ as derived from (4.6) and (4.8). Using $< cos\theta_0 >$ f is independent of c (after ref. 24)

theory using $\bar{J}$ given by (3.9) and

$$f^1 = f_0(1 - p/p_c)/(1 - p) \tag{5.3}$$

with the self consistent value of f_0 derived from (2.32) with $J = J_0 = \bar{J}$. There are considerable discrepancies as $c \to 1$. However with the improved value of $< cos\theta_1 >$ given by (4.5) the agreement with the simulation results if very good across the whole range.

Results for the random trap model have also been obtained from simulation on a square lattice. These are compared in figure 8 for the case $J_A = 2J_B$ with the effective medium result obtained by using $\bar{J}$ as given in (3.18) and the self-consistent formula (2.32) for f. Since $J_0 = J = \bar{J}$ in this case f is independent of the concentration c. However with the improved approximation for $< cos\theta >$ shown in figure 6 the variation of $f(r, c)$ observed in the simulation is reproduced with high accuracy as in figure 9.

6 Conclusions

The theory of atomic diffusion summarised in this article gives a convenient approximate treatment for systems containing a finite concentration of moving atoms such as are found in useful ionic conductors. It also provides an insight into the effects of lattice disorder by considering simplified models with random barriers and random trapping sites. The overall effect of disorder is conveniently and adequately described by effective medium theory to determine an effective hopping rate and hence an effective diffusion constant. The consequences of the interactions between the particles are more complex but are given to a first approximation by the correlation factor f given by (2.34). However these correlation effects are affected by the disorder so that in addition to the inclusion of the effective medium values of J, some modification of $< cos\theta >$ and hence of M is necessary.

These effects are only dramatic if there is a wide variation in the hopping rates due to the disorder, and most especially if some diffusion paths are blocked - either by static atoms giving rise to site percolations as in (2.33) or by impenetrable barriers giving bond percolation as in (3.9).

The real situation in lithium borate glasses is more complicated as is discussed in the article by Barrio. They appear to be described satisfactorily by a model which allows for double occupancy of available sites which leads to a variation of the random trap model discussed above. This model has been treated in more detail by Perondi et al[28].

References

1. A.R. Allnatt and A.B. Lidiard *Atomic Transport in Solids* (CUP Cambridge, 1993)
2. S.W. Lovesey *Theory of Neutron Scattering from Condensed Matter* Vol. 1 (OUP Oxford 1984)
3. T. Springer *Tracts in Modern Physics* Vol. **64** (Springer-Verlag Berlin 1972)
4. L. Onsager and R.M. Fuoss *J. Phys. Chem.* **36** (1962) 2689
5. C.P. Flynn *Point Defects and Diffusion* (OUP, Oxford 1972)
6. R.A. Tahir-Kheli and R.J. Elliott *Phys. Rev.* **B27** (1983) 844
7. R.A. Tahir-Kheli and R.J. Elliott *J. Phys.* **C15** (198) 5751
8. J. Bardeen and C. Herring *Imperfections in Nearly Perfect Crystals* (Wiley,New York 1952)

9. A.D. Le Claire and A.B. Lidiard *Phil. Mag.***1** (1956) 518
10. O.F. Sankey and P.A. Fedders *Phys. Rev.* **B22** (1980) 5135
11. L. Perondi Thesis (Oxford University 1993)
12. R.A. Tahir-Kheli *Phys. Rev.* **B27** (1983) 7229
13. J.W. Essam in *Phase Transitions and Critical Phenomena* Vol. **2** (Academic Press New York 1972) Eds. C. Domb and M.S. Green
14. R.A. Tahir-Kheli and N. El-Meshad *Phys. Rev.* **B27** (1983) 7759
15. P.C.W. Holdsworth and R.J. Elliott *Phil. Mag.* **A54** (1986) 601
16. J.P. Bouchard and A. Georges *Phys. Rep.***195** (1990) 126
17. R.J. Elliott, J.A. Krumhansl and P.L. Leath *Rev. Mod. Phys.* **46** (1974) 465
18. I. Webman *Phys. Rev. Lett.* **47** (1980) 1496
19. R. Brak and R.J. Elliott *J. Phys. Cond. Matt.* **1** (1989) 10299
20. J.W. Haus and K.W. Kehr *Phys. Rev.* **150** (1987) 265
21. L. Perondi, K.K. Kaski and R.J. Elliott *Phys. Rev. B* (to be published)
22. B. Velicky *Phys. Rev.* **184** (1969) 614
23. N.F. Schwabe and R.J. Elliott *Phys. Rev. B* (to be published)
24. L. Perondi, K.K. Kaski and R.J. Elliott *Phys. Rev. B* (to be published).
25. K.W. Kehr, R. Kutner and K. Binder *Phys. Rev.* **B23** (1981) 4943; **B26** (1982) 2967; **B28** (1983) 1846; R.A. Tahir-Kheli and N. El Meshad *Phys. Rev.* **B32** (1985) 6166
26. H. van Beijeren, K.W. Kehr and R. Kustner *Phys. Rev.* **B28** (1983) 5711
27. C.H. Mak, H.C. Anderson and S.M. George *J. Chem. Phys.* **88** (1988) 4052 T. Wichmann and K.W. Kehr *J. Phys. Condens. Matt.* **I** (1995) 717
28. L. Perondi, R.A. Barrio and R.J. Elliott *J. Non. Cryst. Solids.* **168** (1994) 167

Chapter 2
A Theory of Glass Formation

Richard Kerner

1 Introduction

It does not seem that a simple rule exists, regulating the relationship between the theory and the experiment in physics. The importance of a sound experimental basis for any physical theory has been often stressed since Galileo Galilei[1], and during the seventeenth and eighteenth centuries scientists usually gathered the experimental data first, elaborating the theoretical explanations later, like it happened with classical thermodynamics, electricity, optics and chemistry.

During the nineteenth century the mathematical methods used by physicists became more various and more powerful, enabling them to predict new and yet unknown observational facts. After the famous Leverrier's prediction of the existence of Uranus, followed the predictions concerning the existence of the electromagnetic waves made by Maxwell, of the unknown new elements by Mendeleev, and many other surprising theoretical discoveries.

Although the Twentieth Century has been the arena of a spectacular upsurge of experimental observations and new techniques, leading to the discovery of the X-rays, radioactivity, superconductivity and nuclear energy, still few things seem to be considered as rewarding as the ability to predict such fine phenomena as the exact bending angle of a light ray passing close to the Sun (computed by Einstein) or the existence of the positron (deduced by Dirac).

In many branches of modern physics the theory and the experiment coexisted fruitfully, alternating successful discoveries and predictions. However, in the second half of this century, many domains of human activity were pushed to an unprecedented excess, and to the extremes that often transformed them into the absurd opposite of the intended goal. It seems that physics has become subjected to the same kind of infirmity: in many domains, the hiatus between the experiment and theory has grown so large, that a new and unexpected situation has emerged: on the one hand, many very elaborate and sophisticated theories are being created, discussed and abandoned, all of them describing the objects whose real existence is not proved at all, and very dubious in most of the cases (e.g. the superparticles, magnetic monopoles, heterotic strings etc.); on the other hand, the abundant and various experimental data are being stored in many domains without a well-structured mathematical model that would organize and interpret them in a simple way. By a *mathemati-*

cal model we mean a well-defined set of rules applied to a small number of well-defined parameters, which inserted into a set of equations, enable one to compute from a given input the set of unique solutions predicting the values of measured quantities.

One of the most spectacular examples of this situation is offered by the glass science. Glass has become familiar to many civilizations since the ancient Egypt, and is widely used in all its varieties. Last century has brought the development of sophisticated glass technologies, both in production and applications. In parallel, many new methods of experimental investigation of glasses have been discovered and applied, among which the most important are the Raman spectroscopy and Nuclear Magnetic Resonance (NMR) [2,3,4].

The amount of the experimental data obtained is tremendous indeed; new compounds are being tested and described, bringing to our possession the comparative data for glasses obtained via different vitrification techniques, such as annealing or rapid quenching. Yet, with all this information at hand, it seems that the very wealth of the empirical data enhances the birth of a theory that would not only unify and order them, but also provide the physicists with a mathematical tool for computing and predicting the most important physical characteristics of glass given its composition and the preparation method.

In order to be able to do so, such a theory should provide a good description of short and medium range structure of local neighborhoods containing dozens of atoms; this will be possible if the theory contains also a good model of the non-equilibrium process of glass transition, taking into account its main features, including its temperature and time dependence.

The attempts to develop such a theory have been made earlier, some of them with certain success. We should mention the pioneering works of J.C.Phillips [5,6], the models of R.Collins [7], J.M.Kosterlitz and D.J.Thouless [8], the papers of P.J.Bray [9,10], R.A.Barrio and F.Galeener [11,12,13], M.Thorpe [14,15], and many others. Here we describe our attempt to produce a model of glass formation conceived as an irreversible process based on the agglomeration and local growth of clusters containing a few atoms at the beginning. This theoretical model has been developed during past decade, and applied to covalent binary glasses with some success. It is exposed in articles that have appeared recently, [16,17], some of them written in collaboration with R.A.Barrio, M.Micoulaut and D.M. dos Santos-Loff [18,19,20,21].

A theory of glass formation exposed here is based on the mathematical description of agglomeration and growth at microscopic level. This theoretical model, which is an elaborate version of a model of growth of two-dimensional networks [(22,23)], and its derivatives published since then in other papers [(11,18,19,20,24,25)], displays a non-negligible predictive power when applied to the description of the medium-range order in binary covalent glasses such as $B_2O_3 + nLi_2O$ or $SiS_2 + nLi_2S$.

After a careful counting of all types of clusters containing up to twelve atoms of parent glass-former compound (B or Si) and evaluation of the number of different pathways leading to their formation from the elementary building blocks, their probabilities are computed, taking into account both the multiplicities (statistical factors) and the Boltzmann factors related to the different energetic costs of different elementary agglomeration processes. Then one can evaluate new probabilities of characteristic features that are repeatedly observed in a given glass (e.g. boroxol rings, edge-sharing or vertex sharing doublets of rings); finally, a system of differential equations can be produced, by requiring the time derivatives of these probabilities to vanish.

This can be interpreted as the expression of the *saturation* or the *stationary régime* of the agglomeration process, attained at a relatively early stage of the process. Mathematically, the derivatives are approximated by the finite differences between the expression for the probabilities of the characteristic features evaluated at the $(n+1)-th$ and $n-th$ steps, i.e. averaged on all possible clusters containing $n+1$ or n elementary units.

These finite differences are quite complicated functions of the original probabilities evaluated on the "singlets" $(n = 1)$. The simultaneous vanishing of these functions yields many solutions, which may be classified as attractive, repulsive or saddle-type points, according to the eigenvalues of the system linearized in the vicinity of these stationary solutions. Among these, the solutions that display vanishing of almost all probabilities except one or two, correspond to the local crystallization tendency, because they privilege the repetition of the same configuration, while the solutions displaying the maximal amount of non-vanishing probabilities (and therefore the maximum configurational entropy $S_{conf} = -k_B \sum P_i \ln P_i$) are interpreted as representing the local glass-forming tendency.

Information can be drawn from the model concerning the medium-range structure, and therefore the density of the glass as function of the modifier concentration, the behaviour of the glass transition temperature, relative frequency of different structural groups, etc., many of which can be compared with the Raman, Infrared or NMR experimental data. The time dependence of the agglomeration and glass-formation processes can also be taken into account by including into the time derivatives of the configuration probabilities P_i the terms $\frac{\partial P_i}{\partial T}\frac{dT}{dt}$, depending directly on the time variation of the temperature. We show how the addition of this term to the right-hand side of our system of differential equations describing the agglomeration process leads to an acceptable model of rapid quenching.

Finally, we discuss a possible interpretation of our model in terms of a non-standard formulation of statistical mechanics, which consists in replacing the ensemble of states of the clusters by the ensemble of all possible *pathways* leading to the formation of a cluster of given type.

The paper is organized as follows:

1) General Introduction
2) Description of the Model: the mathematical analysis of the growth and agglomeration processes; derivation of the differential system
3) Illustration of performances of the model on the simplest case of star-like building blocks associating into doublets and triplets.
4) Application to a well-known binary glass: $B_2O_3 + nLi_2O$
5) A model of pure B_2O_3 - glass; analysis of ring structures, also in the amorphous SiO_2.
6) Inclusion of time dependence on the temperature: a model of rapid quenching and its applications to the edge-sharing systems (e.g. $B_2S_3, SiS_2 + nLi_2S$)
7) Discussion and interpretation of the description of irreversible processes by our model in terms of Statistical Physics.
8) Conclusion

2 Preliminary Theoretical Considerations

We shall take for the starting point the considerațions formulated by Zachariassen in his famous paper written in 1932[26]. Essentially, his point of view is that the structure of covalent glass can be characterized as a *continuous random network* made out of atoms of the glass former (usually an element from the III-rd or IV-th group of the periodic table), with oxygen bridges between them; in many cases, oxygen atoms can be replaced by sulfur or selenium. The general formula of the most frequently used covalent glass former is then X_2Y_3 (for $X = B$ or As for the compounds of the III - VI periodical table groups), and XY_2 for $X = Si$ or Ge (for the compounds of the IV - VI groups);in both cases, $Y = O$, or S, or Se.

While the three or four $X - Y$ bonds can be considered as stiff *planar tripods* or *tetrapods* spanning an almost perfect stiff tetrahedron, the oxygen (resp. sulfur, or selenium) bridges $X - Y - X$ are quite elastic, and the angles they form can easily take on the values contained in some non-negligible range around their mean value; this meekness and elasticity is responsible for the breakdown of the short- and medium-range order in the network resulting from the solidification of the melt during the glass transition, when most of the bridges are definitively fixed.

However, the glass-forming property strongly depends on the form of the first micro-clusters, containing two or three atoms of the glass former and a few bridges, which depends on chemical differences between the particular glass formers. Zachariasen [26] pointed out that in the case when the *edge-sharing* units are easily formed, the network becomes so highly organized at a local level, that it does not display anymore the "randomness" that would qualify it as a glass. It can be expressed by saying that the *edge-sharing oxides are bad glass formers.* In contrast to that, the compounds that do not produce the the edge-sharing units (which can be also described as the smallest possible *rings* made of only two atoms of the glass-former), and form bigger rings, (like the boroxol rings in B_2O_3 containing three boron atoms), or even bigger ones (4, 5 and 6-membered rings in the case of SiO_2), display a pronounced glass-forming tendency [27, 28, 29, 30, 31].

Of course, the glass-tendency forming may be favored by the addition of a compound that would act destructively on any local attempt to start an organized network. The addition of certain amount of *modifiers* often plays this rôle. These are usually the corresponding alkali oxides or sulphides, like Li_2O or Na_2S, etc.), whose influence at the local level can be seen as the modification of the coordination number of the glass-former atoms - e.g. transforming the three-coordinated boron atoms into B_4 ions in B_2O_3, or lowering the coordination from 4 to 3, or even 2 in the case of SiO_2 or SiS_2 based glasses. This of course tends to rise the disorder and to strenghten the glass-forming ability.

In the late seventies J.C.Phillips [5, 6] has found an elegant expression of this observations by encoding them in the notion of local constraints; this approach has been also developed by M.Thorpe [14, 15]. The glass-forming tendency in the constraint theory is directly related to the difference between the average number of degrees of freedom and the average number of constraints on the local level; when this difference vanishes, the glass-forming tendency attains its maximum.

The statistical model of glass formation that we expose here is based on a detailed description of agglomeration and growth of micro-clusters that already exist in the hot melt about to undergo the glass transition. In order to predict the structural features of the resulting random network, we must have a reasonable model indicating in what way these clusters grow and agglomerate together, as well as their own structure and relative statistical weight. To do so, we must find first a description of the irreversible process of agglomeration in terms of variables directly related to the configuration space of the system.

Our approach is inspired by the image displayed in Fig.1. Let us mentally divide the configuration space containing all possible micro-clusters that are present in the hot melt into a series of sheets, numbered from 1 to k_{max} , where k_{max} is the maximal size (i.e. the number of atoms of the glass-former involved) of a cluster observed in the liquid at the time we are observing it. Each of these sheets contains only the clusters of the size k.

Obviously, k_{max} is a function of temperature T, therefore, of the time t when $T = T(t)$ varies during the glass transition. A hot melt at constant temperature contains a lot of clusters of different sizes and shapes, although negligibly few ones that are bigger than k_{max}. Also the statistical distribution of the clusters depends on the temperature. At a given constant temperature some of the clusters split into smaller units, while some of the smaller ones agglomerate growing into bigger entities; however, on the average the balance is maintained and the mean distribution of clusters on the sheets of the configuration space remains constant.

But let the temperature vary, and the distribution of probabilities of finding a given number of clusters on the k-th sheet will also evolve; the higher the temperature, the greater the "population" of clusters concentrated on the sheets corresponding to lower values of k, and *vice versa*. It is quite obvious that when the temperature slowly goes down, the "center of gravity" of the overall distribution also displaces itself towards the sheets of greater k; at the same time, k_{max} is growing, too.

We believe that the best way to get more insight into the structural properties of a glassy network is to reconstruct as precisely as possible the very process of creation of such a network, which occurs during the glass transition and stops at the temperature T_g consists essentially in the saturation of all the bridges between the micro-clusters existing already in the hot melt; the whole phenomenon is an off-equilibrium thermodynamical process, during which, starting from some moment, clusters agglomerate with the exponentially growing speed, which gives rise to an exponential growth of viscosity and a sharp variation of the heat capacity.

We do not know the exact dependence on the temperature (and therefore, on time, because usually $T = T(t)$ is imposed by the experiment), neither that of the probability densities corresponding to various clusters of a given size k, nor of the k_{max}. Still, we can follow the variations of the probabilities from one sheet to another by means of the parameter s ("*step*"), defined as follows:

$$s := \left\langle \frac{N_1 + 2N_2 + 3N_3 + \ldots + kN_k + \ldots + k_{max}N_{k_{max}}}{N_1 + N_2 + \ldots + k_{max}N_{k_{max}}} \right\rangle - 1 \qquad (1)$$

where the averaging symbol $\langle\ \rangle$ refers to samples which are representative enough, so that the local fluctuations do not matter anymore (in practice such

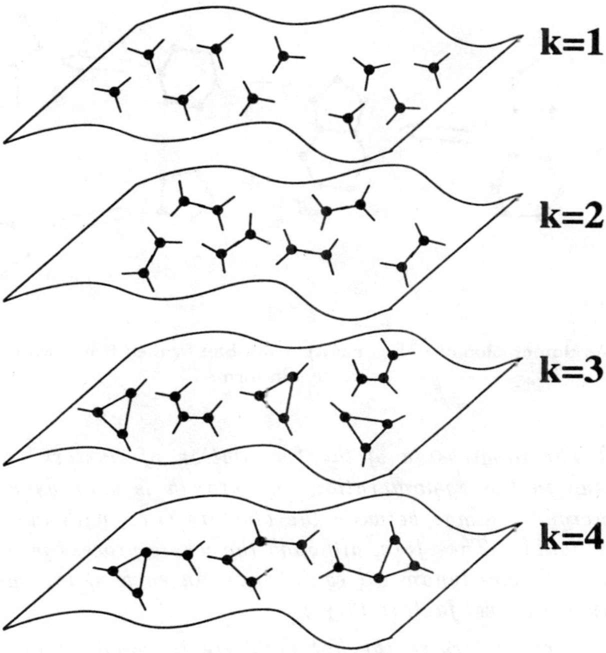

Figure 1: The configurations space of the liquid melt divided into separate sheets containing all clusters of a given size k

samples should contain at least hundreds of clusters of all possible sizes and types); N_k is the total number of clusters with k atoms in the sample.

It is easy to check that if all the atoms were isolated (i.e. the whole system located on the sheet of "singlets", $k = 1$, then all N_k would vanish except N_1, in which case $s = 0$; if all the atoms were in the state of "doublets", i.e. in clusters containing just two atoms, then $s = 1$, etc. During the process of formation of the glassy network in a freezing liquid the parameter s varies in a quasi continuous manner when evaluated on sufficiently large samples.

Our model will become operational and able to give predictions if we can define a way to compute the distribution of probabilities of all geometrically possible non-redundant clusters of a given size $k+1$ as a function of the probability distributions on the sheets numbered 1, 2, ... k. In order to define such recurrent algorithm, we need the following physical assumptions.

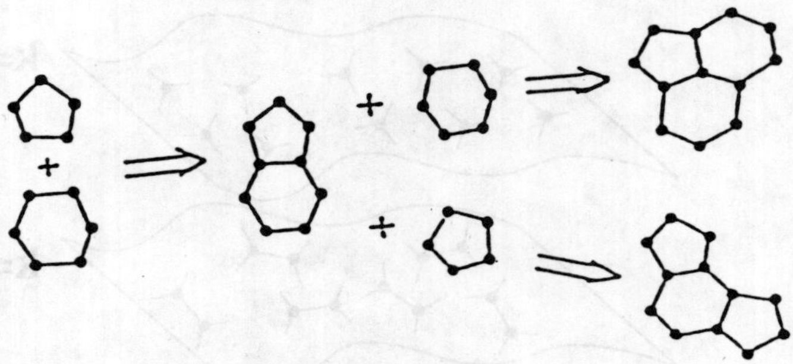

Figure 2: Agglomeration of polygons: a) A doublet formed from two singlets; b) a cell c) a chain formed

1) The progression of the "population of clusters" from one sheet to another due to the agglomeration and growth is slow as compared with the kinetic energy exchange between the clusters (i.e. with the speed of phonons in the hot liquid). Therefore, although the whole process is off equilibrium, we may assume thermodynamical equilibrium on each of the sheets, and use the Boltzmann statistical factors there;

2) The clusters in the hot melt are in constant proximity, and their movements are not as rapid as the molecular motion in a gas; consequently, we shall assume that the probability of collisions is roughly constant, i.e. depends very weakly on temperature, so that the most important contribution to the probability of a given agglomeration step is given by the statistical weight (i.e. the number of ways in which this step can be performed) and the Boltzmann factor depending on the energy cost of a given process;

3) We shall consider the energies stored in a given cluster during its agglomeration as additive quantities, which is equivalent to suppose that the energy is locally conserved. This means that we can evaluate the probability of observing a given cluster by summing up all the agglomeration pathways that lead to its creation from elementary "singlets", by multiplying the probabilities of each transition by corresponding statistical weights and Boltzmann factors.

The best way to illustrate the algorithm for the evolving probabilities is to consider the simplest example elaborated already in [23]. Let us consider 5, 6 and 7-sided equilateral, but not equiangular polygons on the plane, that can stick together side by side, as shown in Fig. 2.

If we suppose that the polygons are endowed with some amount of stiffness that draws them naturally to the most regular (*i.e. equiangular*) shape, then at first approximation we can attribute the energy cost of deformation proportionally to the absolute value of the curvature introduced in a flat tri-coordinate network by a given polygon. Thus, hexagons will be considered as zero-curvature quantities, whereas each pentagon or heptagon locally creates an equal amount of local *positive* (resp.*negative*) curvature.

Let the probability of finding a k-sided polygon in the reservoir be $P^{(0)}{}_k$, with $k = 5, 6, 7$; the probabilities of six different *doublets* that can be formed with these entities are:

$$P^{(1)}_{km} = \frac{1}{Q}(2 - \delta_{km}) P^{(0)}_k P^{(1)}_m e^{-E_{km}/kT} \tag{2}$$

with $k \leq m$, and Q denoting the normalizing factor assuring that the sum of all the probabilities of doublets remains equal to 1. The probability of finding a k-sided polygon in a sample containing many such doublets is given by:

$$P^{(1)}_k = \frac{1}{2}(2P^{(1)}_{kk} + \sum_{i \neq k} P^{(1)}_{ik}), \quad k = 5, 6, 7; \tag{3}$$

Of course, if we were considering the *entire amount* of the melt, without any interaction with the exterior except a thermal exchange, then $P^{(1)}_k$ would be strictly equal to $P^{(0)}_k$; however, we are interested in a local analysis, which means that we are averaging over the relatively small samples containing several dozens of elementary entities, and which can exchange them with the rest of the reservoir. These samples of "subcritical size" should be big enough in order to assure that the quantity ds/dt evaluated inside them remains always positive during glass transition, but small enough in order to display noticeable fluctuations of the probabilities $P^{(n)}_k$ evaluated on the consecutive "*sheets*" of the configuration space defined above.

When we compare the probabilities $P^{(n)}_k$ and $P^{(n+1)}_k$ evaluated on the subsequent sheets n and $n+1$, they converge to the common limit P_k; the label n for which the difference of the subsequent averages becomes negligible will be called *critical size*. Although such saturation is certainly attained at the stages of agglomeration corresponding to the values of n no less than a dozen, we can show how the system of non-linear differential equations can be deduced from the simple requirement of *maximal local homogeneity* of the network, which is expressed by setting $P^{(n+1)}_k - P^{(n)}_k = 0$, even for $n = 1$ or 2.

Consider, for example, just the first two sheets of the configuration space containing respectively the "singlets" and the "doublets". At a given time the probability P_k evaluated on subcritical samples taken from both sheets with equal frequency will be given by the expression:

$$P_k(t) = (1-s)P_k^{(0)} + sP_k^{(0)} \tag{4}$$

where $s = s(t)$ is the agglomeration "step" parameter defined above. Then, if the time derivative of the temperature can be neglected (the slow annealing case), then the time derivative of P_k is given by:

$$\frac{dP_k}{dt} = \frac{ds}{dt}\frac{dP_k}{ds} = (\frac{ds}{dt})[P_k^{(1)} - P_k^{(0)}] \tag{5}$$

Of course, only $N-1$ equations are independent, consecutive to the normalization condition $\sum_{k=1}^{N} P_k^{(n)} = 1$. The resulting set of differential equations can be studied as a dynamical system, whose validity is restrained to the close vicinity of its *singular solutions* ($P_k = Const$). The character of the singular points is determined by the eigenvalues of the differential system *linearized* at these points; it obviously depends on the energies E_{km} we attribute to each elementary agglomeration of polygons.

The simplest choice consists in making these energies proportional to the absolute value of the local curvature produced by each of the polygons, which means that we set:

$$E_{55} = 2\epsilon, \quad E_{56} = \epsilon, \quad E_{57} = 0, \quad E_{66} = 0, \quad E_{67} = \epsilon, \quad E_{77} = 2\epsilon$$

With these values of energies the phase diagram of the corresponding dynamical system looks as shown in Fig. 3.

In the triangle ABC containing the probabilities P_5, P_6 and P_7 constrained by $P_5 + P_6 + P_7 = 1$ one can observe *five* singular solutions represented by the points A, B, C, D and E. The points A and C represent unstable solutions, because all the trajectories starting there point outwards (a repulsive behaviour); the points B and D represent stable solutions (attractive behaviour on the phase diagram); finally, the point E inside the triangle is *metastable* (often called a *"saddle point"*, although this denomination makes sense only in two dimensional phase diagram).

Quite obviously, the dynamical character of these singular solutions corresponds to the behaviour of a real tri-coordinate network on the plane: the trajectories converge towards the solution B because it represents the most natural tiling of the plane by means of pure hexagonal lattice, which according to our assumptions, does not require any effort (no local curvature, therefore no energy cost involved, $E_{66} = 0$). Another attractive point (D) represents the solution $P_5 = P_7 = \frac{1}{2}, P_6 = 0$, corresponding to the possibility of tiling the

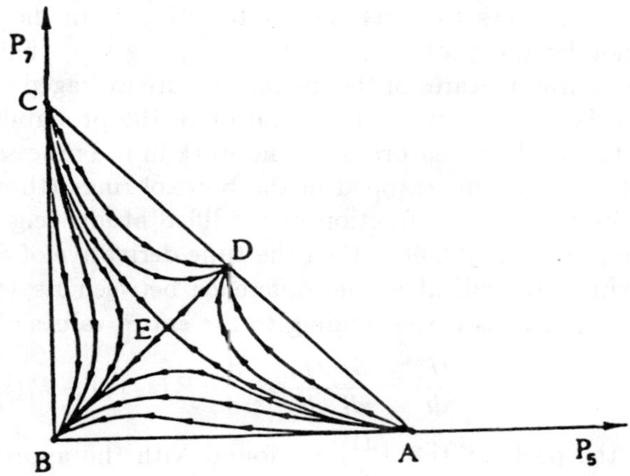

Figure 3: The phase portrait of the differential system (5)

plane with equilateral pentagons and heptagons in equal number.

The *metastable* point E, whose position inside the triangle depends on the temperature, clearly represents the tendency of the system to create the amorphous network which is a mixture of all the three kinds of polygons; the configurational entropy defined as

$$S_{conf} = -k_B \sum_{k,m} P_{km}^{(1)} ln P_{km}^{(1)}$$

attains its maximum exactly at this point.

The above description of agglomeration and growth process corresponds at this stage to the first *linear* approximation; in order to get more insight into the agglomeration and clustering process, we should follow the construction of bigger clusters, step by step, and compute in a similar way the probability distributions on each of the "sheets" of the configuration space represented in Fig.1. This approach can be generalized for the investigation of agglomeration and growth processes during glass transition in more complicated systems.

The essential features of the toy model discussed above are reproduced, with minor variations, in any model conceived in this way and based on the analysis of the agglomeration process. The differences come from the choice of the elementary configurations, their ability to form well-determined clusters, therefore, the multiplicities of various pathways of agglomeration and the energies determining the Boltzmann factors.

The variables that are most interesting from the physical point of view need not be the probabilities of finding a given singlet, but may represent some essential features of the microstructure averaged over the samples of sub-critical size. If we follow the evolution of the probability of finding a certain characteristic local feature of the network in the process of formation, e.g. the rate of boron atoms trapped in the boroxol rings (three borons + three oxygens), or the relative fraction of modified atoms (e.g. B4 ions) in a binary glass $B_2O_3 + nLi_2O$, etc., then the time derivative of such probability can be approximately defined as the difference between its values evaluated on the consecutive sheets corresponding to the entire values of the parameter s:

$$\frac{dP_i}{dt} = \frac{ds}{dt}\frac{dP_i}{ds} \approx \frac{ds}{dt}(P_i(s_{k+1}) - P_i(s_k)), \tag{6}$$

where the probabilities $P_{ij}^{(1)}$ are found with the appropriate Boltzmann and statistical factors, evaluated accordingly to the particular situation. (Note that in order for this approximation to be valid, one must suppose that the logarithmic time derivative of the temperature, $\frac{1}{T}\frac{dT}{dt}$, can be neglected when compared with ds/dt ; if it is not the case, the time derivative of P_i should also take into account the extra term of the form $[\partial P_i/\partial T][dT/dt]$):

$$\frac{dP_i}{dt} = \frac{ds}{dt}(P_i(s_{k+1}) - P_i(s_k)) + \frac{dT}{dt}(\frac{\partial P_i}{\partial T}) \tag{7}$$

We shall discuss this possibility in the section devoted to the mathematical description of rapid quenching).

The derivative ds/dt has a clear physical meaning. As a matter of fact, its inverse, $\tau = dt/ds$, may be interpreted as the average time needed for an elementary agglomeration step to be completed, i.e. the time it takes to create a new bridge in the liquid during the cooling process.

In what follows, we shall neglect the eventual dependence of τ on the temperature; the only important thing for the analysis of the solutions of the differential system produced is that during the process, $ds/dt > 0$, which we obviously assume.

By definition, P_i's are the probabilities of a certain number, say M, of characteristic features or local configurations whose evolution we are following; therefore, they are subjected to the obvious constraints:

$$\sum_{i=1}^{M} P_i = 1, \quad 0 \leq P_i \leq 1.$$

This means that the only physically relevant *constant* solutions of the differential system above, which are also called *stationary solutions* or *singular points* of the differential system, must be found inside the $(M-1)$-dimensional simplex defined by these constraints, as shown in the Fig. 4.

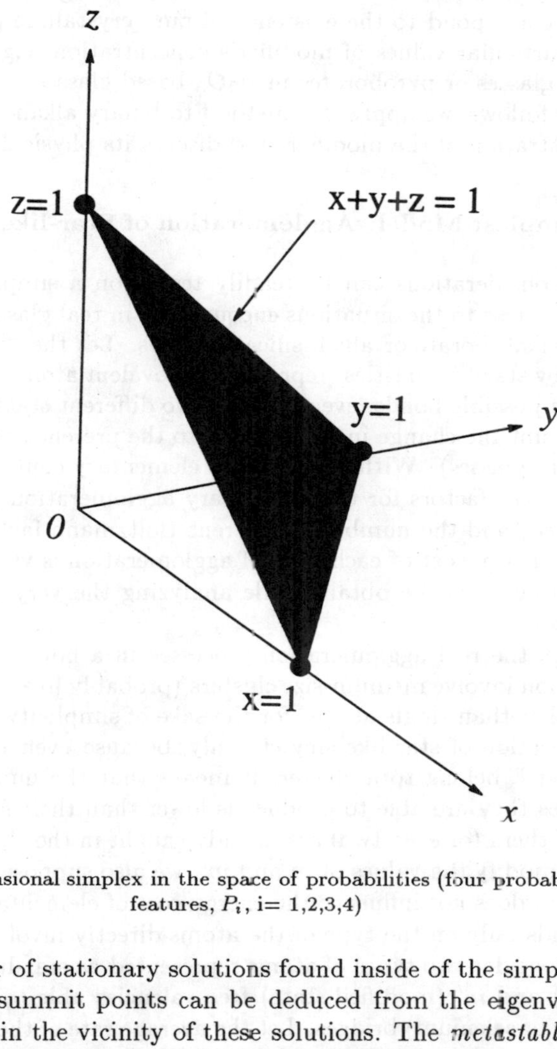

Figure 4: A 3-dimensional simplex in the space of probabilities (four probabilities of local features, P_i, i= 1,2,3,4)

The character of stationary solutions found inside of the simplex or on its faces, edges and summit points can be deduced from the eigenvalues of the *linearized system* in the vicinity of these solutions. The *metastable* solution is found always inside the simplex, and the configirational entropy $-k_B \sum P_k ln P_k$ always attains a local maximum there. In most of the systems investigated by this method until now, only one metastable solution was found; in some more complex cases one can not exclude the existence of two or even more metastable solutions, corresponding to different choices of local statistics in amorphous structures that could eventually coexist during the glass transition. In the case of binary glasses, the solutions found on the edges or in the corners of the simplex correspond to the existence of pure crystalline phases which may appear at particular values of modifier's concentration, e.g. metasilicates in SiO_2-based glasses or pyroborates in B_2O_3-based glasses.

In what follows, we apply this method to binary alkali-oxyde glasses with a low concentration of the modifier, and discuss its physical consequences.

3 The Simplest Model: Agglomeration of Star-like Entities

The above considerations can be readily tested on a simple model which is already quite close to the situations encountered in real glass-forming systems, like binary alkali-borate or alkali-silicate glasses. Let the "singlets" be represented now by star-like entities (representing covalent atoms of the glass former with their m possible bonds, eventually in two different states, m and m', taking into account the change in valency due to the presence of a modifier in the case of binary glasses). With two kinds of elementary configurations only, the statistical weight factors for the elementary agglomeration precesses are very easily deduced, and the number of different Boltzmann factors that take into account the energy cost of each kind of agglomeration is very low, so that the significant results can be obtain while analyzing the very first stages of the growth process.

Although the real agglomeration processes in a hot melt undergoing the glass transition involve medium-size clusters (probably five to ten atoms on the average) rather than single atoms, for the sake of simplicity we shall represent the agglomeration of star-like singlets only, because even if in real situation these "singlets" belong to a cluster, it means that the number of oxygen or sulfur bridges they are able to produce is lower than their formal valency, because one of them (or even two) are already caught in the cluster behind. This amounts to modify the values of m and m'; we also suppose that the presence of the clusters does not influence the energy cost of elementary agglomeration, which depends only on the type of the atoms directly involved.

Let us consider two types of atoms represented as star-like entities with m (basic state) and m' (modified state) free valencies able to form new oxygen (resp. sulfur or selenium) bridges. Let the average rate of the basic state atoms in a micro-sample be x, and that of the modified ones $1 - x$. We shall not exclude *a priori* the possibility of creation of two bridges at once, thus forming a *two-membered* ring; later on, depending on chemical properties of the system considered, such tendency may be either totally excluded or dominant.

The probabilities of six possible elementary agglomeration processes are displayed in the Fig. 5 below:

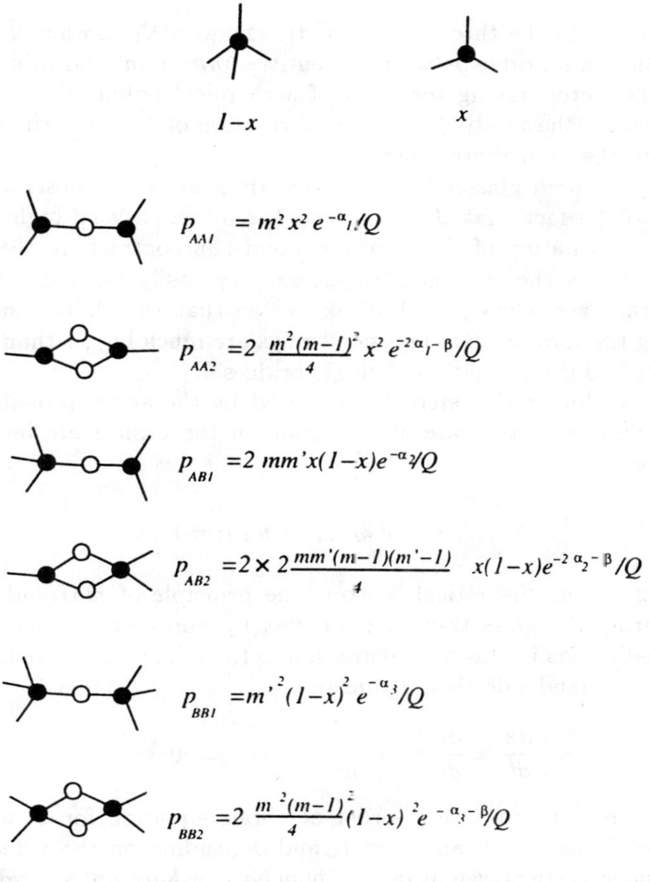

Figure 5: Six possible elementary agglomerations of two types of atoms with their explicit probabilities

The normalizing factor Q is given by the following formula:

$$Q = m^2x^2e^{-\alpha_1} + \frac{m^2(m-1)^2}{4}x^2e^{-2\alpha_1-\beta} + 2mm'x(1-x)e^{-\alpha_2}$$
$$+mm'(1-m)(1-m')x(1-x)e^{-2\alpha_2-\beta} + m'^2(1-x)^2e^{-\alpha_3}$$

$$+\frac{m'^2(m'-1)^2}{2}(1-x)^2 e^{-\alpha_3-\beta}$$

Note that we have introduced *three* different Boltzmann factors,

$$e^{-\alpha_1}=e^{-\frac{E_1}{kT}},\quad e^{-\alpha_2}=e^{-\frac{E_2}{kT}},\quad and\quad e^{-\alpha_3}=e^{-\frac{E_3}{kT}}$$

corresponding to the three characteristic energies E_1, E_2 and E_3 related to the formation of a bridge between the entities mm, mm' and $m'm'$, respectively; another factor having the form of a chemical potential $e^{-\beta}=e^{-\frac{\mu}{kT}}$ takes into account the relative facility of formation of two-side rings (double bonds between the immediate neighbors).

In the B_2O_3-based glasses the two-sided rings are never observed, which corresponds to the fact that $\beta \gg \alpha_i$, $i=1,2,3$, i.e. a very high potential barrier for the formation of simultaneous bonds; in contrast, in the SiS_2 or $SiSe_2$-based glasses the two-sided rings are very easily formed, which corresponds to the case when $(2\alpha_i+\beta) \ll \alpha_i$, so that the Boltzmann factors accompanying the formation of two-sided rings are much bigger than the ones that are related to the formation of single bridges.

After the agglomeration step characterized by the above probability distribution is completed, the rate of the atoms in the basic state in a typical sample will be:

$$x^{(1)}=\frac{1}{2}(2p_{AA1}+2p_{AA2}+p_{AB1}+p_{AB2}) \tag{8}$$

According to our theoretical scheme, the principle of *maximal local homogeneity* during the glass transition process by *annealing* , when dT/dt is negligibly small, leads in this approximation to the following differential equation, whose right-hand side should vanish:

$$\frac{dx}{dt}=\frac{ds}{dt}\frac{dx}{ds}=\frac{ds}{dt}(x^{(1)}-x)=0; \tag{9}$$

This leads in turn to a third-order algebraic equation for x, which has always two solutions $x=0$ and $x=1$, and depending on the values of the Boltzmann and statistical weight factors *may have* or *have not* a third solution contained between 0 and 1, that we shall denote by x_{am} . The two situations are displayed in the Fig. 6 below:

Figure 6: Two diagrams representing the behaviour of the system: a) with two solutions only, b) with a third solution x_{am}

We should note that the solution $x = 1$ does always exist, and in the case a), when the corresponding point is *attractive* it describes the local tendency of the system to remain "pure", i.e. to expel the modified atoms from local clusters, which should correspond to the beginning of crystallization, or at least a phase separation; the solution $x = 0$ is always *repulsive*, whereas the solution x_{am} inside the segment $[0, 1]$, when it appears, is *attractive*, while the solution $x = 1$ becomes automatically repulsive; this corresponds to a local tendency of maximal mixing and homogenization of both kinds of entities, which is characteristic for an *amorphous* configuration.

The detailed analysis of all possible situations as a function of the parameters $m, m', \alpha_1, \alpha_2, \alpha_3$ and β is very complicated; moreover, it seems precarious to be seriously involved in such an enterprise knowing that the equation we are analyzing is a very rough approximation of a greater variety of simultaneous agglomeration processes, e.g. forming three-sided rings in the next step of agglomeration.

That is why we shall restrain the analysis to the extreme cases corresponding either to the total absence of the two-sided rings, or to the situtation in which the growth process is dominated by the formation of such rings. Even with such simplified assumptions, our model has an impressive predictive power.

The condition for vanishing of the right-hand side of the equation (9) can be multiplied by the normalizing factor Q, and it reads then as:

$$x(1 - x)(\Phi x + \Psi) = 0 \tag{10}$$

with Φ and Ψ functions of $m, m', \alpha_i, i = 1, 2, 3$, and β.

The solution corresponding to the local amorphization tendency exists when

$$0 < x_{am} = -\frac{\Psi}{\Phi} < 1 \tag{11}$$

Let us now check two important limit cases:

1) The case when $\beta > 0$ and great enough to ensure that $e^{-2\alpha_i - \beta} \ll e^{-\alpha_i}$ for $i = 1, 2, 3$; then, neglecting these terms, we get the following simplified expression for x_{am}:

$$x_{am} = \frac{mm'e^{-\alpha_2} - m'^2e^{-\alpha_3}}{2mm'e^{-\alpha_2} - m'^2e^{-\alpha_3} - m^2e^{-\alpha_1}} \tag{12}$$

The condition for x_{am} to be contained in the interval $[0, 1]$ is equivalent to:

$$me^{-\alpha_2} > m'e^{-\alpha_3} \quad \text{and} \quad m'e^{-\alpha_2} > me^{-\alpha_1} \tag{13}$$

(Note that this condition will remain valid if we exchange simultaneously m with m' and α_1 with α_3; but this amounts to re-define the roles of the "basis" and "modified" states, which is one of the symmetries of our approach in the case of binary systems.)

A good glass former should display the amorphisation tendency independently of the modifier's rate; this means that the solution x_{am} has a continuous limit when $x \to 1$. In this case, the condition reduces to

$$m'e^{-\alpha_2} = me^{-\alpha_1} \tag{14}$$

Introducing explicitly the Boltzmann factors, $e^{-\alpha_i} = e^{-\frac{E_i}{kT}}$, we obtain:

$$\frac{E_2 - E_1}{kT} = ln(\frac{m'}{m}) \tag{15}$$

This equation defines the relation between the entropic and energetic factors and the glass transition temperature T, that are the most important for the formation of binary glasses. If $m' > m$, the logarithm of $\frac{m'}{m}$ is positive while the energy E_2 involved in the formation of a bond between the atoms $A - B$ is greater than the energy E_1 needed to form a bond $A - A$; and vice versa, when $m' < m$, in a good glass former one must have $E_2 < E_1$.

This is what should be intuitively expected from a good glass former, which even on a miscroscopic scale behaves in a "frustrated" way: as our simple formula shows, the two main contributions to the probability of formations of new clusters always act in opposite directions: if the modifier raises the number of possible bonds, thus creating more entropy and raising the probability of agglomeration ($m' > m$), the corresponding energy cost is higher, thus *reducing* the probability, and vice versa.

We believe that this "undecided" character of the agglomeration process is responsible for the tendency to choose the amorphous rather than the crystalline state.

We can test the predictive power of our model using it for the evaluation of the derivative of the glass transition temperature T_g with respect to the concentration of the modifier n at $n = 0$.

In the vicinity of $x = 1$, we have $rn = 1 - x$, where r is the ratio of the number of modified molecules of the glass former X_pY_s to the number of the molecules of the modifier M_qY, i.e. $r = q/p$.

For example, $r = 1$ for the alkali-borate glass $B_2O_3+nLi_2O$, because each molecule of the modifier transforms one molecule of the glass former, as in the case of the mixture $SiO_2 + nCaO$, while we must put $r = 2$ for the mixture SiO_2+nNa_2O, because here one molecule of Na_2O transforms *two* molecules of the glass former SiO_2. Therefore, for the concentrations n close to 0 we have:

$$[\frac{dT_g}{dn}]_{n=0} = -[r\frac{dT_g}{dx}]_{x=1} = -[r(\frac{dx}{dT})^{-1}]_{\substack{x=1\\T=T_0}} \tag{16}$$

where T_0 is the glass transition temperature observed at $n = 0$, i.e. in the absence of the modifier.

With the expression of $x = x_{am}$ defined above, and it explicit dependence on the temperature (which we identify here with the glass transition temperature T_g), we can easily calculate the above expression, which yields the following simple formula:

$$\left[\frac{dT_g}{dn}\right]_{n=0} = \left[\frac{(2\frac{m'}{m}\frac{E_2}{E_1} - 1)^2}{\frac{m'}{m}\frac{(E_2-E_1)}{kT^2}}\right]_{T=T_0} = r\frac{T_0}{ln(\frac{m'}{m})} \tag{17}$$

because at $T = T_0$ we have $n = 0$ and therefore $x_{am} \to 1$, so that $\frac{m'}{m}\frac{E_2}{E_1} = 1$ and $E_2 - E_1 = kT_0 ln(\frac{m'}{m})$.

The last formula is very significant and may be announced as a universal rule for binary glasses (*without edge-sharing elementary units!*):

The modifier that increases the average coordination number of the glass former tends to raise the glass transition temperature (like in e.g. $B_2O_3 + nLi_2O$-glass), whereas the modifier that decreases the average coordination number tends to lower the glass transition temperature (like in e.g. $SiO_2 + nCaO$-glass).

The two typical curves $T_g(n)$ for the opposite types of behavior are displayed in Fig.7 below:

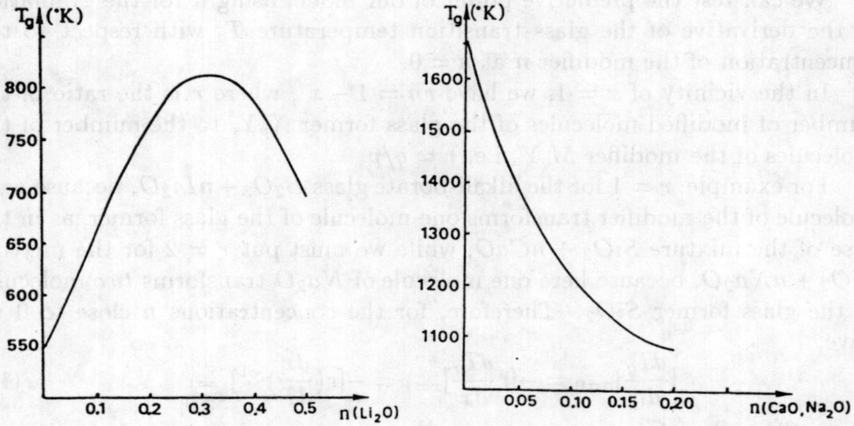

Figure 7: Two curves $T_g(n)$ for binary glasses: a) $T_g(n)$ for $B_2O_3 + nLi_2O$ b) $T_g(n)$ for $SiO_2 + n(CaO, Na_2O)$

Our simple formula can be tested now by comparison with the experimental data: the average values of the derivative (dT_g/dn) at $T = T_0$ are:

1200° K/mole and $T_0 \sim 550°$K for the alkali-borate glass $B_2O_3 + nLi_2O$;

which yields $ln(\frac{m'}{m}) = 0.458$, giving $\frac{m'}{m} = 1.58$, (a value quite close to $\frac{3}{2}$);

$-1900°$ K/mole and $T_0 \sim 1700°$ K for the ordinary silicate glass $SiO_2 + n(CaO, Na_2O)$;

which yields $\frac{1}{r} ln(\frac{m'}{m}) = -0.895$ and $\frac{m'}{m} = 0.409$ for $r = 1$, and $\frac{m'}{m} = 0.167$ for $r = 2$.

However, in real silicate glasses, one ususally has 70% of Si_2O, 20% of CaO and 10% of Na_2O; the normalized ratio between the compounds CaO and Na_2O is as $\frac{2}{3}$ to $\frac{1}{3}$, which means that we should ponderate the parameter r as follows: $r = 1 \times \frac{2}{3} + 2 \times \frac{1}{3} = \frac{4}{3}$. Then we get $\frac{m'}{m} = 0.303$, the value quite close to $\frac{1}{3}$.

The energy difference $E_2 - E_1$ for each of these cases is:

$$E_2 - E_1 = 0.024eV \text{ for the alkali-borate glass,}$$

$$\text{and } E_2 - E_1 = -0.142eV \text{ for the alkali-silicate glass}$$

One should keep in mind that our model is based on a very simplified analysis, averaging the contributions coming from many different elementary processes, especially in the case of the ternary silicate glass; that is why the above energy differences can not be taken very seriously, but regarded as an approximation, in which only the sign and the order of magnitude are relevant.

The obtained estimations of the fraction $\frac{m'}{m}$ suggest that the most frequent elementary agglomerations in the aforementioned binary glasses should be as represented in the Fig. 8 below :

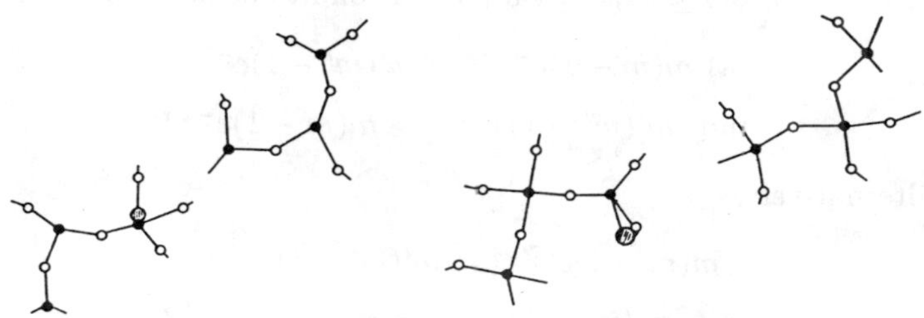

a b

$B_4 + B_3 \ (m' = 3, m = 2) \qquad Si_2 + Si_4 \ (m' = 1, m = 3)$

Figure 8: Two elementary agglomeration processes: a) in a $B_2O_3 + nLi_2O$ glass, b) in a $SiO_2 + n(CaO, Na_2O)$ glass

A third example confirming our formula is provided by the germanate-selenium glass $Ge_{1-x}Se_x$ [33]. In the limit of $x = 1$ (pure selenium) the glass transition temperature tends to $T_{g0} = 318K$; at $x = 0.95$, i.e. with 5% of germanium, the glass transition temperature raises to about $340K$ to $342K$, which corresponds to the slope $[dT/dn]_{n=0}$ contained in the interval $440K/mole - 460K/mole$. Obviously, in this case $r = 1$, so the the formula yields the following result:

$$\frac{318K}{460K.mole^{-1}} \leq ln(\frac{m'}{m}) \leq \frac{318K}{440K.mole^{-1}}, \text{ i.e. } 1.996 \leq \frac{m'}{m} \leq 2.06$$

which means that $\frac{m'}{m} = 2$. This is in perfect agreement with what is known about the local structure of $Ge_{1-x}Se_x$ at low concentrations of Ge, namely, that selenium has a very strong chain-forming tendency, whereas the germanium atoms get trapped in the selenium chains, thus creating the four-valenced entities instead of two-valenced ones (see Fig. 9).

2) Another interesting limit corresponds to the preferential formation of the edge-sharing units (lozenges formed by two simultaneous bridges between

the glass-former atoms). Such tendency is observed indeed in the sulfides, e.g. in $B_2S_3 + nLi_2S$ or in $SiS_2 + nLi_2S$. In our model such a situation prevails if $\beta < 0$, so that the factor $e^{-\beta}$ becomes great enough in order to make all other factors $e^{-\alpha_i}, i = 1, 2, 3$ negligible. Then the "amorphous" solution is given by:

$$x_{am} = \frac{mm'(m-1)(m'-1)e^{-2\alpha_2} - (m')^2(m'-1)^2e^{-2\alpha_3}}{2mm'(m-1)(m'-1)e^{-2\alpha_2} - (m')^2(m'-1)^2e^{-2\alpha_3} - m^2(m-1)^2e^{-\alpha_1}} \tag{18}$$

In order to have $0 \leq x_{am} \leq 1$ a pair of conditions have to be satisfied:

$$\text{a) } m(m-1)e^{-2\alpha_2} > m'(m'-1)e^{-2\alpha_3}$$

$$\underline{\text{and}} \quad m'(m'-1)e^{-2\alpha_2} > m(m-1)e^{-2\alpha_1}$$

or alternatively,

$$m(m-1)e^{-2\alpha_2} < m'(m'-1)e^{-2\alpha_3}$$

$$\underline{\text{and}} \quad m'(m'-1)e^{-2\alpha_2} < m(m-1)e^{-2\alpha_1}$$

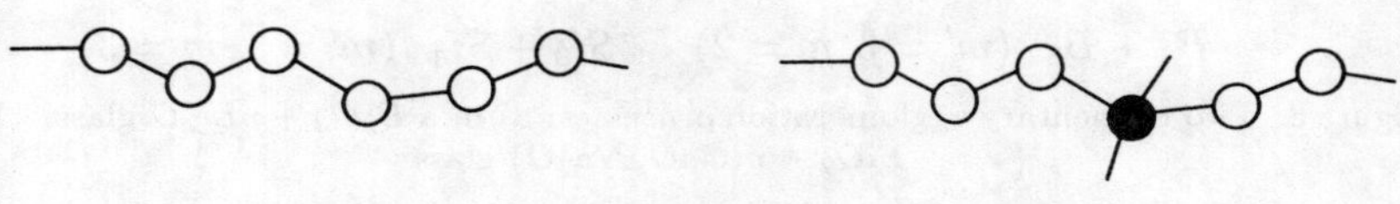

Figure 9: a) A pure selenium chain, $m = 2$ b) A selenium chain with a Ge atom, $m' = 4$

In contrast with the previous case when $\beta \gg \alpha_k$, the conditions involve three different Boltzmann factors corresponding to the energies of formation of two simultaneous bonds: $2E_1$ for two bonds of the type AA, AA, $2E_2$ for two bonds AA, AB and E_3 for two bonds of type BB, BB.

Moreover, these conditions are much more difficult to satisfy: as a matter of fact, in both cases the energy E_2 has to be either greater than both E_1 and E_3, or smaller than both these energies, which is quite hard to believe in the case of oxydes, where the repulsive forces between the ions of the modifier should rather produce the effect $E_1 < E_2 < E_3$ found in e.g. alkali-borate glass.

This simple qualitative result conforms well with Zachariasen's observations that the *oxydes in which the edge-sharing units are easily produced are always bad glass formers*[26].

Although the predictive power displayed is quite good for a model so simple, it will not bring any more relevant information if we don't continue the analysis, extending it to next steps of agglomeration producing bigger clusters.

The number of such clusters grows quasi-exponentially, and many new situations do appear at farther steps, namely, the formation of rings of different number of sides, then of double rings sharing either one atom or one bridge, etc.

It is not difficult to evaluate the statistical weight of each of these structures; however, the energies involved can be evaluated only roughly, and their number is growing, so that the model depends on a too great number of parameters.

We shall describe now how the model can be adapted to this new situation, when it becomes too difficult to follow the agglomeration of very primitive entities, which can be replaced by certain typical small clusters.

4 The Alkali-Borate $B_2O_3 + nLi_2O$ Glass

Our model has been applied to the real glass-forming systems, such as $B_2O_3 + nLi_2O$ [18, 53], which turned out to be the easiest example to be treated in this way, and to the pure B_2O_3 [21, 49], in which the only significant feature that can be analyzed on the medium-range scale is the presence of boroxol rings; finally, a few attempts have been made in order to describe ring structures in the amorphous SiO_2 [17], by far the most complicated case.

Let us first recall the results obtained for the $B_2O_3 + nLi_2O$ glass; the details can be found in [18, 53]. One of the main features of the borate glass, with or without alkali modifier, is the total absence of edge-sharing units and the easiness with which three-sided rings are formed.

The presence of the boroxol rings is strongly evidenced by the typical breathing modes in the vibrational Raman and infrared spectra; the problem of the real abundance of these structures remains unsettled, although many authors believe that the rate of boron atoms trapped in the boroxol rings is as high as 80 to 83% [4, 43, 44, 42, 40]. We shall give strong arguments in favour of this value when we shall consider in the next section the model adapted to the pure B_2O_3 glass.

It is believed that the most important structural change provoked by the presence of the Li_2O modifier is the change of coordination number of boron atoms, which are transformed from the state of valence three borons (B_3) into four-coordinated boron ions (B_4), corresponding to $\frac{m'}{m} = \frac{4}{3}$.

In order to investigate the formation of the boroxol rings in the alkali-borate glass, we have to follow as far as possible the agglomeration of simple units, including the steps that lead to the formation of many rings and more complex local structures. The number of configurations grows exponentially, so that it is very hard to keep the trace of them all; we show in Fig. 10 ten non-redundant configurations resulting from the third agglomeration step, during which the first boroxol rings start to appear; it is quite obvious that in order to see the first correlations between the neighboring rings we must continue beyond the sixth step, which is a difficult task.

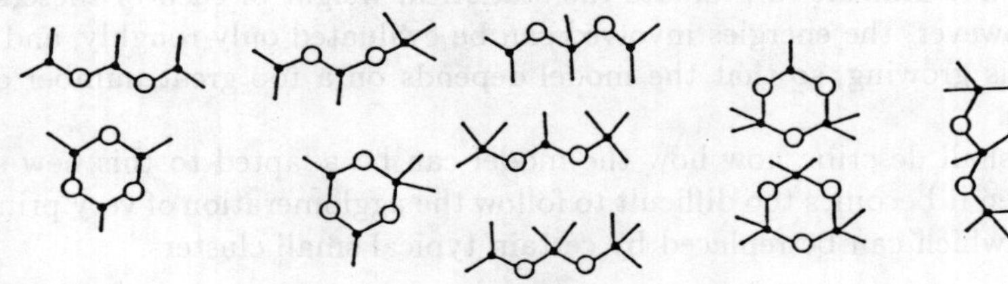

Figure 10: The ten non-redundant configurations of the third step of agglomeration of B_3 atoms and the B_4 ions in the $B_2O_3 + nLi_2O$ glass

Although the controversy concerning the abundance of boroxol rings in borate glasses is still quite vivid, and some authors maintain that these rings do not contain more than 10% of boron atoms forming the network, we believe that the contrary is true, i.e. that conformally with the estimations of Jellison *et al.*[4], Johnson *et al.*[42], Hannon *et al.*[43, 44, 45], and Walrafen *et al.*[40, 41], the boroxol rings contain more than 80% of all boron atoms in the glassy network.

In that case, especially at low concentrations of the modifier, we can replace the agglomeration of star-like B_3 and B_4 singlets with the agglomeration of boroxol rings, containing either pure B_3 atoms, or one or two B_4 ions. We shall give new arguments, based on our model, in favour of the abundance of boroxol rings close to 80% in the next section.

The results of this approach have been described in the papers[16, 34], and are also the subject of the article by R.A.Barrio in this volume. The general idea remains exactly the same, with *four* elementary configurations displayed in Fig. 11 below:

We shall be concerned with the structural changes of borate glasses with low alkali content ($n \leq 0.3$). According to the interpretation of the experimental NMR curves the only structures present in that range are boroxol rings, diborate and tetraborate groups. We shall assume that the basic units present already in the melt are six-membered rings with zero, one, two, and three B4 atoms in them. We give them the probabilities x, y, z, and t, respectively; they are sketched in Fig.11, where we named them A, B, C, and D, respectively. The normalization condition leaves only three independent variables, since $t = 1 - x - y - z$, and therefore, the range in which these variables are defined as probabilities forms a unitary simplex in the space (x, y, z), that is bounded by the planes $x = 0$, $y = 0$, $z = 0$, and $x + y + z = 1$, like in Fig. 4. If one excludes the possibility of forming 2-fold and 4-fold rings in the assembly process, one finds 36 distinct pairs of rings, and according to the method exposed above, the probabilities associated with each configuration can be calculated by counting the number of ways in which one can get a particular cluster. Observing that there are only three types of bonds formed, one associates energy costs for the creation of new bonds, which are included in the corresponding Boltzmann factors as $e_i = \exp(-E_i/k_B T), i = 1, 2, 3$.

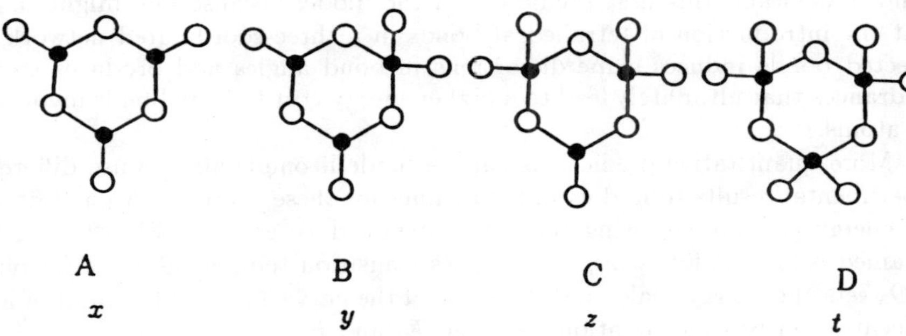

Figure. 11: The four elementary building blocks: boroxol rings with 0,1,2 and 3 B_4-ions

Once the pair probabilities are obtained, the new probabilities for the rings are evaluated, for instance, the new probability of finding a ring of type A in the pairs (or the new value of x) is

$$P_A^{(2)} = x_1 = \frac{1}{Q} \sum_{k=1}^{36} N_A(k) P_k \tag{19}$$

where k labels one of the 36 pair configurations and $N_A(k) = 0, 1, 2$ is the number of rings of type A found in the cluster k (actually some N_D are equal to 3). The normalizing factor Q is the sum of all pair probabilities and takes automatically into account the multiplicity factors γ. Near the stationary solutions one could substitute in the equations $P_A^{(2)} = P_A^{(1)} = x$, and so on.

At this point an analytical solving of Eq.(7) is not possible, so a numerical program was developed for this purpose, and all the stationary solutions have been found.

The character of the solutions depends on their position in the simplex. For any temperature there are always two solutions, one at $x = 1$, and other at $t = 1$. These kind of solutions can be interpreted as extreme values, probably associated with pure orderly crystals, and are of no interest here, since the glassy state must maximize the configurational entropy. It is clear that only solutions in the interior of the simplex (not in corners or faces) are likely to be pertinent to the amorphous state. We have numerically checked that the solutions in the interior have a greater entropy. It turns out that the existence of such a (saddle point) solution is not guaranteed for all values of the parameters E_i, and T. We have scanned the parameter space numerically and found that for a reasonably large range of temperatures no solution of this kind exists, unless $E_1 < E_2 < E_3$, which is sound. Phenomenologically one could understand this first prediction of the model because one might think that the introduction of tetrahedral bonds in a three-coordinated network of directed bonds induces some distortions in bond angles and produces steric hindrances that ultimately lead to a higher energy cost to form bonds involving B4 atoms.

More quantitative predictions can be made if one is able to use different experimental results to find estimated values for these energies. A good fit for the energy parameters using the experimental data given in Fig.12. can be obtained using the following[16]: the glass transition temperature T_{g0} for pure B_2O_3 sets the energy scale, and the slope of the curve $T_g(n)$ in the limit of low concentration provides relations between E_1 and E_2.

A more involved consideration is needed to set the value of E_3, since in the low concentration limit B4-B4 bonds are scarce. A thorough numerical examination of the parameter space was performed here to give an accurate fitting, and the final results do not differ much from the ones published in reference[67]. They are

$$E_1 = 0.078eV, \qquad E_2 = 0.123eV, \qquad E_3 = 0.213eV.$$

which are close to the estimates of Uchida et al[46].

We can note that the value of $E_2 - E_1 = 0.043eV$ is of the same order as the one obtained via the simplified model with star-like entities. With these values one can get the probabilities x, y, z corresponding to the amorphous solution, and the concentration of B4 atoms should be

$$n = \frac{1}{3}(y + 2z + 3t) = \frac{1}{3}(3 - 3x - 2y - z)$$

As in reference[16], let us define P_{tetra} and P_{dibor} as the proportion of corner-sharing and edge-sharing rings, respectively. These quantities are functions of the pair probabilities that contain these configurations in the group of 36 pair diagrams. The calculations for different temperatures give different values of n, and one can get them as functions of Li_2O concentration. The resulting theoretical curves are displayed in Fig.12 below, as compared with the Raman spectroscopy data (M.Balkanski et al.,[2,3]).

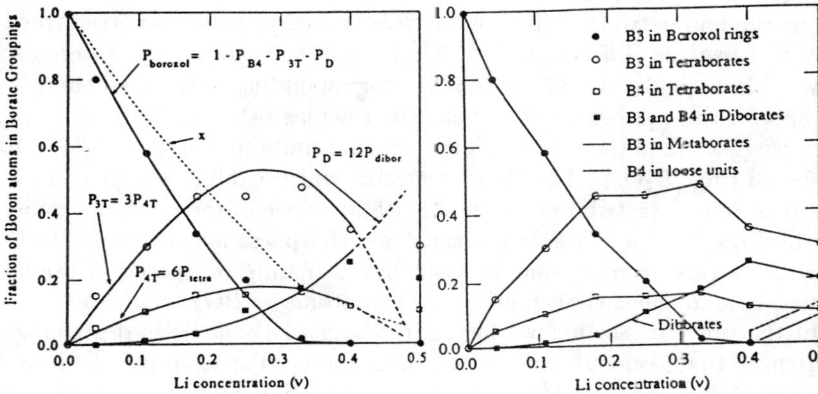

Figure. 12: Comparison of relative probabilities of boroxol rings, diborates and tetraborates as predicted by our model with the Raman spectroscopy data

The slope of the theoretical curve $T_g(n)$ for the values of n below 0.3 is also very close to the experimental one, displayed in Fig. 7a. As a matter of fact, the value of $\frac{m'}{m}$ is the same if we consider the agglomeration of the three-valenced boron atoms BO_3 and the B_4 ions, or the boroxol rings with 0 and 1 B_4-ions; in both cases, supposing that one of the valences is already occupied in the cluster, the remaining ones are $m = 2$ and $m' = 3$ in both cases.

The breakdown of the experimental curve $T_g(n)$ in the vicinity of $n = 0.35$ can be explained by the massive creation of the *metaborates* at this stage, which represent long chains of borons connected by oxygen bonds, in which the mean value of m' per boron tends to 1, which gives the negative slope.

In the article by Rafael A. Barrio in this volume other consequences of the model are discussed, namely the density and electric conductivity of the alakali-borate glasses.

5 Structural Model of Pure B_2O_3 Glass

Now we shall turn to another application of the model, i.e. a detailed study of pure B_2O_3 , which will enable us to expose a more general statistical mechanical approach to the glass transition. At the same time, it will give us *a posteriori* argument in favor of the great abundance of boroxol rings in this glass, justifying the boroxol ring approximation used in the above description of the alkali-borate glass.

The stœchiometry of vitreous boron oxide (a III-VI compound) suggested long ago that its structure could be represented as a three-dimensional continuous random network (CRN) with BO_3 triangles, which was confirmed by X-ray and neutron diffraction [56, 57, 69]. Indeed, the diffraction patterns show two very sharp peaks at 1.37 and 2.37 Å , corresponding to the $B-O$ and $O-O$ distances in the equilateral BO_3 triangle, whereas other much broader peaks clearly suggest that there are no edge-sharing units in the glass (like a four-membered ring, B_2O_4). The Raman spectra which exhibit a high vibrational peak at $808cm^{-1}$ (attributed to the breathing mode of the oxygens inside the boroxol rings [58, 11], and whose intensity and sharpness is comparable to those of usual crystals, suggest that the network is mainly composed of randomly oriented boroxol rings connected by oxygen atoms or BO_3 triangles.

In the previous section we have already cited the articles describing the experiences that seem to confirm this hypothesis. For example, Jellison and co-workers [4] who applied $^{17}O - NMR$ spectroscopy to B_2O_3, constructed a model of oxygen sites inside and outside boroxol rings, which explained that the best agreement with experience can be obtained assuming that as much as 83% of the borons are trapped inside the boroxol rings

^{18}O enriched compounds have been also analyzed in order to show that the breathing mode of the oxygens inside the boroxol rings at $808cm^{-1}$ is strongly modified by the isotopic substitution and follows the rate of the isotope[59]. Similar work has been done on vitreous and molten B_2O_3 by Walrafen et al. who investigated the temperature dependence of the integrated Raman intensity of the $808cm^{-1}$ peak [40]. They have observed that the sharp peak disappears with increasing temperature, suggesting a general breakdown of the boroxol groups in the liquid. These authors have also shown that the formation of a boroxol ring might be energetically favoured and evaluated a corresponding formation energy of 5 $kcal.mol^{-1}$ [41].Also neutron and inelastic neutron diffraction results can be satisfactorily interpreted by assuming a high fraction of boroxol rings; Johnson et al. have concluded the rate being of 60% [42]; Hannon et al. give the evaluation of $80 \pm 5\%$ [43, 44].

Controversy around boroxol rings has started with the publication of numerous Molecular Dynamics (MD) simulations in ten recent years [60, 62, 68]. Various modelling methods of MD have been used, such as two-body potentials, three-particle interactions with the bond angle maintained at 120° or 130°. All of them, except one, predict the absence of boroxol units in the network, or at least, a very small number.

Our aim is to evaluate the relative amount of boron atoms trapped in the boroxol rings during the glass transition. Besides, the boroxol rings are the only reliable characteristic structure whose statistics can be followed during the agglomeration of BO_3 triangles. We shall follow the geometrical construction of all possible small and medium-size clusters containing a given number of atoms (up to 15 borons and 20 oxygens), starting from the most elementary configuration for which there exists clear and unambiguous experimental evidence, i.e. the BO_3 triangle and which will be called a "*singlet*" in the forthcoming construction. During the agglomeration of the singlets, the boron atom and its three oxygen bonds form oxygen bridges with next borons.

Although we can safely assume that the network can be entirely tiled with these singlets, the vitreous matrix of B_2O_3 which has undergone a glass transition can be tiled also with larger agglomerates produced by our construction, each step of agglomeration yielding larger clusters and a better approximation, because the final medium-range structures we shall produce contain obviously more significant information about the typical structure of B_2O_3 than the initial doublets and triplets.

Whatever the construction step, contribution to the configurational entropy coming from a given cluster can be evaluated with a satisfactory approximation by computing the number of independent agglomeration pathways leading to that particular cluster.

Besides the entropy contribution, we can assume that the energy stored in a given cluster can be treated as an additive quantity and can be evaluated by summing the energies involved in each elementary step of agglomeration consisting in the formation of a new single oxygen bridge or of two simultaneous bridges (corresponding to the creation of a boroxol ring).

In this construction we don't exclude the formation of a random network with BO_3 triangles or a boroxol linked network. The processes leading to such structures are characterized by two unknown binding energies we shall evaluate later on. Each single oxygen bridge creation should represent the energy cost of E_1 and each boroxol creation, the energy E_2.

Starting from a BO_3 singlets, we can create a doublet (Fig. 13a), excluding the possibility of formation of a two-membered ring (i.e. edge-sharing triangles) for which there is no experimental evidence in X-ray results,[56]. Therefore, the probability of the doublet after normalization is $p_{2A} = 1$, and we can tile the entire network with such doublets.

During the second step ($l = 2$), two possibilities must be taken into account, i.e. formation of a pure chain or of a boroxol ring (Fig. 13b).

Besides the energies E_1 and E_2 that characterize these two processes, the two triplets have a different multiplicity, which corresponds to the number of ways in which a given triplet can be constructed. A boroxol group has the multiplicity $2 \times 3 \times 4 = 24$, whereas for a chain we count $3 \times 4 = 12$ (Fig. 13).

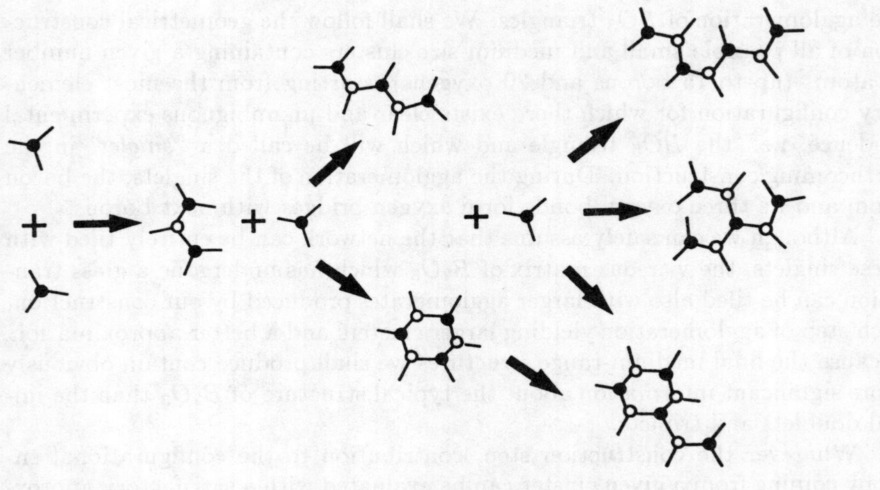

Figure. 13 The first three steps of agglomeration of BO_3 singlets: a) a doublet; b) two triplets, a chain and a boroxol ring; c) three different quadruplets

We evaluate the probabilities of these two structures exactly as before, with respective binding energies E_1 for the creation of a simple oxygen bond, and E_2 for the creation of a boroxol ring.

$$p_{3A} = \frac{12e_1}{12e_1 + 24e_2} p_{2A} = \frac{1}{1 + 2e_2/e_1} \tag{20}$$

$$p_{3B} = \frac{24e_2}{12e_1 + 24e_2} p_{2A} = \frac{2e_2/e_1}{1 + 2e_2/e_1} \tag{21}$$

where $e_1 = e^{-E_1/kT]}$ and $e_2 = e^{-E_2/kT}$.

The next step leads to three different clusters: their multiplicities and the Boltzmann factors e_1 and e_2 of each pathway of production are shown in figure 4 and the normalized probabilities are given below:

$$p_{4AA} = \frac{12e_1}{15e_1 + 24e_2} p_{3A} = \frac{4}{5 + 8e_2/e_1} p_{3A} \tag{22}$$

$$p_{4A3} = \frac{3e_1}{15e_1 + 24e_2} p_{3A} = \frac{1}{5 + 8e_2/e_1} p_{3A} \tag{23}$$

$$p_{4AB} = \frac{24e_2}{15e_1 + 24e_2} p_{3A} + \frac{9e_1}{9e_1} p_{3B} = \frac{8e_2/e_1}{5 + 8e_2/e_1} p_{3A} + p_{3B} \tag{24}$$

We can go on with this construction and for each new step produce a set of clusters with the corresponding probabilities. Each new step l costs at least one new binding energy, so that the stored energies contained in a given cluster are given by the following accessible states:

$$E_j^{(l)} = (l - j)\, E_1 + j\, E_2 \quad \text{with} \quad j = 0, 1, .. \frac{l+1}{3} \tag{25}$$

Note that these construction depends only on the energy difference $E_1 - E_2$ (via the factor e_2/e_1), and that the quadruplets displayed in figure 13 involve only two characteristic energies, $E_I^{(3)} = 3E_1$ and $E_{II}^{(3)} = 2E_1 + E_2$. We have followed the agglomeration up to the 10th step yielding 288 non-redundant multiplets with eleven boron atoms each (Table I):

Step l	Number of multiplets	Number of multiplets with: 1 boroxol	2 boroxols	3 boroxols
0	1	-	-	-
1	1	-	-	-
2	2	1	-	-
3	3	1	-	-
4	4	2	-	-
5	9	4	1	-
6	14	6	2	-
7	31	14	6	-
8	62	30	13	1
9	135	60	34	4
10	288	129	74	15

Table I: The number of multiplets at each step of the agglomeration process, with number of configurations containing 1, 2 *or* 3 *boroxols.*

Some of them contain one, two or three boroxol rings; some other contain no rings at all and are composed of pure or ramified chains. Rings with more than four boron atoms should also be expected in our construction but we have neglected them for the sake of simplicity. Nevertheless, such rings can be represented by a pure chain, because it seems reasonable to assume that the energetic cost of formation of such rings is approximatively twice the energy E_1 (during the lth step, a pure chain with $(l-2)$ borons is "*closed*" by a BO_3 triangle).

Such an assumption becomes more and more realistic if one considers rings of growing size (the effect of the ring decreases with its size, i.e. the influence of a 3-membered ring is more significant than the one produced by a 12-membered ring). Therefore, the main structural influence should be produced by the boroxol rings represented by energy E_2; two examples of clusters with eleven boron atoms are given in Fig. 14.

The first one is a typical example of an artificial ring (or pure chain), whereas the second one contains three boroxol groups.

In order to evaluate the rate of boron atoms contained in the boroxol rings among all the configurations obtained after l steps of agglomeration, we need to substitute the right values of Boltzmann factors $e1$ and $e2$ into all the probabilities. Therefore, we must know both the glass transition temperature T_g and the energies E_1 and E_2 (in fact, only their difference appears in the cluster probabilities).

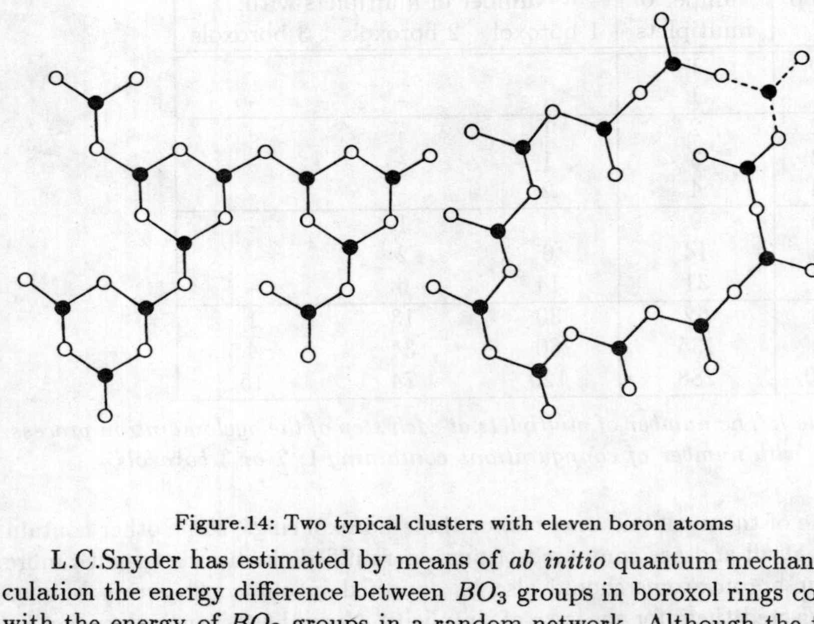

Figure.14: Two typical clusters with eleven boron atoms

L.C.Snyder has estimated by means of *ab initio* quantum mechanical calculation the energy difference between BO_3 groups in boroxol rings compared with the energy of BO_3 groups in a random network. Although the first approximation gave a relatively low value of about $\Delta E = 1.5\ kcal.mol^{-1}$, the corrections taking into account the electron correlation energy raised this value up to 6 $kcal.mol^{-1}$ [71].

Krogh-Moe has proposed the stabilization energy of a boroxol unit to be roughly 8 $kcal.mol^{-1}$ after discussing the reorganization of BO_3 units from boroxol rings into a random network of BO_3 triangles [50]. Walrafen and coworkers have proposed the value 6.4 ± 0.4 $kcal.mol^{-1}$, which has been deduced from the investigation of vitreous and molten B_2O_3 via computation of the dependence of the integrated Raman intensities on the temperature [40].

The glass transition temperature is a function of thermal history of the melt which is is highly influenced by the cooling (or quenching) rate. The usual range of T_g for B_2O_3 is known to be approximatively $470 - 530^oK$. Trying to evaluate the factors $e1$ and $e2$, we get different values following the choice of T_g.

We have chosen the mean value of (5.3 $kcal.mol^{-1}$), close to the most recent result which has been obtained by Walrafen *et al.* Indeed, they have estimated the energy of formation of a boroxol unit as being roughly equal to 5.0 $kcal.mol^{-1}$ [41, 19].

With the energy difference established and the glass transition temperature fixed, we can compute the fraction $F_{(l)}$ of boron atoms trapped inside the boroxol rings by summing the probabilities of clusters containing this characteristic structural group, with an appropriate weight, e.g. 3/4 for the configuration p_{4AB} of Fig. 13 or 9/11 for the structure displayed in figure 14, etc. The results show the convergence to the limit value of 80%; the continuous curve extrapolating the behaviour of $F_{(l)}$ versus l (from 1 to 10) is plotted in the region I of Fig. 15. The obtained rate is consistent with various theoretical or experimental results cited above, especially with the models of Jellison et al. who suggested that 83% of boron atoms are contained in boroxol rings [4] and Hannon et al.(80%) [43, 44].

The evaluation of the boroxol rate and its dependence on T shows also that for $T = T_g$ the variation of $F_{(l)}$ is very low; note that this happens independently of the agglomeration step we are considering. When the liquid freezes at T_g, all structural recombinations of clusters are stopped (or, at least become negligibly slow); this is confirmed by our computation, in which our functions $F_{(l)}$ remain constant from $T = -200^oC$ up to T_g consistently with the experimental data of Walrafen [40, 41].

However, at this stage of agglomeration, the amplitude of the variations in the boroxol rate computed for different steps, even for the last ones (step 9-10 in figure 15) remains quite large. Therefore, it seems necessary to go farther in the agglomeration process, in order to evaluate the boroxol rate from larger clusters in a more precise and conclusive manner.

For the ultimate step of agglomeration ($l = 10$) that we were able to follow exactly, we have got 288 non-redundant multiplets, and it seems quite impossible to perform the analysis of the agglomeration process further on by direct joining of BO_3 triangles "by hand". Indeed, the forthcoming step $l = 11$ would have brought at least thousand non-redundant multiplets with twelve boron atoms which seem very difficult to follow without many errors. But at this stage of the agglomeration ($l = 10$), we can already compute all the probabilities of 288 different clusters by inserting the average value of T_g value and the best fit for the difference $E_1 - E_2$ based on the estimations cited above.

Fortunately enough, among the 288 clusters of the $10 - th$ agglomeration step, only very few are really significant, i.e. have a non-negligible probability. For example, the clusters displayed in Fig. 14 have very different probabilities. The pure chain shown in 14*a* has the probability of 10^{-19} whereas the second structure in 14*b* appears with the 37.8% rate !

This is why we can keep only the most representative clusters of 10-th step and start a new agglomeration by direct joining of BO_3 singlets to them. Of course, the sum of the probabilities of thus reduced number of clusters should be close to 99% in order to minimize the error produced by the approximation.

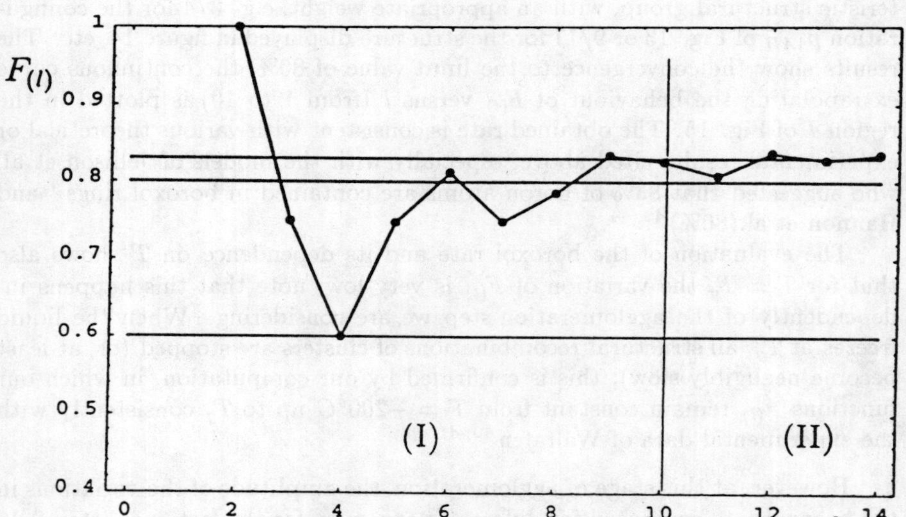

Figure. 15: The boroxol rate $F(l)$ in clusters with l boron atoms

In order to test the validity of such an approximation, we have started the multiplet approximation at step $l = 9$ which has a population of 135 multiplets (Table I). With the value $E_1 - E_2$ obtained in [4, 40], only five of these clusters represent already 99.11% of all the probabilities.

We can now realize a new step of agglomeration (called 10*), computing the probability of the twelve structures with eleven borons, obtained from the five most important structures of previous step, and compare them with the probabilities of the direct computation based on the 288 multiplets of step 10 (Table II).

Step l	Number of multiplets	Number of multiplets with: 2 boroxols	3 boroxols
9	135	34	4
	5	3	2
10*	12	7	5
10	288	74	15

Table II: The number of different structures obtained by the multiplet approximation as compared with the directly computed ones

We have checked that the differences between the probabilities of new clusters obtained via the approximate and exact way of agglomeration (10 and 10*) are infinitesimal indeed, which confirms the validity of such an approximation, as long as one keeps the multiplets that totalize about 99% of all the probabilities.

Having tested the validity of this approximation, we can now confidently apply it further on and continue the agglomeration process by constructing the multiplets with 12 boron atoms, starting from the six most representative multiplets at step 10.

A second multiplet approximation can be realized in order to obtain the structures with fifteen boron atoms during the steps labelled 13** and 14** (Table III) if we start from the 11 most representative clusters with 13 boron atoms (which totalize 99.10% of all the probabilities).

With these new steps of agglomeration, we are able to compute again the fraction of borons trapped inside the boroxol rings, following the same scheme as explained above. The result is plotted in the region II of Fig. 15 and shows for the steps 11* to 14** a much better convergence to approximately 84%, remaining in good agreement with the value established by Jellison et al.[4] and remains inside the evaluation range of Hannon et al.[43, 44].

The variation of $F_{(l)}$ with respect to l is also decreasing, as shown in the Fig.15.

In section 7, where we develop a non-standard statistical theory based on the *agglomeration pathways* instead of *states* in the statistical sum, we shall give an independent argument in favour of this result.

Step l	Number of multiplets	Number of multiplets with: 2 boroxols	3 boroxols	4 boroxols	5 boroxols
10	288	74	15	-	-
	6	1	5	-	-
11*	17	4	11	2	-
12*	39	8	26	5	-
	11	-	8	3	-
13**	32	-	23	10	-
14**	93	-	62	29	2

Table III: The number of structures sharing from 11 to 15 boron atoms, using two consecutive multiplet approximations

Before we proceed further, we would like to mention briefly another pure glass with a pronounced ring structure [66, 17], which is the amorphous SiO_2. Although the ring structure of SiO_2 is very rich, and the significant rings range from the smallest ones, containing three Si atoms, to the biggest significant entities composed of seven or even eight Si atoms. Moreover, rings can join into various clusters displaying different geometries, because the elementary building block is now a $Si(O_{\frac{1}{2}})_4$ tetrahedron (Fig. 16). F.L.Galeener [66] has estimated on the basis of different experimental data that the smallest 3-rings in the amorphous SiO_2 contain about 3% of all the Si atoms. The first such rings start to appear during the second step of agglomeration (Fig.16). If the rate of the Si atoms contained in these smallest (and very stiff, therefore energetically costly) rings is lower than 3%, then certainly it cannot exceed this value at the very beginning, i.e. during the first agglomeration step. This enables us to set a simple equation from which we can evaluate the energy cost of formation of a 3-ring in the amorphous SiO_2. Let the energy related to a simple $Si-O-Si$-bond formation be E_1, and the energy cost for a 3-ring creation be E_3. Denoting the corresponding Boltzmann factors by $e_1 = e^{-\frac{E_1}{kT}}$ and $e_3 = e^{\frac{E_3}{kT}}$, we can easily find the probabilities of two elementary processes shown in Fig.16, p_{3A} for the chain, and p_{3B} for the 3-ring:

$$p_{3A} = \frac{24e_1}{24e_1 + 108e_3}, \quad p_{3B} = \frac{108e_3}{24e_1 + 108e_3} \tag{26}$$

Now, the condition of having no more than 3% of atoms in 3-rings reads:

$$p_{3B} = \frac{108e_3}{24e_1 + 108e_3} = \frac{9e_3}{2e_1 + 9e_3} \leq 0.03 \tag{27}$$

which leads to $e_3/e_1 \leq (2/291)$, and finally, to the evaluation

$$\frac{(E_3 - E_1)}{kT} \geq -ln(\frac{2}{291}) = 4.98 \tag{28}$$

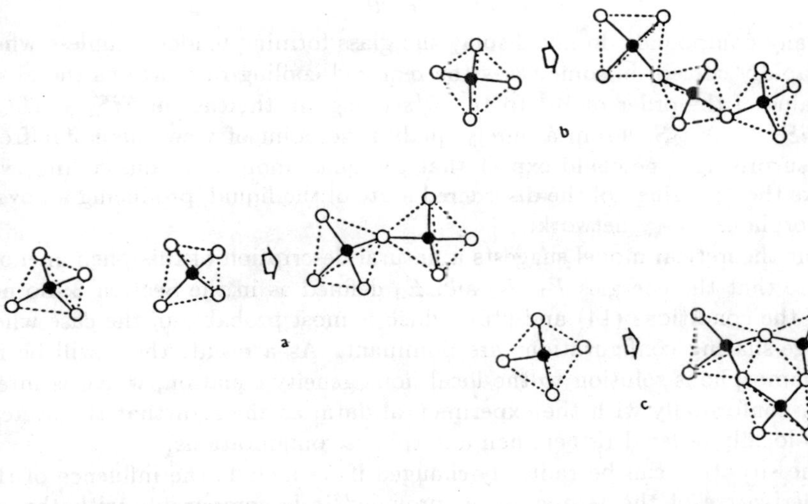

Figure. 16: The first two steps of agglomeration of SiO_2: a) the doublet; b) a chain; c) a three-ring

At $T = T_g \sim 1500K$, this gives $E_3 - E_1 \geq 0.698eV$. One should keep in mind that the energy E_3 contains already the two contributions E_1 corresponding to the formation of two $Si - O - Si$ bridges, plus the extra energetic price for the 3-ring creation. This energy is relatively high because of a pronounced *planar* character of such a ring, in which case the angle of oxygen bridges is close to 120°, while it is known that the energy is minimal in such bridges at the angle of about $135^o - 140^o$ [66].

In section 7 we shall derive this result in an alternative manner, confirming the above estimation quite independently.

6 Theoretical Model of Rapid Quenching

As mentioned already in Sections 2 and 3, the evaluation of the time derivative of the probabilities as

$$\frac{dP_k}{dt} = \frac{ds}{dt}[P_k^{(1)} - P_k^{(0)}] \tag{29}$$

is valid *only* if the temperature (on which the expressions $P_k^{(1)}$ depend via the Boltzmann factors contained in each partial probability) can be considered as constant during the agglomeration process, which means that we must have

$$\frac{ds}{dt} \gg \frac{1}{T}\frac{dT}{dt} \tag{30}$$

Many compounds do not display the glass-forming tendency unless when very rapidly cooled; in some cases the required cooling rate attains the absolute value of the order of 10^4 to 10^5 K/sec, e.g. in the case of $SiS_2 + nLi_2S$ and $SiS_2 + nNa_2S$. From a purely qualitative point of view this is far from being surprising: one could expect that the quasi momentaneous cooling will provoke the "freezing" of the disordered state of the liquid, producing an overall amorphous glassy network.

Our theoretical model suggests a natural description of this phenomenon. Suppose that the energies E_1, E_2 and E_3 defined as in the Section 3, do not satisfy the conditions (14) and (15), which is most probable in the case when the edge-sharing configurations are dominant. As a result, there will be no stable amorphous solution to the local homogeneity equation, which is interpreted, conformally with the experimental data, as the sign that the system prefers locally ordered rather than amorphous configurations.

The situation can be radically changed if we include the influence of the time derivative of the temperature, provided it is comparable with the inverse of the characteristic time of elementary agglomeration process. Then our equation should take on the following form:

$$\frac{dP_i}{dt} = \frac{ds}{dt}(P_i(s_{k+1}) - P_i(s_k)) + \frac{dT}{dt}(\frac{\partial P_i}{\partial T}) \tag{31}$$

Let us apply this formula to the simplest model with the star-like entities of two kinds, with coordination numbers m for the glass-former and m' for its modified atoms; their respective probabilities at each agglomeration step (k) will be denoted by $P_A^{(k)}$ and $P_B^{(k)}$.

We shall restraint our analysis for low values of modifier's concentration, which makes the formulae much simpler. We shall also consider the two limit cases of pure corner-sharing or pure edge-sharing tendencies, corresponding to the extremal choices of the parameter β, as discussed in Section 3.

In what follows, we keep the same notation as before, i.e. we use the Boltzmann factors $e^{-\alpha_1}, e^{-\alpha_2}, e^{-\alpha_3}$ for the agglomeration of the type $A-A$, $A-B$ and $B-B$, respectively, and we introduce the extra Boltzmann factor $e^{-\beta}$ whenever an edge-sharing configuration appears.

Let us recall that the expression for the time derivative of probabilities $P_k, k = A, B$, without taking into account dT/dt was of the form:

$$\frac{dP_k}{dt} = \frac{ds}{dt}\left(P_k^{(1)} - P_k^{(0)}\right) = \frac{ds}{dt}\frac{1}{Q}\left[x\,(1-x)\,(\Phi x + \Psi)\right], \tag{32}$$

with Φ and Ψ functions of m, m' and the Boltzmann factors e^{α_i} ; Q is the normalizing factor. Recalling that the expressions for $P_k^{(1)}$ contain Q as the denominator, and the Boltzmann factors, one can easily see that the time derivative of $P_k^{(1)}$ will contain both T and Q_2 in the denominator; our "master equation" expressing maximal local homogeneity can be reduced to:

$$x\,(1-x)\,[\Phi x + \Psi]\,Q + \Lambda\,F(x; m, m', e^{\alpha_i}) = 0 \tag{33}$$

where the function F is computed separately for each particular situation, (i.e. corner-sharing or edge-sharing dominated process, modifier's concentration n close to 0, etc.), the coefficients Φ and Ψ are the same functions of the Boltzmann and statistical factors as in the examples considered above (with $dT/dt = 0$, and where the characteristic parameter Λ is defined as:

$$\Lambda := \frac{1}{T}\frac{dT}{dt}\left(\frac{ds}{dt}\right)^{-1} = \frac{d\,(ln\,T)}{ds} \tag{34}$$

and gives the logarithmic time derivative of the temperature measured in the time units $\tau = dt/ds$.

Here are the explicit forms of "master equation" in the case of a low modifier concentration, when we can neglect the terms containing $(1-x)^2$:

$$x(1-x)[(\Phi x + \Psi)Q + \Lambda(\alpha_1 - \alpha_2)m^3 m' x^2 e^{-\alpha_1 - \alpha_2}] = 0 \tag{35}$$

for the pure *corner-sharing* configurations, and

$$x(1-x)[(\Phi x + \Psi)Q + \frac{1}{2}\Lambda(\alpha_1 - \alpha_2)m^3(m-1)^3 m'^3(m'-1)^3 x^2 e^{-2(\alpha_1+\alpha_2+\beta)} = 0 \tag{36}$$

for the case when the *edge-sharing* configurations are dominant.

The coefficients Φ and Ψ are different in both cases, and depend on the parameters α_1, α_2 and β; we don't write them explicitly here for the sake of simplicity. As before, there are always two solutions, $x = 0$ and $x = 1$, when we require these expressions to vanish. In many cases however there is no solution in between, or it turns out to be unstable, unless $\Lambda \neq 0$. The expression for Λ which makes the stable solution $0 \leq x_{am} \leq 1$ possible is particularly simple in the limit of low modifier concentration, i.e. when $x_{am} \to 1$. We have then:

$$\Lambda = -\frac{m\,e^{\alpha_2 - \alpha_1} - m'}{(\alpha_1 - \alpha_2)\,m'} \tag{37}$$

for pure corner-sharing agglomeration, and

$$\Lambda = -\frac{m(m-1)\,e^{2(\alpha_2-\alpha_1)} - 2\,m'(m'-1)}{(\alpha_1-\alpha_2)\,m'\,(m'-1)} \tag{38}$$

for pure edge-sharing agglomeration.

Now we can apply our model to various glasses, some of which can not be obtained without rapid quenching. First of all, we can easily determine under which conditions on the parameters ($\alpha_1, \alpha_2, \alpha_3, m, m'$ and β), and without any quenching term (i.e. $\Lambda = 0$) the crystallisation spontaneously occurs ($P_A^{(0)} = 0$ or $P_A^{(0)} = 1$ attractive) , or a disordered structure is preferred ($P_A^{(0)} = 0$ and $P_A^{(0)} = 1$ repulsive).

These conditions depend very strongly on the possibility of formation of edge-sharing doublets, i.e. on different limit values of the parameter β which seems to play a key role here (Tab.IV).

	System III-VI (m=3, m'=4)	System IV-VI (m=4, m'=3)	typical network former satisfying the condition
$\beta \ll 0$	no glass	formation	SiS_2, $SiSe_2$
$\beta \sim \alpha_k$	$\alpha_3-\alpha_2 > -0.062$ $\alpha_1-\alpha_2 > 0.604$	$\alpha_3-\alpha_2 > 0.604$ $\alpha_1-\alpha_2 > -0.062$	B_2S_3, GeS_2 $GeSe_2$,...
$\beta \gg \alpha_k$	$\alpha_3-\alpha_2 > 0.285$ $\alpha_1-\alpha_2 > -0.288$	$\alpha_3-\alpha_2 > -0.288$ $\alpha_1-\alpha_2 > 0.285$	B_2O_3, SiO_2 GeO_2,...

Table IV: Conditions on $\alpha_3-\alpha_2$ and $\alpha_1-\alpha_2$ for amorphisation (the glass formers in the table are put into the corresponding category following the experimental data concerning the existence of edge-sharing doublets)

It appears that the possibility of amorphisation is mainly favoured by the absence of edge-sharing doublets (four-membered rings). For great amounts of these rings ($\beta \ll 0$), one notes that is is impossible to favorize the intermediate singular point $P_A^{(0)} = x_{am}$. This corresponds to the systems whose network former is SiX_2 ($X = S, Se$) with which a glass can not be obtained without high quenching rates[52] and whose network structure is almost exclusively composed of very long parallel chains of edge-sharing tetrahedra SiX_4, (this is especially true for the thiosilicate glasses [51] (identified in our model with the case $\beta \ll 0$). The model predicts strong crystallisation tendency of these systems.

There are other binary systems in which the four-membered rings can be found, but not with the same abundance as in silicon sulfide and silicon selenide based glasses. We put them in the intermediate category ($\beta \sim \alpha_k$). The structure of B_2S_3 and GeS_2 also seem to match with such a condition on β because they are not only composed of four-membered rings but also of chains and 6-membered rings [36], especially the thiogermanate and the $GeSe_2$ based systems whose structure is believed to be composed of parallel chains which are cross-linked by edge-sharing tetrahedra and which contain also six-membered rings, forming even bigger clusters [37, 35].

In principle, these glasses could be formed, although it is very difficult to obtain them in the experiments with low quenching speed at low modifier concentrations. This probably means that the factors α_1, α_2 and α_3 do not satisfy the amorphisation conditions displayed in Table IV ($\beta \sim \alpha_k$) in GeX_2 ($X = S, Se$) based glasses. For the B_2S_3 based systems, the glass formation for low concentrations n depends on the nature of the modifier. The conditions displayed in Table IV could therefore be satisfied for $B_2S_3 - nNa_2S$ systems which form glasses quite easily for $0 \leq n \leq 1$, whereas the $B_2S_3 - nLi_3S$ system produces separate phases and do not form glass with any concentration of modifier (which means that the conditions on α_1, α_2 and α_3 are not satisfied).

For the compounds in which the creation of four-sided rings is excluded ($\beta \gg \alpha_k$), the formation of a disordered network seems easy and the energetic constraints (Tab.IV) are not too restrictive (typically, such glasses are the borate glasses for $III - VI$ systems obtained very easily, and the silicate and germanate glasses for $IV - VI$ system [31, 39]). In order to include all the above cases, we must go one step farther, when the six-sided rings (like the boroxol rings) start to appear.

The second step of the agglomeration yields 22 triplets of four different types, constructed with the previously obtained doublets and singlets (Fig.17).

The triplet 4 corresponds here to a six-sided ring which is observed in several glasses of the III-VI and IV-VI systems. We denote by γ the the energy difference taking into account the creation of such a triplet, similarly to the definition of the factor β. Some of the new explicit expressions of probabilities are displayed below (we show an example of probability expression for each of four particular kinds of triplets):

$$p^{(2)}_{AAA1} = \frac{2\,(m-1)m}{Q_2}\, P^{(0)}_A\, p^{(1)}_{AA1}\, e^{-\alpha_1}$$

$$p^{(2)}_{AAB2} = \frac{(m'(m'-1)(m-1)(m-2)}{Q_2}(1 - P^{(0)}_A)\, p^{(1)}_{AA1}\, e^{-\beta} + \frac{(m-2)m}{Q_2}\, P^{(0)}_A\, p^{(1)}_{AB2} e^{-\alpha_1}$$

$$p^{(2)}_{BBB3} = \frac{m'(m'-1)(m'-2)(m'-3)}{Q_2}(1 - P^{(0)}_A) p^{(1)}_{BB2} e^{-\beta}$$

$$p^{(2)}_{AAB4} = \frac{m'(m'-1)(m-1)^2}{Q_2}\,(1 - P^{(0)}_A)\, p^{(1)}_{AA1} e^{-\gamma} + \frac{m(m-1)^2(m'-1)}{Q_2}\, P^{(0)}_A\, p^{(1)}_{AB1} e^{-\gamma}$$

where Q_2 is the new overall normalizing factor. The probability of finding an A-configuration among the doublets is now constructed with the relative probabilities $P^{(1)}_A$ and $P^{(2)}_A$:

$$P_A(t) = (s(t)-1)P^{(2)}_A + 2(1 - \tfrac{s(t)}{2})P^{(1)}_A$$

leading to:

$$\frac{dP_A}{ds} = [P^{(2)}_A - P^{(1)}_A] \tag{39}$$

Of course, if we put $P^{(2)}_A - P^{(1)}_A) = 0$, this means that the former equation $P^{(1)}_A - P^{(0)}_A = 0$ is not valid anymore; the stationary régime is now supposed to occur at the second step, and not at the first. The singular boundary solutions of this equation are the same as for the first step: $P^{(0)}_A = 0$ or $P^{(0)}_A = 1$, whereas the intermediate solution is too complicated to be written down explicitly. Here again, we study the dominant agglomeration tendency of the system by performing the linearization of equation in the vicinity of its singular solutions [19].

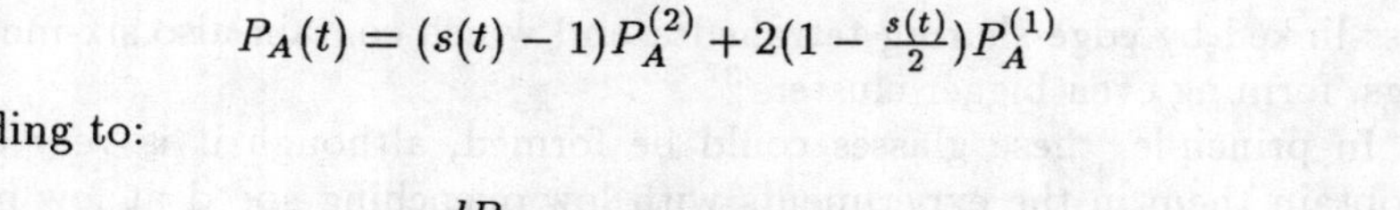

Figure. 17: The four types of triplets in III-VI systems; the corresponding probabilities are: $p^{(2)}_{AAA1}$, $p^{(2)}_{AAB2}$, $p^{(2)}_{BBB3}$ and $p^{(2)}_{AAB4}$

The energy constraints of the first two steps, explicited above, show that the intermediate solution (corresponding to the glass-forming tendency) sometimes can not be found, especially for the compounds having low formation energy of four-membered rings ($\beta \ll 0$ and $\beta \sim \alpha_k$). This is confirmed by the experimental data on the SiS_2 and $SiSe_2$ based glasses, and more generally, the silicium and germanium chalcogenides are very difficult or impossible to obtain without high quenching rates (sometimes reaching values as high as $10^6 K.s^{-1}$), especially at low concentrations of the modifier.

As explained in the beginning of this section, the use of such quenching rates imposes a new term in the equations (5). We must now take into account the derivative dT/dt which has become non negligible. Consequently, the equations defining singular points are modified[19]:

$$\frac{dP_A}{dt} = \frac{\partial P_A}{\partial s}\frac{ds}{dt} + \frac{\partial P_A}{\partial T}\frac{dT}{dt} = 0 \tag{40}$$

leading to:

$$\frac{dP_A}{dt} = \frac{ds}{dt}[P_A^{(i)} - P_A^{(i-1)}] + \frac{dT}{dt}\frac{\partial P_A^{(i)}}{\partial T} = 0 \tag{41}$$

where $i = 1, 2$ in order to analyze the first and the second step alike. The equations determining the singular solutions $dP/dt = 0$, can be interpreted as:

$$\frac{ds}{dt}\frac{\partial P_A}{\partial s} = -\frac{dT}{dt}\frac{\partial P_A}{\partial T} \tag{42}$$

with the right-hand side acting as an external "force" that is needed in order to drive the system into an otherwise impossible equilibrium. In this equation, the term dT/dt is the time derivative of T and it represents the rapidity of the temperature decrease during the quenching. If $|dT/dt|$ is very large (fast cooling which incidentally supposes a high initial value of T), $\partial P_A/\partial T$ might be relatively small, or, *vice versa*, the same quenching effect can be obtained with slower cooling rates if $\partial P_A/\partial T$ is bigger. Because $\partial P_A/\partial T$ depends indirectly on formation energies E_{AA}, E_{AB} and E_{BB} and the particular geometrical properties of the assembling rules, the approximate evaluation of the term $\partial P_A/\partial T$ can be made to throw some light on the minimal values of dT/dt that would assure *amorphisation*, i.e. the emergence of an intermediate solution for P_A contained between 0 and 1.

In this equation, where $\frac{ds}{dt}\frac{\partial P_A}{\partial s}$ is supposedly of the same order that $\frac{dT}{dt}\frac{\partial P_A}{\partial T}$, we have defined two important functions: $a(t)$ represents the average speed of agglomeration in the melt, i.e. the derivative of the step function $s(t)$ with respect to time t; $\mu(t)$ represents the quenching rate which must be *negative* (the amorphous solution should be favoured by *cooling* the melt).

At the first step, the equation (41) has always the solutions $P_A^{(0)} = 0$ and $P_A^{(0)} = 1$, whereas the intermediate solution $P_A^{(0)} = x_{am}$ may also appear, but depending on the value of the ratio:

$$\Lambda(T,t) = \frac{\mu(t)}{Ta(t)} \tag{43}$$

The Table V shows typical values of the quenching parameter Λ obtained for different choices of the parameters α_k, $(k = 1, 2, 3)$, and β, that allow the intermediate (amorphous) solution for P_A to appear.

α_1 α_2 α_3	$\beta \ll 0$			$\beta \sim \alpha_k$		
	$\Lambda(T,t) =$			$\Lambda(T,t) =$		
	0	-10^2	-10^4	0	-10^2	-10^4
0.5 0.1 0.5	-	.396	.427	-	.640	.643
0.5 0.1 1.0	-	.489	.512	.290	.722	.724
0.5 0.1 1.5	-	.463	.487	.467	.728	.729
0.5 0.1 2.0	-	.386	.414	.538	.708	.710
0.5 0.1 2.5	-	.271	.306	.572	.676	.677
0.5 0.2 0.5	-	.243	.300	-	.639	.643
0.5 0.2 1.0	-	.419	.454	.232	.753	.755
0.5 0.2 1.5	-	.378	.415	.462	.764	.765
0.5 0.2 2.0	-	.236	.285	.544	.747	.749

Table V: Solutions $P_A^{(0)}(x_{am})$ of the first step for $\beta \ll 0$ and $\beta \sim \alpha_k$ and different values of α_1, α_2 and α_3 in the system IV-VI

For the second step, the equation (41) also yields the solutions $P_A^{(0)} = 0$ and $P_A^{(0)} = 1$, and an intermediate solution which is a function of β, γ and $\Lambda(T,t)$. The results for both steps are shown in the Table VI.

It is easy to observe that for the network formers with ($\beta \sim \alpha_k$), the quenching term $\Lambda(T,t)$ of the order of -10 is sufficient to obtain the formation of a disordered structure, if only $\alpha_3 \leq \alpha_2$, whereas in systems displaying important number of four-membered rings ($\beta \ll 0$), the quenching term which is necessary to obtain glass attains the values close to -10^4.

This situation seems to describe well the experimental production of glasses with different rates of four-membered rings [55]. The first type of glass ($\beta \sim \alpha_k$) which can be identified with B_2S_3 or GeX_2 ($X = S, Se$) based glasses, is usually obtained with air or water quenching [48, 37]; the second type, with ($\beta \ll 0, \gamma \gg \alpha_k$, or $\gamma \gg \alpha_k$), identified with SiX_2 ($X = S, Se$) based glasses, can be obtained only with twin-roller quenching with cooling rates of the order of those proposed by our model. The domains ($\alpha_1 - \alpha_2 > 0$) and ($\alpha_3 - \alpha_2 > 0$) are reached already at $\Lambda(T,t) = -10^{-2}$ for systems without edge-sharing units ($\beta \gg \alpha_k$). The identification with borate and germanate glasses proposed in the Tables IV and VI is therefore quite realistic, because the cooling rates of these glasses are of the same order [27].

The linearized equations can be used in order to define the critical cooling rate Λ_{cr}. It corresponds to the value at which the right-hand sides of the equations (40) or (41) vanish. The boundary singular point (0 or 1) which was attractive (favouring the crystallisation tendency) becomes repulsive for cooling rates satisfying $|\Lambda(T,t)| > |\Lambda_{cr}|$.

The critical cooling rate depends on the factors (α_1,α_2) or on (α_3,α_2)

respectively, depending on the initial definition of the attractive point ($P_A = 0$ or $P_A = 1$). Its expression for the first step becomes very simple when $\beta \gg \alpha_k$:

$$\Lambda^1_{cr} = \frac{1 - \frac{4}{3}e^{(\alpha_2 - \alpha_3)}}{6(\alpha_2 - \alpha_3)} \tag{44}$$

for III-VI based systems, and similarly for the IV-VI based systems,

$$\Lambda^1_{cr} = \frac{1 - \frac{3}{4}e^{(\alpha_2 - \alpha_1)}}{6(\alpha_2 - \alpha_1)} \tag{45}$$

		First	step		Second	step	
α_1	α_2	$\beta \ll 0$ $[SiS_2]$ $[SiSe_2]$	$\beta \gg \alpha_k$ $[GeO_2]$ $[SiO_2]$	$\beta \ll 0$ $\gamma \sim \alpha_k$ $[SiSe_2]$	$\beta \ll 0$ $\gamma \gg \alpha_k$ $[SiO_2]$	$\beta \gg \alpha_k$ $\gamma \sim \alpha_k$ $[GeO_2]$	$\beta \gg \alpha_k$ $\gamma \gg \alpha_k$ [?]
4	5	1.0×10^2	1.3	1.1×10^1	1.1×10^1	6.6×10^{-2}	5.3
4	6	4.5×10^1	2.2	4.2	4.3	8.9×10^{-3}	6.7
5	6	2.0×10^2	1.3	2.0×10^1	2.0×10^1	6.6×10^{-2}	3.7×10^1
5	7	9.4×10^1	2.2	8.6	8.7	8.9×10^{-3}	4.8×10^1
6	7	4.3×10^2	1.3	4.3×10^1	4.3×10^1	6.6×10^{-2}	2.6×10^2
6	8	2.1×10^2	2.2	1.9×10^1	1.9×10^1	8.9×10^{-3}	3.5×10^2
7	8	9.5×10^2	1.3	9.5×10^1	9.6×10^1	6.6×10^{-2}	1.9×10^3
7	9	4.8×10^2	2.2	4.3×10^1	4.4×10^1	8.9×10^{-3}	2.6×10^3
8	9	2.2×10^3	1.3	2.2×10^2	2.2×10^2	6.6×10^{-2}	1.4×10^4
9	10	5.2×10^3	1.3	5.2×10^2	5.2×10^2	6.6×10^{-2}	1.0×10^5

Table VI: The critical cooling rate $|\Lambda_{cr}|$ for the first step and second step in the IV-VI based systems [Network formers are put in the corresponding category (α_k,β) following the experimental data concerning the existence of edge-sharing doublets and six-membered rings]

One can observe that the values of Λ_{cr} displayed in the Table VI for different (α_1,α_2) show a behaviour which is in close correspondence with the experimental data measured in the glass preparation techniques. The formation energy of a doublet is of the order of $0.2 - 0.6$ eV, whereas the process is carried out for temperature ranges of $0.08 - 0.1$ eV (i.e. 860K - 1100K), giving the (α_1,α_2) values displayed in Table IV.

The most significant result is that in the domain covering the usual values of α_1 and α_2 the three categories of glasses corresponding to the typical values of the parameter β: $\beta \ll 0$, $\sim \alpha_k$ or $\beta \gg \alpha_k$, need the critical cooling rates Λ_{cr} whose orders of magnitude are radically different: 5×10^3 for $\beta \ll 0$, 10^1

for $\beta \sim \alpha_k$ and 10^{-2} for $\beta \gg \alpha_k$. This is consistent with the critical cooling rates which are necessary to form the silicate, germanate and borate glasses $[\beta \gg \alpha_k]$ $(10^{-4} - 10^{-2} K.s^{-1})$ [27] and those which are necessary to form the germanium and boron chalcogenides $[\beta \sim \alpha_k]$ and silicon chalcogenides [$\beta \ll 0$; usual cooling rates are of the order of $10^6 K s^{-1}$]

Let us remind that the cooling rate term Λ_{cr} displayed in the Table VI, which can vary from one type of glass to another as much as by the factor of 10^6, contains the product of three essential ingredients: the real cooling rate $\mu(t) = \frac{dT}{dt}$ expressed in $K s^{-1}$, the temperature T in expressed in Kelvins, and the rapidity of agglomeration $a(t) = \frac{ds}{dt}$ expressed in number of elementary agglomeration steps per second.

It seems reasonable to assume that the agglomeration rapidity is roughly of the same order of magnitude in all the aforementioned types of glasses during the annealing or quenching alike. As for the temperature T, it cannot vary during the glass formation process more than between 1000 K and 300 K, which does not bypass the factor of 3. This means that the variations of many orders of magnitude in the Λ_{cr} factor do reflect mostly the variations of the real cooling rate $\frac{dT}{dt}$ as observed in the experiments.

7 Statistical Physics of Non-Equilbrium Growth Process

The use of the normalizing factor of the cluster distribution at each step of agglomeration as the statistical sum, which enabled us to evaluate the energies involved in elementary agglomeration processes with and without boroxol rings being formed, can be generalized and incorporated into a more general theory. The interpretation of the statistical sum has to be modified in a drastic way, because it refers no more to a system in thermodynamical equilibrium, but to an off-equilibrium process, which might be irreversible.

With the physical assumptions enumerated in the Section 2, we can describe a hot melt at constant temperature as a "soup" of clusters of different sizes and shapes, although one can safely assume that before the glass transition, when the temperature is still above the value T_g and the viscosity is still not too high, there is a sharp cut-off in the size of clusters at a given $k_{max}(T)$, so that the number of clusters containing more than k_{max} atoms is negligibly small in the hot liquid. Of course, in the state of quasi-equilibrium at constant temperature some of the clusters split into smaller units, while some other agglomerate giving bigger entities, but on the average the balance is kept constant, and the mean distribution of clusters on the sheets of the configuration space (Fig.1) remains unchanged.

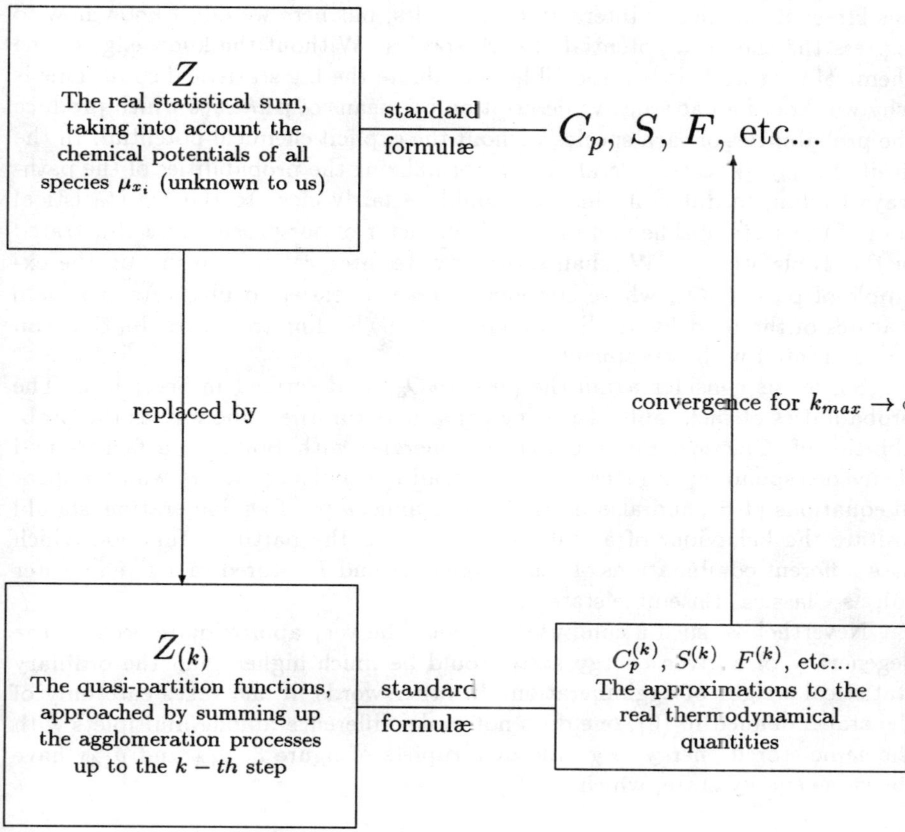

Table VII: The alternative ways of evaluating C_p, S, F, and other thermodynamical functions

When the temperature goes down slowly, the distribution of probabilities of finding the proportions of clusters of given type on the k-th sheet also evolves: the lower the temperature, the greater the "population" of sheets corresponding to higher values of k. It is quite obvious that the "centre of gravity" of the overall distribution will drift towards the sheets with greater k when the temperature continues to go down; parallely, the cut-off size k_{max} is growing, too.

The situation is therefore similar to the one observed in a mixture of various kinds of chemically interacting molecules; but here we don't know how to express the chemical potential of each species. Without the knowledge of the chemical potentials it is impossible to evaluate the big statistical sum. This is why we choose an alternative description by means of *pathways* which produce the probabilities of each species without the explicit chemical potential. In the limit of $k_{max} \to \infty$ the overall factor normalizing the probabilities of the pathways leading to different clusters should be fairly close to the big statistical sum of the system. The non-standard character of our procedure is illustrated in the Table VII: We shall show how we interpret this theory on the example of pure B_2O_3, whose simplicity makes it easier to illustrate the main features of the model with direct computations leading to the results that can be confronted with experiment.

So, let us consider again the pure B_2O_3, as described in Section 5. The probabilities of each multiplet being constructed in the same way as the probabilities of 'Classical Ensemble states' energies with Boltzmann factors and their corresponding degeneracy, the overall normalizing factors which appear in equations (1-5), and also in the forthcoming steps of agglomeration, should imitate the behaviour of a statistical sum, i.e. the partition function which uses different combinations of the energies E_1 and E_2 stored in a given cluster (6), as Classical Ensemble state energies.

Nevertheless, such a computation would be very approximate because the degeneracy of a given energy state would be much higher than the ordinary statistical weight of agglomeration. In other words, if one takes care only of the states defined in (6), one does not make difference among multiplets with the same stored energy (e.g. the quadruplets of figure 3, p_{4A} and p_{4A1} have the same energy state, which is $3E_1$).

That is why the "*partition function*" $Z_{(m)}$ should not refer to an ensemble of momentaneous states of the same system, but rather to an Ensemble of "*Evolution processes*" (or pathways) that lead to the set of all clusters of given size m.

In this sense, the obtained statistical sum does not refer to a system in an equilibrium thermodynamical state, but rather enables us to use a similar technique for a non-equilibrium process such as the glass transition we want to describe.

This means that for the second step (which yields the triplets p_{3A} and p_{3B}), the partition function has the simple form (see equations (26)- (27)):

$$Z_{(2)} \sim 12e^{-E_1/kT} + 24e^{-E_2/kT} \tag{46}$$

For the third step, the construction is quite similar:

$$Z_{(3)} \sim \prod_{\alpha=1}^{2} Q_\alpha^{(3)} = 9e_1(15e_1 + 24e_2) \tag{47}$$

For further steps, the partition function taking into account also the kinetic energy contribution, can be written on the same way:

$$Z_{(l)} \sim \left[2\pi M_{(l+1)} k_B T\right]^{3N_{(l+1)}/2} \prod_{\alpha=1}^{N_{(l)}} Q_\alpha^{(l)} \tag{48}$$

where $N_{(l)}$ is the initial number of multiplets (clusters) of the step l and $N_{(l+1)}$ the number of created multiplets during this step, $M_{(l+1)}$ is the atomic mass of a structure with $l+1$ boron atoms and $Q_\alpha^{(l)}$ is the normalizing factor of the pathway labelled α [$\alpha = 0, \ldots, N_{(l)}$]; finally, k_B denotes the Boltzmann constant. The first factor results from the usual integration of the kinetic energy term of the partition function [contribution of classical harmonic oscillators]. One can now evaluate the normalized internal energy of a given step l:

$$U_{(l)} = \frac{3}{2} k_B T + \frac{1}{N_{(l+1)}} k_B T^2 \frac{\partial}{\partial T} \sum_{\alpha=1}^{N_{(l)}} ln\ Q_\alpha^{(l)} \tag{49}$$

and the heat capacity C_p by the derivation with respect to T

$$C_p^{(l)} = \frac{3}{2} k_B + \frac{1}{N_{(l+1)}} \left(2k_B T \frac{\partial}{\partial T} + k_B T^2 \frac{\partial^2}{\partial T^2}\right) \sum_{\alpha=1}^{N_{(l)}} ln\ Q_\alpha^{(l)} \tag{50}$$

Now we can check these formulae on the example of pure B_2O_3. To do this, we shall study the set of clusters constructed above and its thermodynamical functions $U_{(l)}$ and $C_p^{(l)}$ in the vicinity of glass transition, although it seems probable that the heat capacity should take into account other contributions (e.g. the rotational or vibrational modes' contribution to the kinetic energy terms in the partition function); but we shall assume that the most important contribution to the behaviour of $C_p^{(l)}(T)$ is fairly approximated by the second and third terms of equation (50).

Although the characteristic quantities of a liquid such as α_T, C_p or κ_T (bulk compressibility) might be discontinuous at T_g suggesting a manifestation of a second-order or a more complex phase transition, we can nevertheless expect the presence of an inflexion point in the curve of the heat capacity at this temperature, which should reflect what is generally observed in the $C_p(T)$ curve for various glass-forming systems:

$$\frac{\partial^2}{\partial T^2}\left[C_p^{(l)}\right]_{T=T_g} = 0 \tag{51}$$

This simple requirement will be used in order to fit the parameter $(E_1 - E_2)$ of our model and to satisfy the equation (51). Nevertheless, the glass transition temperature is a function of the thermal history of the melt which is is highly influenced by the cooling (or quenching) rate. The usual glass transition's temperature range for B_2O_3 is known to be approximatively $470 - 530^o K$.

Trying to fit the energy difference $E_1 - E_2$ in order to satisfy the equation (51), we obtain different values following the choice varying from $T_g = 470^o K$ to $530^o K$, which suggests that our value $E_1 - E_2$ might need a correction whose value should depend on the cooling rate, but which can be neglected as far as the cooling rate is low enough in order to influence in the same way all particular agglomerations processes taken into account.

For $T_g = 470^o K$, we obtain the best fit for the steps (3-10) for:

$$E_1 - E_2 = 0.200 \; eV = 4.605 \; kcal.mol^{-1} \tag{52}$$

and for $T_g = 530^o K$:

$$E_1 - E_2 = 0.260 \; eV = 5.986 \; kcal.mol^{-1} \tag{53}$$

The normalized internal energy and heat-capacity curves based on the above numeric values of $E_1 - E_2$ are plotted in figures 18 and 19, corresponding to the agglomeration steps $k = 8$ and $k = 9$. The energy difference $E_1 - E_2$ obtained here is in fair qualitative agreement with previously cited theoretical and experimental estimates.

The same method applied to the first steps of agglomeration of pure SiO_2 yields the energy cost of creating a 3-ring equal to $0.614eV$ if we fix the inflexion point of the curve $C_p(T)$ at $T = T_g = 1500K$, which is very close to the value $0.698eV$ obtained in the previous Section by requiring that the amount of Si atoms contained in the 3-rings be lower that 3%. Here we obtain this result in a completely independent manner.

The last point we would like to investigate here is the rèal time evolution of the glass transition. To do this, let us return to the analysis of the agglomeration process, trying now to evaluate its time dependence, i.e. the way in which the characteristic features that can be evaluated statistically on the micro-clusters considered above, approach their limit values when evaluated in the resulting network after the glass transition has been completed. This will enable us to analyze the evolution of the glass transition process as function of time and temperature, assuming that the latter may be considered as a given function of time.

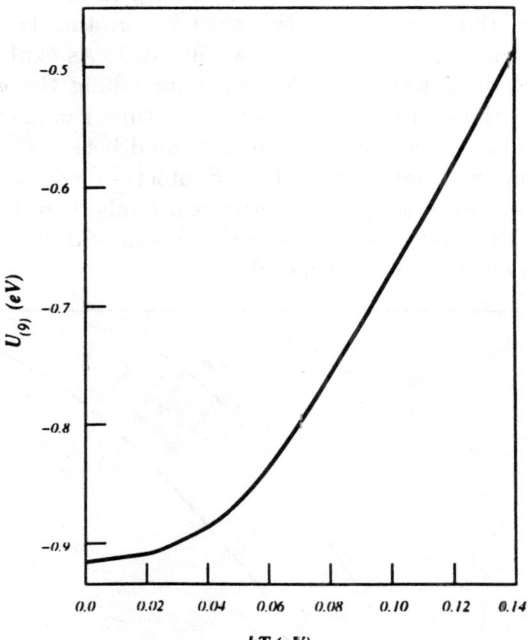

Figure. 18: The normalized internal energy $U_{(9)}(T)$, $E_1 - E_2 = 0.26eV$

In the models of glass transition based on the statistical analysis of local growth and agglomeration that are supposed to occur when the temperature of a melt decreases, we have used simple differential systems, or even algebraic systems, reflecting the properties of local homogeneity of the network. In both cases, the real time dependence of the agglomeration process remained undetermined, hidden in the unknown function $s(t)$ (the *"step"* variable).

There are many direct and indirect indications that this function must be of exponential character, i.e. behaving roughly as $s(t) \sim e^{\lambda t}$. It is well known that during the glass transition many important parameters such as density and viscosity display the Arrhenius-like behavior, i.e. their variation becomes exponentially accelerated at the approach of the glass transition, before they stabilize around the values characteristic for the glass; in some cases, more subtle effects like the Kohlrausch relaxation, can be measured [73], which lead to the functional dependence known as "stretched exponential", $e^{\lambda t^{\beta}}$, but we are not discussing them here.

One of the simplest arguments is provided by the following mechanical analogy. Suppose that we want to evaluate the average "velocity" (measured in number of steps per second), when a single step is equivalent to joining a singlet to a cluster, thus transferring this cluster from the sheet k to the next sheet $k+1$ of the configuration space shown in Fig.1. As explained in section 2, we believe that during the agglomeration modelling the glass transition process, it takes on the average the characteristic time $\tau = [ds/dt]^{-1}$ to create a new bridge; each time such a new bridge is created between one of the atoms on the cluster's rim and another singlet (or doublet, or even a bigger cluster), the cluster which belonged to the k-th sheet is propulsed to the sheet number $k+m$, where m is the number of atoms in the cluster with which a new bridge has been created (which took the time τ).

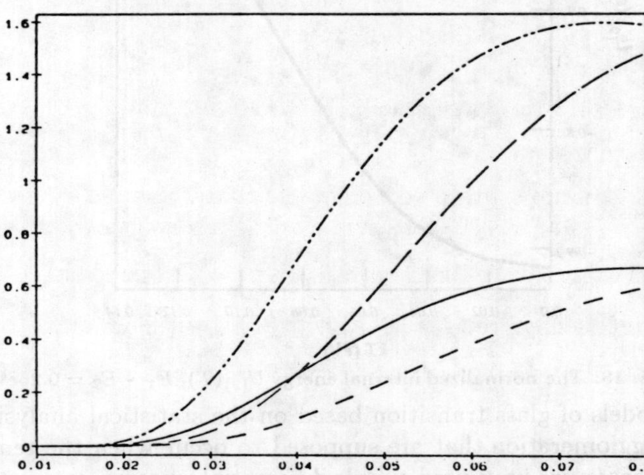

Figure. 19: Normalized heat capacity curves (arbitrary units) for two choices of ΔE : $E_1 - E_2 = 0.20eV$ and $E_1 - E_2 = 0.26eV$ a), b): the 8-th step; c), d): the 9-th step of agglomeration

If we denote by N_k the number of clusters containing k atoms, and by $N = \sum_{k=1}^{k_{max}} N_k$, then, if the average mobility of the clusters is roughly the same, and they are still small enough, so that we can neglect the steric hindrances that may later exclude many agglomeration pathways, then the probability of agglomerating with a cluster of the size m is proportional to N_m/N; this is also the probability that the cluster we are considering is propulsed towards the sheet number $k+m$, corresponding to the "velocity" of m steps per time τ.

Then the *average "velocity"* of any cluster travelling towards the sheets with higher k is easily obtained as:

$$\langle v \rangle = \frac{ds}{dt} = \lambda \frac{1}{N}(N_1 + 2.N_2 + 3.N_3 ... + k_{max}.N_{k_{max}}) = \lambda(s+1), \quad (54)$$

following the definition of the parameter s given in Section 2; λ is a proportionality coefficient depending on the nature of the melt and on the average mobility of clusters.

The differential equation thus obtained leads to the exponential behavior of the "step function" $s(t)$:

$$\frac{ds}{dt} = \lambda\,(s+1) \quad \Rightarrow \quad s(t) = A\, e^{\lambda\, t} + B \quad (55)$$

Another argument concerns any variables that are being averaged over a big set of clusters whose size is steadily growing. Suppose that we are averaging certain physical parameter, denoted by P, over the clusters containing about N atoms. Following the agglomeration and growth, during the time Δt the average size of a typical cluster changes by ΔN, and the average value $< P >$ varies by ΔP.

Let us evaluate the new average $< P > (t + \Delta t) =< P > (N + \Delta N)$:

$$< P(t+\Delta t) >= \frac{N\, P(t) + (P(t) + \frac{\Delta P}{\Delta t})\, \Delta N}{N + \Delta N} = P + \frac{(\Delta P/\Delta t)}{N + \Delta N}\Delta N; \quad (56)$$

The time variable t is dimensionless here, i.e. measured in number of units τ. After developing and neglecting ΔN before N, yields

$$\frac{\Delta P}{\Delta N} \simeq \frac{1}{N}\frac{\Delta P}{\Delta t}, \quad \text{which means that} \quad \frac{d}{dt} = N\frac{d}{dN}, \quad (57)$$

so that again we have $t \sim ln\, N$ or equivalently, $N \sim e^{\lambda t}$.

Let us show how this simple intuition is confirmed by the micro-analysis of glass transition process, as described by our model. The best worked out example of glass transition viewed as an agglomeration process that we have at our disposal is the pure B_2O_3 glass. In this case we may use the statistics of micro-clusters of all possible types, containing up to eleven boron atoms, and a good approximation up to 15 boron atoms per cluster.

The main question to be answered now is: what is the relative weight of the full ensemble of clusters containing a given number of boron atoms among all the clusters present in the melt, and how this distribution of probabilities evolves with time (or, what is roughly the same, with temperature).

In other words, we would like to know how the entire population of multiplets travels from one sheet of the configurational space to another, as suggested in our statistical description of the hot melt. We shall show how the quasi-geometric progression of the number of different types of multiplets with growing l finds its image in the exponential behaviour of the average cluster size as function of temperature, and in the exponential approach of the boroxol ring statistics towards its limit value. Incidentally, it will give us an idea of the *critical size* of clusters, at which all the significant statistics are almost identical with the overall statistics that characterize the "frozen" network resulting from the glass transition that has definitively taken place.

Although in the pure B_2O_3 glass we have only one characteristic parameter at our disposal, i.e. the rate of boron atoms contained in the boroxol rings, it can be used in order to decipher the unknown statictics of clusters of given type.

Walrafen *et al.* give the experimental curve describing the temperature dependence of the integrated intensity of the sharp $808cm^{-1}$ peak (attributed to the boroxol breathing mode) which is proportional to the boroxol rate $F(T)$ [41] and has been modelized by these authors with the following function:

$$ln\left(\frac{F(T)}{B_0 - F(T)}\right) = \frac{3237.66}{T} - 2.58893 \qquad (58)$$

where B_0 is the boroxol rate at T_g. With the characteristic energy differences established above, our model enables us to compute the boroxol rate $F_{(k)}$ averaged over clusters of a given size k; this quantity is also function of temperature T which enters via the Boltzmann factors contained in the expression of $F_{(k)}$.

Now, in order to produce a more realistic picture, we must take into account that the melt contains a mixture of clusters of all sizes up to some k_{max} that grows with time as the temperature goes down. The problem is how to evaluate the relative part of the clusters of the size k, for $k = 1, 2, ...k_{max}$. What is helpful here is the following simple observation: if the agglomeration pace is roughly the same at any point of the melt, it means that it takes the same characteristic time for a cluster of the size k to transform into a cluster of the size $k+1$. This means in turn that the ratio of the number of clusters of size $k+1$ to the number of clusters of size k is roughly given by the ratio of the numbers of possible multiplets given in the Table I. The numbers N_k of different multiplets containing k boron atoms grow as $1-1-2-3-4-9-14-31-62-135-288-...$, which is very close to the geometric progression with the constant ratio close to 2.13. If we ponderate the numbers of multiplets with the numbers of boron atoms contained in them, i.e. if we consider the series kN_k, the progression is even more impressive.

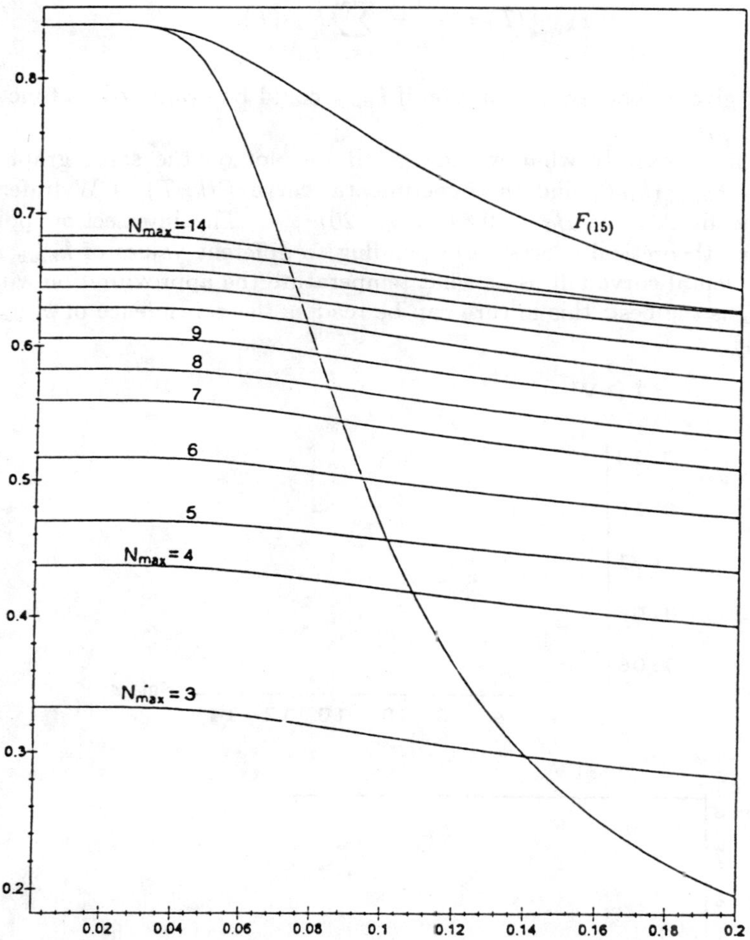

Figure. 20: The experimental curve of Walrafen et al.[41] and the curves $F_{k_{max}}(k_BT)$; $E_1 - E_2 = 0.20eV$

This means that at first approximation at least, the most important contribution to any statistics will come from the last and biggest type of clusters. That is the reason why we propose to evaluate theoretically the boroxol rate with a simple formula

$$F_{k_{max}}(T) = \frac{1}{k_{max}} \sum_{k=1}^{k_{max}} F_{(k)}(T). \tag{59}$$

which will give a good approximation if k_{max} could be evaluated as function of temperature.

But this is exactly what we can get if we plot on the same graph all the curves $F_{k_{max}}(k_B T)$ and the experimental curve $F(k_B T)$ of Walrafen *et al* with the limit value $B_0 = 0.84$ (Fig. 20). The intersection points between the theoretical curves corresponding to different values of k_{max} and the experimental curve tell us at what temperature the approximation with a given k_{max} is the best; this in turn can be read as the dependence of k_{max} on T (Fig. 21).

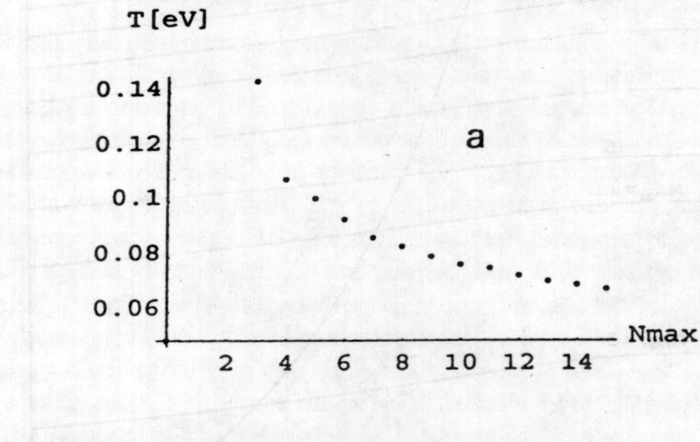

Figure. 21: a) The dependence of k_{max} on T, with $E_1 - E_2 = 0.20eV$ b) The boroxol rate as a function of k_{max} (with the same ΔE).

The points displayed in Fig. 21 are very well iterated by an exponential with a constant:

$$k_B T = A + B e^{-\lambda k_{max}}$$

with $A = 0.0628\ eV$, $B = 0.0966\ eV$, $\lambda = 0.1897$ for the choice $E_1 - E_2 = 0.20eV$. We see that when the maximal size of considered clusters attains 15 boron atoms, the corresponding temperature is close to $0.07eV$, i.e. still above the real glass transition temperature.

Parallelly, we can draw the development of the boroxol rate with temperature by projecting the same intersection points on the vertical axis of the fig. 11; this gives another exponential approach to the limiting value of 84% (fig. 21). We believe that this exponential behaviour is also reflected in the exponential (i.e. very rapid) behaviour of other interesting parameters, above all, of the viscosity, which should have a typical Arrhenius behaviour as similar strong glass-forming systems [72].

The fact that with a simple thermostatistical model, based on the topological construction of large clusters, we can obtain the usual shape for $C_p(T)$ and $U(T)$ curves, the realistic values for the energy of creation of a boroxol ring (5 $kcal.mol^{-1}$) and a satisfying fraction of boroxol rings (84%) is very encouraging, and incites us to apply it in the future to other glass-forming systems.

8 Conclusion

The model of glass formation exposed above has been elaborated during past ten years, in collaboration with Dina Maria dos Santos, Frank L. Galeener, Matthieu Micoulaut and Rafael A. Barrio; we should also mention numerous enlightening discussions during all these years with Maurice Kléman, Nick Rivier, Minko Balkanski, Michel Massot, and with J.C.Phillips. The essential features of this model has been published in many different versions, either general, or applied to specific situations. This article summarizes most of the results in a synthetic way, enabling the reader to have an overall view of the possibilities and limitations of this approach.

Here we have insisted on showing the universality of this approach and to display its applications to various physical systems. It seems to have a non-negligible predictive power combined with the simplicity of the equations; the only practical problem remaining is the difficulty of producing all the redundant clusters, whose number grows exponentially with each step of agglomeration. It is possible that an algorithm producing and identifying the clusters could be found, in which case the model would become more powerful and better suited for numerical computer methods.

Above all, the model provides us with the statistical distribution of various local structures, which is necessary for the modelling of all the important physical phenomena in glasses, such as thermal and electric conductivity, ionic diffusion, propagation of sound etc. This is why we hope that our model will become a common tool for physicists working on glasses and their properties.

9 Acknowledgements

The author is greatly indebted to F.L.Galeener (deceased in June 1993) for encouragement and help during the first years of work on this model.
Innumerable discussions and common work with D.M.dos Santos, M.Micoulaut and R.A.Barrio, as well as constant encouragement by J.C.Phillips, have influenced and improved the results of this work.

References

1. Galileo Galilei, Dialogo di Galileo Galilei Linceo, Firenze, MCDXXXII.
2. *"Microionics, Solid State Integrable Batteries"*, ed. M. Balkanski (North Holland, Amsterdam, 1991).
3. M. Massot, M. Balkanski, and A. Levasseur, in *"Microionics, Solid State Integrable Batteries"*, ed. M. Balkanski (North Holland, Amsterdam, 1991) p. 139.
4. G.E.Jellison Jr, L.W.Panek, P.J.Bray and G.B.Rouse Jr, J. Chem. Phys. **66** (1977) 802.
5. J. C. Phillips, Journ.Non-Cryst.Solids, **34** (1979), 153.
6. J. C. Phillips, Journ.Non-Cryst.Solids, **43** (1981), 37.
7. R.Collins, Proc.Phys.Soc. **83** (1964) 553.
8. J.M.Kosterlitz and D.J.Thouless, J.Phys. C **6** (1973), 1181.
9. P.J. Bray, J. Non-Cryst. Solids **95&96** (1987) 45.
10. *The structure of Non-Crystalline Materials*, Proc. of the Symp. in Cambridge 1976, Taylor and Francis (London) 1977.
11. R.A.Barrio, R.Kerner, in "Lectures on Thermodynamics and Statistical Physics", ed. M.Lopez de Haro and C.Varea, World Scientific, (1992).
12. R.A. Barrio, and F.L. Castillo-Alvarado, Phys. Rev. B **46** (1992) 3779.
13. F.L.Galeener, R.A.Barrio, E.Martinez and R.J.Elliott, Phys.Rev.Letters **53** (1984) 2429.
14. M.F.Thorpe, J.Non-Cryst.Solids, **57** (1985), 355.
15. M.F.Thorpe, J.Non-Cryst.Solids, **182** (1995), 135.
16. R. Kerner and M. Micoulaut, J. Non-Cryst. Solids **176** (1994) 271.
17. R. Kerner, J. Non-Cryst. Solids **182** (1995) 9.

18. R.A.Barrio, J.P.Duruisseau and R.Kerner, Phil.Mag. B, **72** (5), (1995) 535.
19. R.Kerner and M.Micoulaut, C.R.Acad.Sci.Paris, **315**, Sér.II (1992), 1307.
20. D.M. dos Santos-Loff and R.Kerner, Journal de Physique (France), **4** (1994), 1491.
21. D.M. Dos Santos-Loff, M. Micoulaut and R. Kerner, Europhys. Letters **28**(8) (1994) 573.
22. R.Kerner, Phys.Rev. B, **28** (1983), 756.
23. R.Kerner and D.M.Dos Santos, Phys. Rev. B **37** (1988) 3881.
24. R.Kerner, J.Non.Cryst.Solids, **135** (1991) 155.
25. R.Kerner, Physica B, **215**, (1995), 267.
26. W.H.Zachariasen, J.Chem.Phys **54** (1932) 3841.
27. J.Zarzycki, *Les verres et l'état vitreux*, Ed.Masson, Paris 1982.
28. S.R. Elliott, *Physics of amorphous materials*, Longman Scientifical & Technical (1990).
29. A.F.Wells, Acta Cryst. **7** (1954) 535.
30. A.Smekal J.Soc.Glass Techn. **35** (1951) 411T.
31. *The Physics and Technology of amorphous* SiO_2, Ed. R.A.B.Devine, Plenum Press (New York), 1987.
32. A.H.Silver and P.J.Bray, J.Chem.Phys. **29** (1958) 984.
33. A.Feltz, H.Aust and A.Blayer, Journ; of Non-Cryst. Sol. **55** (1983), 179.
34. R.A. Barrio, F.L. Castillo-Alvarado and F.L Galeener, Phys. Rev. B**44** (1991) 7313.
35. P.M.Bridenbaugh, G.P.Espinosa, J.E.Griffiths, J.C.Philipps and J.P.Remeika, Phys.Rev. B **20** (1979) 4140.
36. M.Rubinstein, Phys. Rev. B **14** (1976) 2778.
37. J.E.Griffiths, G.P.Espinosa, J.C.Philipps and J.P.Remeika, Phys. Rev. B **28** (1983) 4444.
38. J. Krogh-Moe, J. Non-Cryst. Solids, **1** (1969) 269.
39. G.E.Henderson and M.E.Fleet, J.Non-Cryst.Solids **134** (1991) 259.
40. G.E. Walrafen, S.R. Samanta and P.N. Krishnan, J. Chem. Phys. **72** (1980) 113.
41. G.E. Walrafen, M.S. Hokmabadi, P.N. Krishnan and S. Guha, J. Chem. Phys. **79** (1983) 3609.
42. A.V.Johnson, A.C.Wright and R.N.Sinclair, J.Non-Cryst.Solids **50** (1982) 281.
43. A.C.Hannon, R.N.Sinclair and A.C.Wright, Physica A **201** (1993) 375.
44. A.C. Hannon, D.I. Grimley, R.A. Hulme, A.C. Wright and R.N. Sinclair, Proc. of the Pacific Rim Meeting, Hawaii, J. Non-Cryst. Solids **175**

(1994).
45. A.C.Hannon, R.N.Sinclair and A.C.Wright Physica A, **201** (1993) 375.
46. N. Uchida, T. Maekawa, and T. Yokokawa, J. Non-Cryst. Solids **74** (1985) 25; *ibid*, **85** (1986) 290.
47. R.A. Barrio, F.L. Castillo-Alvarado, and F.L. Galeener, Phys. Rev. B **44** (1991) 1744.
48. A.Levasseur, J.C.Brethous, J.M.Réau and P.Hagenmüller, Mat.Res.Bull. **14** (1979) 921.
49. M.Micoulaut, R.Kerner and D.M.dos Santos-Loff, Journ.of Phys.C **7** (42), (1995) 8035.
50. J.Krogh-Moe, Phys.Chem.Glasses **6** (1966) 46.
51. M.Tenhover, M.A.Hazle and R.K.Grasselli, Phys. Rev. B **29** (1984) 6732.
52. A.Pradel, M.Ribes and M.Maurin, Solid St.Ionics **28-30** (1988) 762.
53. R.Kerner, J.Non.Cryst.Solids, **135** (1991) 155.
54. M.Micoulaut, Physica B, **212** (1995) 43.
55. W.Müller-Wahrmuth and H.Eckert, Phys. Reports **88** (1982) 91.
56. R.L. Mozzi and B.E. Warren, J. Appl. Crystallogr. **3** (1970) 251.
57. J. Goubeau and H. Keller, Z. Anorg. Allg. Chem. **272** (1953) 303.
58. F.L. Galeener, G. Lucovski and J.C. Mikkelsen Jr., Phys.Rev.B**22** (1980) 3983.
59. C.F. Windish and W.M. Risen, J. Non-Cryst. Solids **48** (1982) 307.
60. T.F.Soules and A.K.Varshneya, J.Amer.Ceram.Soc **64** (1981) 145.
61. T.S. Soules, in *"Glass Science and Technology" Vol. 4A Structure, Microstructure and Properties*, ed. D.R. Uhlmann and N.J. Kreidl, (Academic Press, Boston 1990) p. 276.
62. W.Soppe and H.W. den Hartog, J.Phys.C: Solid State Phys. **21** (1989) L689.
63. M.P.Teter, Amer.Ceram.Soc.Bull. **66** (1987) 562.
64. J. Keizer, *Statistical Thermodynamics of Nonequilibrium Processes*, Springer Verlag (1987).
65. R.J. Elliott, L. Perondi, and R.A. Barrio, J. Non-Cryst. Solids, **168** (1994) 167.
66. F.L. Galeener, G. Lucovski, and J.C. Mikkelsen Jr., Phys. Rev. B **22** (1980) 3983.
67. R.A. Barrio, F.L. Castillo-Alvarado, and F.L. Galeener, Phys. Rev. B **44** (1991) 1744.
68. M.P. Teter, Amer. Ceram. Soc. Bull. **66** (1987) 562.
69. R.N. Sinclair, J.A.E. Desa, G. Etherington, P.A.V. Johnson, and A.C. Wright, J. Non-Cryst. Solids, **42** (1980) 107.

70. P.J. Bray in *"Defects in Glasses"*, ed. F. L. Galeener, D.L. Griscom, and M.J. Weber (Material Research Society Symposia Proceedings Vol. 61, Pittsburgh1986) p. 121.
71. L.C. Snyder, *Quantum Mechanical studies of the structure and Thermochemistry of Species postulated to exist in Borate Glasses*, Am. Ceram. Soc. Mtg., Bedford Springs, 1978.
72. C.A. Angell, J. Non-Cryst. Solids **102** (1988) 205.
73. J.C.Phillips, J. Non-Cryst.Solids **182** (1995) 155.

Chapter 3
Structural and Vibrational Properties of B_2O_3 and Related Glasses

R.A. Barrio

1 Background

Boron oxide and related borate glasses are currently of great interest due to their potential use in microelectronic devices. Particularly important is the fact that the insulating compound B_2O_3 becomes a fast ionic conductor when an alkali metal oxide is added [1]. The modifier introduces important structural changes in the glass that are supposed to be responsible for the increase in the conductivity [2]. In this Chapter we will address this phenomena theoretically.

Our present knowledge of the structure of pure vitreous B_2O_3 allows to say that it is chemically ordered, with the boron atoms sitting in the center of a triangle of oxygen atoms,and the oxygen atoms acting as bridges between borons. This chemical order is short range and very strong. The existence of other type of order involving more distant atoms has been suggested a long time ago, and was invoked to explain X-Rays [3], nuclear magnetic resonance (NMR) [4], and Optical spectroscopy experiments [56]. These are planar six-fold $B3 - O3$ (boroxol) rings, which are supposed to exist in large numbers. Currently, there is a controversy about the existence of boroxol rings, mainly due to the fact that computer simulations are not able to obtain significant numbers of boroxol rings, and to the difficulty of obtaining reasonable values for the density of the material assuming rings. The discussion about the existence of regular small rings in glassy solids will be recurrent throughout this Chapter, since we consider that the structure of the parent compound that forms the fast ionic conductor is essential to determine the properties of the conducting material.

It is also believed that the main effect of the modifier (for instance Li_2O) in the structure of the pure glass is to change the coordination of some boron atoms from three to four and then allowing for the destruction of the boroxol rings and the creation of more complex units found in the crystalline forms of these compounds, as tetraborates and diborates. This assumption is supported by NMR studies [4] where a one to one correspondence has been found between the number of four-coordinated boron atoms and the number of alkali metal atoms, up to a concentration of the latter of 0.4.

There are vibrational spectroscopic studies [5] that are useful to test these assumptions. Of particular interest are the Raman and Infrared spectra of borate glasses, which show extremely narrow features assigned to the presence of boroxol rings and new features that evolve with the modifier concentration.

These are attributed to complex units with four-coordinated boron atoms[7].

We shall start in Section 2 with a general review of the properties of oxide glasses, and we will focus our attention on the types of order one could encounter in these disordered materials. Particular attention will be paid to the presence of small regular rings, and the important effects that they induce in the structural and vibrational properties of the glasses. A good example is vitreous silica ($v - SiO_2$), that shall be treated in detail. In Section 3 we introduce the Bethe Lattice, as a good mathematical model to simulate the effects of a large system, without long range order, on the local properties of the amorphous material. This model has been used extensively in the past for such purposes. We shall use it here as well, and our intention in this Section is to state objectively its virtues and drawbacks.

The study of the vibrational properties of boron oxide and of borate glasses is made in Section 4. These studies use the Bethe lattice to simulate long range disorder, but local and medium range order can also be treated. The assumption that 100% of the boron atoms are in boroxol rings gives a better understanding of the Raman spectrum than the opposite assumption. Section 5 is devoted to the study of the growth processes that take place when preparing boron oxide and borate glasses from the melt. This is an application of a more general theory of growth developed by R. Kerner (see Kerner's Chapter in this book). Again, the question about the reasons for the formation boroxol rings during growth is studied in detail.

Section 6 deals with the problem of atomic diffusion in borate glasses. In here we introduce a phenomenological model based on experimental facts (see Chapter by M. Massot), which is able to reproduce the ionic conductivity dependence on the dopant concentration. A novel ingredient in the theory is that one has to assume that the negative traps in the material may be occupied by more that one positive carrier. Using the theoretical background explained in Elliott's Chapter we deal with the idealized system of atomic diffusion, in a crystalline lattice, allowing double occupancy of sites. The results for the correlated motion are extremely interesting and basically different from the lattice gas model. These results were tested by performing large numerical simulations.

2 General properties of amorphous alloys and related glasses like SiO_2.

Determination of the structure of an amorphous solid is difficult, due to the absence of global translational or rotational symmetries. Since chemical bonding imposes approximate order between nearest neighbours, it is best to attempt first to determine structure on this shortest length scale, and then try to describe additional order on increasingly larger scales of distance. This plan has led to a delineation of four scales of distance [8], namely *short, intermediate, long and global.*

2.1 Ranges of Order

Short-range order (SRO) describes the nearest neighbour bonding environment of each atomic species. The elements of SRO in AX_2 can be visualized with the help of Fig.1(a) where one can see schematically the SRO around the A atom and that around the X atom. Every atom of type A is surrounded by a tetrahedron of X atoms and every X atom bridges between two A atoms. the relative position of two neighbouring tetrahedra is determined by the intertetrahedral angle θ and the dihedral angle δ. Disorder in real glasses, like SiO_2, is associated with variations in *both* angles, although the typical spread in δ is thought to be much larger than in θ [9]. In the 'continuous random network' (CRN) model, δ is taken to be random and uncorrelated with position in the glass structure [10]. In Fig.1(b) one notices that SRO in amorphous B_2O_3 is defined by the boron atom being in the center of an equilateral triangle of oxygen atoms. Therefore, the angle O-B-O is 120° since the triangle is planar. Disorder in this case is associated with the bridging angle θ, and the dihedral angle that fixes the orientation of two bridged triangles.

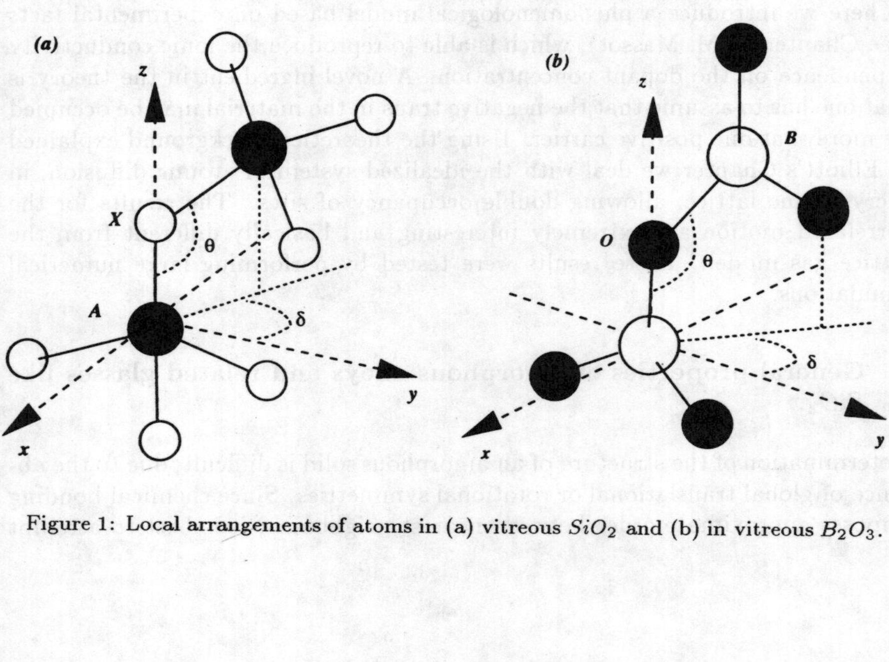

Figure 1: Local arrangements of atoms in (a) vitreous SiO_2 and (b) in vitreous B_2O_3.

Intermediate-range order (IRO) involves specification of relative atomic positions over several nearest-neighbour distances, given the SRO. It may take be specified by a distribution of δ, or distributions of the sizes of completed rings of bonds ('ring statistics'), or properties of network connectivity, etc.

Long-range order (LRO) allows for the possibility of translational repetition of an IRO over several repeat lengths and thus accounts for the possible existence of microcrystals in the structure of amorphous solids.

Global range order (GRO) accounts for structural order defined over macroscopic (entire sample) distances, including specification of network connectivity, chemical order, isotropy, and the like.

Elsewhere, Galeener [11] has described several models for vitreous SiO_2 in terms of all their elements of SRO, IRO, LRO and GRO, providing a correction to the Zachariasen-Warren CRN model [10], by showing that in $a - SiO_2$ the dihedral angle δ is *not* randomly distributed, but at various places in the glass is spatially correlated so as to give small highly regular rings of Si-O bonds at those places.

The same situation might be true in boron oxide, giving rise to IRO in the form of boroxol rings. This turns out to be essential to explain several remarkable features in the Raman spectra of this glass. Therefore let us explain these features more clearly.

2.2 Evidences for IRO in SiO_2 and B_2O_3

Sharp Lines in the Raman Spectra of Glasses

The polarized (HH) Raman spectra of vitreous $(v-)SiO_2$, (reduced by the Bose-Einstein thermal population factor according to the prescription of Shucker and Gammon [12] and Galeener and Sen [13]) is shown in Fig.2(a). One can notice that the HH spectrum consists mainly of broad features which peak at about 450 cm^{-1}, 800 cm^{-1}, 1065 cm^{-1}, and 1200 cm^{-1}. These have been identified [15,16,17] as 'network' modes, accounted for by coupled vibrations of a CRN, assuming random dihedral angles δ produces one high frequency mode,which is thought to be split into a TO-LO pair [18] at 1065 and 1200 cm^{-1}, respectively. The vibrations of this network and the shape of the HH Raman spectrum have been modeled quite well with a Bethe lattice, whose details are reported by Barrio, Galeener and Martínez [14,19].

The two sharp lines D_1 (495 cm^{-1}) and D_2 (606 cm^{-1}) cannot be explained by the same CRN-Bethe lattice model [19]. D_1 and D_2 have full width at half maximum (FWHM) linewidths of 20 cm^{-1} and 30 cm^{-1} respectively, and are highly polarized. Their strength and position in bulk $v - SiO_2$ have been studied variously as a function of neutron bombardment [20,21], sample fictive

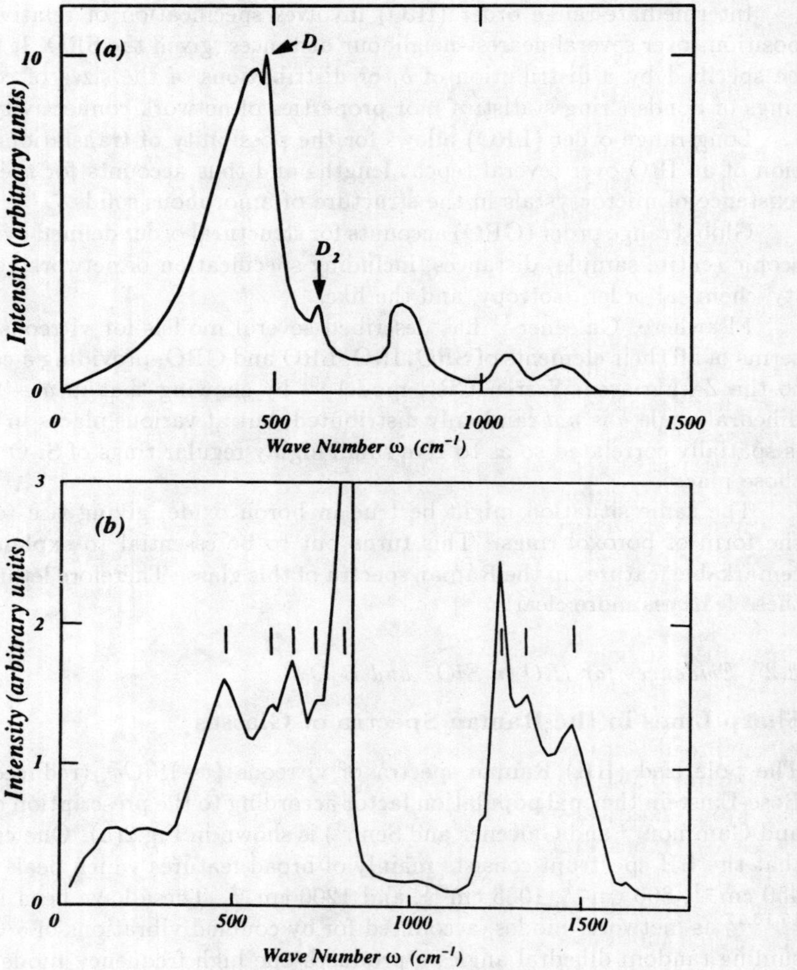

Figure 2: Polarized Raman spectrum of(a) $v - SiO_2$ (taken from Ref. [14]). (b) $v - B_2O_3$.

temperature [22,23,24], trace water content [25,26,27], isotopic substitution on Si [28], isotopic substitution on O [29], tensile stress [30] and pressure compaction [31]. These observations *eliminate* all explanations in terms of wrong bonds or broken bonds [32,33,34].

As a result, Galeener [35] considered regions of *increased* order, IRO in the form of highly regular rings of bonds, connected into the otherwise more disordered network at sites whose possible special nature is unknown. Because D_2 is stronger in porous Vycor glass [36] and in partially desiccated sol-gel glasses [37], Galeener [38] has suggested that the D_2 structures are more numerous in the vicinity of the $v-SiO_2$ surface. These observations and suggestions have been extended by Brinker and coworkers [39].

In Fig.2(b) the HH polarized Raman spectrum of $v - B_2O_3$ presents a very sharp peak at 808 cm^{-1}. This feature is by far the most intense (about 50 times the scale in the figure), suggesting that if the narrow peak is due to the existence of boroxol rings, their abundance is substantial. As we shall see below, an estimation gives $\sim$ 80% of boron atoms in boroxol rings, while only $\sim$ 3% of Si atoms are in regular planar rings.

A Mechanism for Small Rings to be Regular

One of the striking features of D_1 and D_2 is their extreme narrowness, about an order of magnitude smaller than the 200 cm^{-1} width of the dominant broad Raman line (R) which peaks at $\sim$450 cm^{-1}. The Bethe Lattice fit to the width of the broad features of the Raman spectrum is consistent with a distribution of θ having FWHM of about 30° [19].

From empirical selection rules [40] supported by a bond polarizability model [41] Galeener has argued that highly polarized Raman lines in glasses must correspond to symmetric stretch (SS) motions of the kinds shown in Ref. [42]. Since R, D_1 and D_2 are highly polarized and occur at similar frequencies, it is likely their motions are similar, namely SS motions of the bridging oxygen atom along the bisector of θ. This means that the θ for D_1 and D_2 must be much more tightly controlled than that for R. This necessary control is supplied by the regularity of the proposed rings, and regularity can be expected according to the following argument, given in more detail elsewhere [35,44].

Theoretical calculations [43] estimate the energy cost of forming a $Si-O-Si$ bond as a function of the angle θ. It is seen that this curve has a minimum at $<\theta>= 150°$. Therefore, a 3-ring can minimize its energy by making all ist angles as close as possible to $<\theta>$. This is acomplished only if it becomes regular and planar, when $\theta_3 = 130.5°$. Other irregular or puckered configurations should decrease the value of θ_3.

On the other hand, θ_4 for a regular planar 4-ring is 160.5°, and is larger than $< \theta >$. Thus a planar 4-ring, once formed, can reduce its energy by puckering so that some or all of its θ values are less than θ_4 and closer to $< \theta >$. Since θ_4 is not far above $< \theta >$, we assume that a significant number of 4-rings will relax to the lowest value $< \theta >$. Those relaxed 4-rings which are (nearly) regular will exist in otherwise unexpectedly high concentration because they are at an *energetic* 'bottleneck' of lowest possible energy. This provides a rationale for the existence of a significant concentration of regular puckered 4-rings in $v - SiO_2$.

This method of predicting whether or not significant numbers of regular small$-n$ rings are to be expected has been applied to eight rather different AX_2 tetrahedral glasses in Ref. [45]. The number of anomalously sharp Raman lines (like D_1 and D_2) thus predicted agrees with observation for all eight glasses.

The energies of formation of D_1 and D_2 have been measured by Galeener and coworkers [22,24], as follows. A sample of $v - SiO_2$ was annealed at a high temperature T_F until a steady state structure was achieved. the sample was then quenched rapidly to room temperature and the areas under D_1 and D_2 were carefully measured. Those areas exhibit Arrhenius behavior versus inverse T_F. The area under D_1 shows an energy of formation of $0.14eV$ while D_2 shows $0.40eV$. O'Keeffe and Gibbs [46] used first principles calculations to estimate these energies, producing values close to experiment on D_1 and D_2.

No similar studies have been made in boron oxide. However, it should be pointed out that the origin of the polarized Raman line D can be associated to the existence of boroxol rings. The modes of vibration at the frequencies of the D_1 and D_2 lines in silica, and at the 808 cm^{-1} line in boron oxide are similar. Isotopic substitution experiments in both materials show that there is non or very little boron or silicon motion at the Raman active modes, that is, only the oxygen atoms vibrate. This gives a valuable clue to envisage a possible cause of the sharpness of these lines.

2.3 The Vibrational Decoupling of Regular Rings in an Otherwise Disordered Network

Raman spectra of $v - SiO_2$ after isotopic substitution on the silicon atoms [28] and on the oxygen atoms [29] show definitively that the atomic motions for R, D_1 and D_2 in Fig.2(a) are essentially all oxygen motion, with no silicon motion involved.

Since D_1 and D_2 lie in the same frequency range as R, and all these lines are highly polarized (hence represent symmetric stretch of oxygen atoms), one could normally expect the motions of D_1 and D_2 to couple strongly with those of R. This poses a serious technical problem. If D_1 and D_2 are to be narrow band vibrational units, coupled into the more disordered network, then their frequencies will be broadened, and shifted, so as to give much broader features for D_1 and D_2 than are observed.

The resolution of this problem has been found[44], aspects of which were first reported in a letter [47] (GBME). There it was argued that the sharp lines are particular symmetric stretch motions of regular rings that are *fully* connected into the host network; these motions give sharp lines because the rings are regular structures, and because the particular motions are localized to the rings, hence do not couple to the similar motions of the host network. As we shall see, this occurs in a way that automatically also satisfies the isotopic substitution data. We shall repeat the argument of GBME for vibrational decoupling of the breathing mode of the planar 3-rings, as well as appropriate regular but non-planar $n-$rings.

Consider first the planar 3-ring in Fig.3. It is connected to the rest of the network by six Si-O bonds (four of them shown and labeled BL in the figure). In general these oxygen atom (o) displacements will exert a force on the Si atoms, which will communicate their motion to the rest of the network by distorting the BL bonds: This in general will broaden the breathing mode frequency. On the other hand, if there happens to be zero *net* force on the Si atoms, they will remain stationary, and the breathing O motion in the ring will *not* be coupled to the network. It was demonstrated [44] that D_1 and D_2 are consistent with the breathing modes of regular 4-rings and planar 3-rings, respectively, and that these modes are approximately decoupled from the rest of the network, in the manner just described.

It is easy to derive a mathematical condition for such decoupling in terms of the Born force constants α and β, where α measures the 'central' (or length restoring) force and β the 'non-central' (or bond orientation restoring) force. Other more complicated force constant models can be used [48], but the Born forces are the simplest and most convenient and are known to be successful in modern work on the 'optical' vibrations of networks representing disordered solids[48]. (Even so, it should be noted that the Born non-central forces cannot correctly model all of the acoustic modes of a solid[49].)

The individual central ($\vec{f}$) and non-central ($\vec{g}$) forces acting on a Si atom are shown in the lower right of Fig.3. In the absence of Si motion, the magnitudes of these forces are given by $f = \alpha' u_{\|} = \alpha' u \cos(\theta/2)$ and $g = \beta' u \sin(\theta/2)$, where the u are defined in Fig.3 and the prime denotes the force constant for a bond *within* the ring. The magnitudes of the oppositely directed total central ($\vec{F}$) and non-central ($\vec{G}$) forces acting on the Si atom are then given by $F = 2f\cos(\phi/2)$ and $G = 2g\sin(\phi/2)$. The absence of Si motion

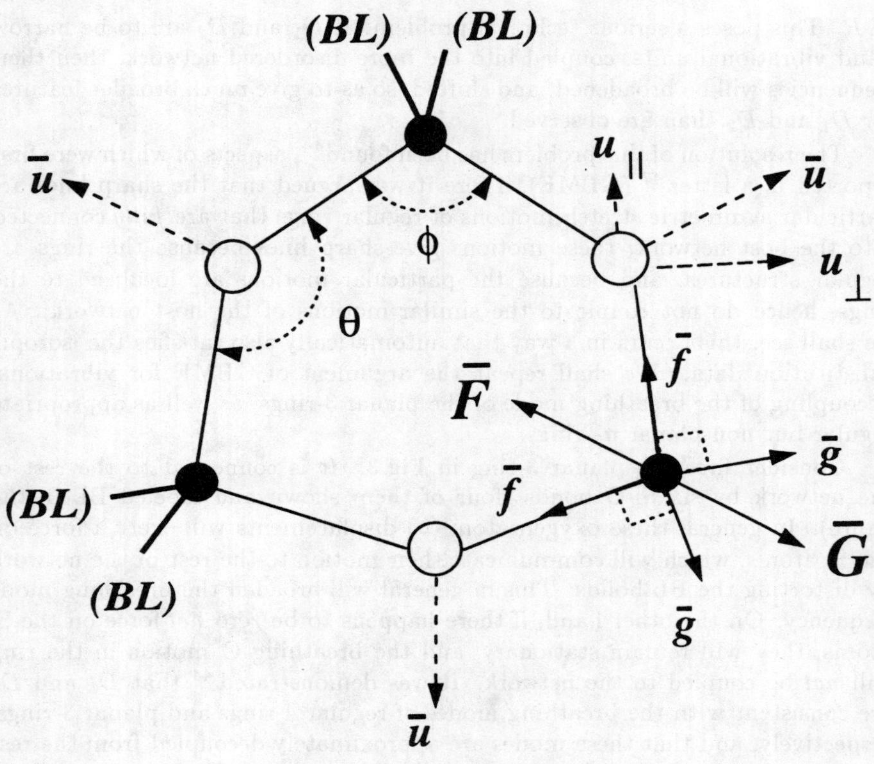

Figure 3: Sketch of a planar 3-ring in silica.

requires $F = G$, so that perfect decoupling occurs when α'/β' has the value

$$(\alpha'/\beta')_d = \tan(\theta/2)\tan(\phi/2). \tag{1}$$

The frequency ω_d of the breathing mode at perfect decoupling follows simply from the fact that the Si atoms are at rest, so that only the oxygen atoms in this ring are in motion. Consider the oxygen atom executing breathing motion at the upper right in Fig.3. The net force inward is $2f\cos(\theta/2) + 2g\sin(\theta/2)$. Application of Newton's second law to the oxygen mass m yields

$$\omega^2 = \frac{2\alpha'}{m}\cos^2(\theta/2) + \frac{2\beta'}{m}\sin^2(\theta/2)$$

or,

$$\omega^2 = \left(\frac{\alpha'}{m}\right)\left[(1+\cos\theta) + \left(\frac{\beta'}{\alpha'}\right)_d (1-\cos\theta)\right], \tag{2}$$

where m is the mass of the O atom. In the case of the planar 3-ring, where $\theta = \theta_3 = 130.5°$, and using $\phi = \phi_T = \cos^{-1}(-1/3) \sim 109.5°$, one has

$$\left(\frac{\beta'}{\alpha'}\right)_d = 0.326, \tag{3}$$

and

$$\omega_d^2 = 0.888(\alpha'/m). \tag{4}$$

Note that Eqns. (1-4) involve ring parameters only: *perfect decoupling of the breathing mode is a property of the regular ring, not of the vibrational systems to which it is attached.* Note also that the perfect decoupling condition [Eq. (1)] is independent of the masses: It is a relationship between force constants and geometry alone.

Perfect decoupling by this mechanism requires that $\vec{F}$ and $\vec{G}$ be antiparallel at all Si atoms in the ring. This is possible for a regular 3-ring (where all θ are equal) only when the ring is planar. It may be verified that there are only three conformations of the three ring that can be regular (always assuming that $\phi \simeq 109.5°$): the planar, the 'crown' and the 'oxygen chair' conformations. The 'crown' can be obtained by puckering the planar ring so that the three O atoms are left in the basal plane while the three Si atoms are tilted upward by a common amount forming a new plane parallel to and above the basal plane. The 'oxygen chair' has three Si atoms in the basal plane, two O atoms tilted up and one O atom tilted down as if the Si atoms formed the 'seat' of a (shallow) chair, the O atoms its 'back' and 'legs'. Similar conformations will be depicted explicitly for a 4-ring later. It can also be readily verified that in the present two non-planar cases, breathing motion along the bisector of θ produces $\vec{F}$ and $\vec{G}$ which *cannot* be antiparallel.

Equations (1) and (2) are also correct for perfect decoupling of oxygen breathing modes of any regular planar ring of order n; in those cases one evaluates the equations for $\phi = \phi_T = \cos^{-1}(-1/3) \simeq 109.5°$ and $\theta = \theta_n = 360(1-n^{-1}) - \phi$ in degrees. The force resolutions at the Si atom are precisely as in Fig.3, so equation (1) follows exactly as before. At perfect decoupling, the force resolutions at the O atoms are also as before, so equation (2) is verified for a planar n–ring.

Equations (1) and (2) are also correct for certain regular non-planar conformations of some n–rings. These require that for oxygen breathing along the bisector of θ the $\vec{F}$ and $\vec{G}$ forces induced at the Si atom are antiparallel.

We note that of the seven regular non-planar conformations of a 4-ring[44], this antiparallel condition is satisfied only for the 'oxygen boat'. The oxygen boat 4-ring is therefore the only regular puckered 4-ring that is capable of perfect decoupling of its oxygen breathing mode, and obeys equations (1) and (2) when decoupled.

This is the explanation for the appearance of both D_1 and D_2 sharp Raman lines in silica. More details and studies of the robustness of this decoupling mechanism can be found in Ref. [44].

The study of the sharp Raman line in boron oxide deserves a separate section. In this material it is very important to calculate the Raman spectrum of the system with and without boroxol rings, in order to discern if the lattice without rings presents sharp peaks. It is therefore very important to include in the calculation the effects of the infinite disordered network on the modes of vibration of local structures determined by SRO or IRO. We have chosen the Bethe lattice for these purposes.

3 The Bethe lattice

3.1 The Bethe lattice as a model for amorphous and glassy solids.

The Bethe Lattice has been extensively used to study amorphous and glassy solids, due to the abundant and easily obtainable analytical results. However, not much attention has been paid to the unphysical and spurious features that make the calculations difficult to interpret. Here we critically review some of the results extracted from Bethe lattice studies and discuss their most serious drawbacks, particularly the effects of an unphysical surface on local and long range correlations.

The most serious difficulty when studying glasses, or amorphous solids in general, is the absence of long range translational symmetry. However, as discussed above, one usually finds local chemical order and some kind of IRO in these, otherwise, disordered solids. For instance, covalent amorphous semiconductors and oxide glasses present nearest neighbour fixed geometry and regular coordination. This fact makes the Bethe lattice an appropriate network to simulate these glassy structures, since it is possible to define a local unit (or effective site) preserving the local SRO or IRO, embedded in an infinite random network without long range order. This method is sometimes more convenient than building up disordered clusters with free ends or periodic boundary conditions, since it is free from unwanted finite size effects and artificial periodicities, although care must be taken when interpreting the results due to the unphysical properties of the Bethe lattice. The main advantage of

Bethe lattice models is that most results are obtained analytically.

3.2 Properties of the Bethe lattice

Suppose one is interested in calculating the magnetic properties of a large system, and one approximates it by a small cluster. The difficulty in doing so is that the magnetization of the atoms on the surface of the cluster is very different to the ones in the interior. Alternatively, one can impose periodic boundary conditions, but one is likely to obtain spurious results due to the artificial periodicity. The idea then is to attach to the surface atoms some effective medium imposing the condition that the magnetization of the surface should be equal to the atoms in the interior of the cluster. This is called the Bethe-Peierls approximation[50].

If one asks whether there is a network in which the Bethe-Peierls approximation is the exact result, the answer is the Bethe lattice [51]. If one studies the vibrations of a tetrahedron with five atoms and devises an imaginary impedance attached to the four surface atoms, imposing the condition that the amplitude of vibration for any atom in the cluster should be the same, one is talking about the "Structural Potential Approximation " [52], which not surprisingly, turns out to be the Bethe lattice again.

Topologically, the Bethe lattice is the Cayley tree, which can be visualized as a simply connected network of identical units with constant coordination Z and without closed loops of paths. From this point of view, the Bethe lattice is a first approximation to any regular network, in which all connected diagrams without closed loops are summed up exactly. There are two important observations, the first is that there is no long range order or periodicity in the Bethe lattice, and second is that there is *only one path* to go from one given site to another.

The first observation has been used as a justification to simulate amorphous infinite networks and calculate electronic[53], vibrational[54], and magnetic [55] properties. The second is responsible for the drawbacks and non-physical features of the results. If there is only one way to transit from one place to another, one should expect that the network is very similar to the linear chain, in fact, some of the properties of the Bethe lattice are as for a one dimensional system, and a linear chain can be considered as a Bethe lattice of coordination $Z = 2$. However, as we shall see below, the linear chain is very peculiar and different from the other Bethe lattices. If one chooses a given site in the Bethe lattice as being the center, the number of steps n to go to another site defines a shell shared by $Z(Z-1)^{n-1}$ sites. Then, as the lattice grows the number of sites in the surface, or the last shell, grows exponentially. The total number of sites in a Bethe lattice with n shells is $1+Z\sum_{i=0}^{n-1}(Z-1)^i = [Z(Z-1)^n-2]/(Z-2)$. Therefore, as the number of shells tends to infinity, the proportion of surface sites tends to $(Z-2)/(Z-1)$.

Except for $Z=2$, in the infinite Bethe lattice the number of sites at the surface is not negligible, as it is in any other physical network. More than half of the sites belong to the surface, and as we shall see, this unphysical fact is responsible for *all* the unwanted properties of the Bethe lattice.

In order to illustrate the ways to solve analytically a problem in the Bethe lattice let us take the simplest possible Hamiltonian

$$H = V\sum_{i,j} |i\rangle\,\langle j| \tag{5}$$

where sites i and j are nearest neigbours. The equations of motion for the Green's function $g=(E-H)^{-1}$ are explicitly

$$\begin{aligned} Eg_{00} &= 1+Vg_{01} \\ Eg_{01} &= Vg_{00}+(Z-1)Vg_{02} \\ &\vdots \\ Eg_{0n} &= Vg_{0(n-1)}+(Z-1)Vg_{0(n+1)} \\ &\vdots \end{aligned} \tag{6}$$

where g_{0n} is the correlation between the central atom and any of the atoms in shell n. The usual way of solving these set of equations is defining a transfer function $g_{0n}=tg_{0(n-1)}$, which is obtained by substituting it in the equations of motion and solving for a second order equation

$$t = \frac{E \pm [E^2-4V^2(Z-1)]^{1/2}}{2V(Z-1)} \tag{7}$$

and

$$g_{00} = (E-ZVt)^{-1} \tag{8}$$

The reason for assuming that t does not depend on n is that one has neglected the surface atoms at shell N, which obey a different equation

$$Eg_{0N} = Vg_{0(N-1)} \tag{9}$$

This is wrong, because there are as many atoms in the surface as there are in the interior. There is an elegant and computationally efficient way of solving the system with surface, which is considering the shell-dependent transfer functions $g_{0n} = T_n g_{0(n-1)}$ and using the Tschevicheff functions of second kind $U(X/2)$, where $X = E/(V\sqrt{Z-1})$ [56]. Equation(9) shows that $T_N = 1/(X\sqrt{Z-1})$ and iterating one finds $T_1 = U_{N-1}(X/2)/(U_N(X/2)\sqrt{Z-1})$. In this case the self correlation function is written as $g_{00} = [E - ZVT_1]^{-1}$. As the lattice tends to infinity

$$\lim_{N\to\infty} T_1 = \frac{(Z-1)^{-1/2}}{X - \frac{1}{X - \frac{1}{X - \dots}}} = \frac{X \pm (X^2-4)^{1/2}}{2\sqrt{Z-1}} = t \tag{10}$$

which means that for a large enough Bethe lattice the effect of the surface on the self correlation is very small.

One could say that properties involving only the density of states $\rho(E)$ are not strongly affected by the surface, since

$$\rho(E) = -\frac{1}{\pi} Im[E - ZVt]^{-1} \tag{11}$$

is the same for an infinite lattice with or without surface, giving a band of width $W = 4V\sqrt{Z-1}$, according to Eq.(7). However, there is a subtle point: If one neglects the surface, one finds that perfectly coherent states (constant and equal amplitude in all sites) at $E = \pm ZV$ exist *out* of the band, since they are eigenstates of Eq. (6), their weight being inversely proportional to the total number of atoms. If one takes into account the surface, $E = \pm ZV$ are *not* eigenstates, thus the presence of an ever growing surface prevents the existence of perfectly coherent states and the band shrinks. One may say that even the very local properties of the Bethe lattice are dominated by the surface.

The case is even worst when one attempts to calculate properties that depend on long range correlations, for instance, the infrared, Raman and Inelastic Neutron cross sections [57] are related to the imaginary part of quantities like

$$\Sigma_N = \sum_{j=1}^{N} g_{0j} \tag{12}$$

which can be calculated in the Bethe lattice [58],

$$\Sigma_N = \frac{1 - ZV(Z-1)^N g_{0N}}{E - ZV} \tag{13}$$

This shows that the result is entirely dominated by the surface for $Z > 2$ because of the very large factor $(Z-1)^N$. Again, if one incorrectly neglects the surface (i.e. $g_{0N} = 0$), there appears a δ-function response at $E = ZV$, but if the surface is taken into account the weight of the δ function is zero because $g_{0N}(E = ZV) = 1/[ZV(Z-1)^N]$.

3.3 Localization of States in the Bethe lattice

The Bethe lattice is good when calculating local properties, as the density of states, but it fails when one is calculating properties that have to do with long-range correlations, as In Infrared, Raman and neutron Scattering cross-sections [57]. In a recent paper [59], the drawbacks using the Bethe lattice as an amorphous network are discussed, and good results we have obtained using it for modeling amorphous and glassy systems are reviewed. In that paper, most of the unwanted features of the Bethe lattice are associated with the enormous surface that cannot be neglected.

One of the most relevant properties of disordered solids is that the electronic states become localized (Anderson localization). This is extremely important to take into account when one is interested in transport properties, since localized states do not contribute to the metallic conductivity. In the Bethe lattice *all* states are localized, since the localization length does not depend on the energy. This can be seen in the following way: Let ϕ_n be the amplitude of the wave function on a site in shell n, and define a transfer matrix

$$\mathbf{T} = \begin{pmatrix} \frac{E}{(Z-1)V} & \frac{1}{(Z-1)} \\ 1 & 0 \end{pmatrix},$$

such that $\mathbf{T}c_{n-1} = c_n$, where $c_n = (\phi_n, \phi_{n-1})$. The eigenvalues of $\mathbf{T}$ are

$$\lambda_\pm = |\lambda|\, e^{\pm i\beta},$$

where $|\lambda| = (Z-1)^{-1/2}$ and $\beta = \arctan\sqrt{4(Z-1) - E^2}/E$. A localization length ξ can be defined through

$$|\phi_n| = |\lambda|^n = e^{-n/\xi},$$

giving

$$\xi = \frac{2}{\ln(Z-1)}.$$

In a linear chain ($Z = 1$) all states are extended, but in any other Bethe lattice all states are strictly localized. As a consequence there is an energy gap Δ between the edge of the band and ZV

$$\Delta/V = (e^{1/\xi} - 1)^2 = Z - 2\sqrt{Z-1}$$

This gap between the theoretical possibility of attaining a totally coherent antiferromagnetic state and the minimum eigenvalue of the system is present in other systems that do not present disorder in the real sense of the word, but that are not periodic. Such is the case of quasicrystals in two dimensions [60].

4 The vibrational properties and structure of pure boron oxide (B_2O_3).

4.1 Structural and vibrational model for vitreous boron oxide

In this Section we shall concentrate in the pure B_2O_3 glass. A model to obtain the density of vibrational states (DOS) of pure boron oxide has been put forward first a long time ago [61]. The theory consists in solving the dynamics of a Bethe lattice of B3-O3 rings (boroxol rings), with nearest neighbour central and non-central forces. This theory was further developed [62] in various aspects: First, the interaction between atoms was extended to take into account the asymmetry of the vibrations in the plane of the rings and perpendicular to it. Then, a model was developed to calculate theoretical Raman and Infrared responses, in order to be able to compare the theoretical spectra with the experimental data.

In Fig.4 there is a sketch of the local arrangement of boroxol rings and oxygen bridges. The interactions between nearest neighbour atoms is represented by a line joining two sites, and taking the z-axis along a given bond, the interaction is a 3×3 diagonal matrix, with elements $D_{xx} = \beta_x$, $D_{yy} = \beta_y$, and $D_{zz} = \alpha$. The dynamics is treated by solving the equations of motion for the displacement-displacement Green's function $\mathbf{G}$ in the Bethe lattice

$$M_\mu(l)\omega^2 G_{\mu\mu'}(l,l') = \delta_{\mu\mu'}\delta(l,l') + \sum_{\mu''}\sum_{l''} D_{\mu\mu''}(l,l'')G_{\mu''\mu'}(l'',l') \tag{14}$$

where M is the mass at site l in the direction of the three Cartesian coordinates μ. The first step to solve these equations is to eliminate the coordinates of the oxygen atoms in them, but care has to be taken to include the rotations by θ in each B-O-B link. This is done by assuming that all the angles are the same and by defining a local frame of reference for each bond, just as it has been made for the SiO_2 network [62]. The Bethe lattice can be solved by defining two transfer matrices, one relating two rings through a bond j

$$\hat{T}_j \hat{G}^{ji}_{nm} = \hat{G}^{ji}_{n+1,m} \tag{15}$$

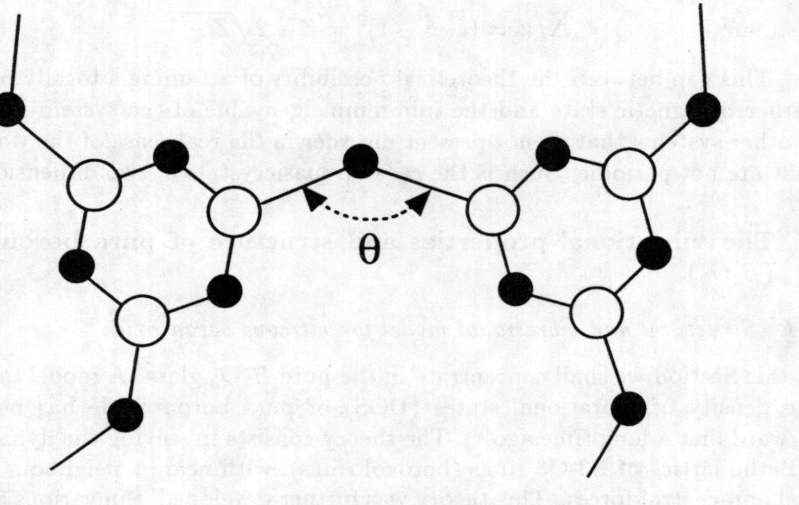

Figure 4: Sketch of the basic geometry of atoms in boron oxide with boroxol rings.

and another transferring information between different boron sites within a given ring n

$$\hat{t}_{jk}\hat{G}^{ki}_{nm} = \hat{G}^{ji}_{nm}. \tag{16}$$

These equations are analytically solved taking advantage of the symmetry of the matrices involved, when one assumes that the angle between the planes of two joint rings is totally random. These matrices allow to define a frequency dependent effective interaction that represents the Bethe lattice. Then the local DOS at any site i is calculated with

$$\rho_i = \frac{-2\omega}{3\pi} Im Tr(M_i \hat{G}_i), \tag{17}$$

and the total DOS is the properly weighted summation of ρ_i at all different sites.

The model to obtain the Raman and infrared responses is based on previous work for SiO_2 [62], where it is shown that the polarized Raman response can be approximated by

$$\omega Im \left[\sum_{\mu,\nu} \sum_{l,l'} \bar{v}^l_\mu G_{\mu\nu}(l,l') \bar{v}^{l'}_\nu \right], \tag{18}$$

and that the Infrared reflectivity at normal incidence by

$$\omega ImTr\left[\sum_{l,l'} e(l)\hat{G}(l,l')e(l')\right] , \tag{19}$$

where $e(l)$ is the dynamical charge associated with site l, and $\bar{v}^l_\mu$ is the μ component of the sum of all the vectors along the bonds arriving to that site. The summations are restricted to sites in a small cluster, an approximation that has proved correct in the Bethe lattice.

These equations have eight parameters, namely the three force constants for the bonds in the rings, another three for the bonds in the bridges and the two angles θ and θ' in Fig.4. These were adjusted to give the best fitting to the experimental Raman and Infrared spectra. The final results for the Raman

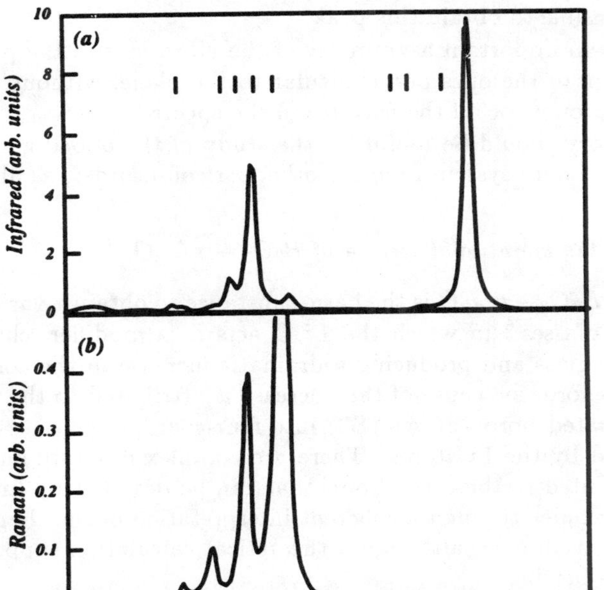

Figure 5: Infrared activity (a) and polarized Raman spectrum (b) calculated using a Bethe lattice of boroxol rings. The vertical bars are at the positions of the main peaks of the experimental data.

and Infrared responses are shown in Fig.5. It is seen that the agreement with the experimental peaks (shown as vertical bars in the figures) is extremely good.

This theory is the simplest one that is able to reproduce *all* the experimental data, but it is complete enough to contain all the features of early theories. In reference [62] a comparison is made with lattice dynamics calculations with central forces only, with a ball and stick hand-made model. The molecular vibrations of isolated rings are also obtained, and we show that former or simpler theories are unable to explain the experiment. The ultimate proof that the calculated dynamics is correct is obtained when one calculates the shifts of the main peaks when one isotopically substitutes ^{10}B or ^{18}O. The comparison with the experimental shifts is shown in Table I of Ref. [62].

There are important conclusions to be drawn from this work:

1) The inclusion of boroxol rings is important to account for the main peak of the polarized Raman spectrum at $\omega = 808$ cm^{-1}. If there are not boroxol rings it is impossible to obtain this peak.

2) There is an important asymmetry of the vibrations in the plane of the ring with respect to the ones perpendicular to the plane, without which it is impossible to account for all the features in the spectra.

3) This theory should be useful for the study of the doped glass, since it provides a simple pure system to apply other sort of disorder, as alloying.

4.2 Model for the vibrational spectra of $B_2O_3 - xLi_2O$

The compound $B_2O_3 - xLi_2O$ is the basic substance to obtain a variety of fast-ion conducting glasses [6] in which the Li_2O acts as a modifier, changing the structure of the glass and producing a dramatic increase in the conductivity. As mentioned before, the cause of this increase is attributed to the conversion of three-coordinated boron atoms (B3) into four-coordinated ones (B4), as a reaction induced by the Li atoms. There are complex diborate and tetraborate units associated to these B4 atoms that can be detected by Infrared and Raman spectroscopies through a coherent interpretation of the changes in the spectra as a function of x, although a theoretical calculation supporting this interpretation is lacking.

In a recent paper [63] we provide such a support by developing a model, both realistic and simple, that incorporates three basic ingredients: (1) The presence of tetrahedral boron sites B4, (2) the disorder, both topological and substitutional, and (3) the activity of the different modes in the experiments. There is an interesting central-force model in a Bethe lattice of alternating B4 and B3 sites to study the changes in the vibrational DOS, and the results are encouraging [64]. We improved that model by taking into account non-central forces, by incorporating substitutional disorder in the Bethe lattice, and by applying the local polarizability model for the Raman and Infrared responses, described above, to the present case. The exceedingly complicated calculations when one includes disorder in the Bethe lattice prevent us to take into account intermediate range order, as diborates, tetraborates, metaborates or even planar boroxol rings, and therefore, all these structures are neglected. This theory is good only if one is interested in studying the influence of substitutional disorder and one is prepared to ignore some rather conspicuous features, like the Raman 808 cm^{-1} peak.

The introduction of B4 atoms at random boron sites in the lattice can be treated using a Coherent Potential Approximation (CPA) theory for bond disorder in a Bethe lattice, which has proved to be rather successful in other amorphous alloys [65]. The basic CPA equations for the displacement-displacement Green's functions are

$$\langle G_{A_i} \rangle = \sum_i P_{ij} \left\langle G_{A_i B_j} \right\rangle , \tag{20}$$

where G_{A_i} is a 3×3 matrix for the displacements of an atom of type A_i embedded in an effective medium, $G_{A_i B_j}$ is the Green's function of an atom with a bond of type ij in the same effective medium, and P_{ij} is the probability of having such a configuration. The triangular brackets mean an average over all possible configurations.

One can use the Hamiltonian used in the last section and eliminate the oxygen coordinates as before, except that in the absence of rings one could label the interaction between an oxygen and a B3 atom as D_3, and the interaction between an oxygen and a B4 atom as D_4. This leaves four possibilities for the renormalized bonds between boron atoms, then the system of equations (20) can be represented by the diagrams in Fig.6.

The quantities σ and ξ stand for the effective medium, and they should be equal after the self-consistent solution is attained. The reason that they are made different is that convergence of the CPA equations is much facilitated by allowing a difference of the effective media bonds attached to B3 or B4 atoms. This is an special case of the more general theory called Nested CPA [66]. The meaning of diagrams in Fig.6 should become clear if we give a couple of examples of the equations they represent, for instance, the first diagram means

$$G_{B4} = \left[M\omega^2 I - \gamma' - \sum_{i=1}^{4} \{\xi_i\} \right]^{-1} , \tag{21}$$

where γ' is the appropriate averaged self energy for phonons that contains the oxygen coordinates and the curly brackets mean a dihedral angle average. The equation represented by the diagram multiplied by P_{43} in the Fig.6 is

$$G_{B4B3} = \left[m\omega^2 I - R_4 - \sum_{i=2}^{4} \{\xi_i\} - \left\{ D_4 (M\omega^2 I - R_3 - \sum_{j=2}^{3} \sigma_j)^{-1} D_4^* \right\} \right]^{-1}, \tag{22}$$

where the B4 atom is linked to the B3 atom through bond 1, and the matrices R_i are appropriate self energies containing γ and γ'.

The diagrams in Fig.6 represent a complete set of equations for ξ or σ, which is solved self-consistently to any desired accuracy. This effective medium, σ or ξ, represents a Bethe lattice containing the disorder, and can be used to calculate any quantity of interest, as the local DOS, with the imaginary part of the Green's functions, or the Raman and Infrared responses, with the local polarizability model described above, applied to a suitable cluster.

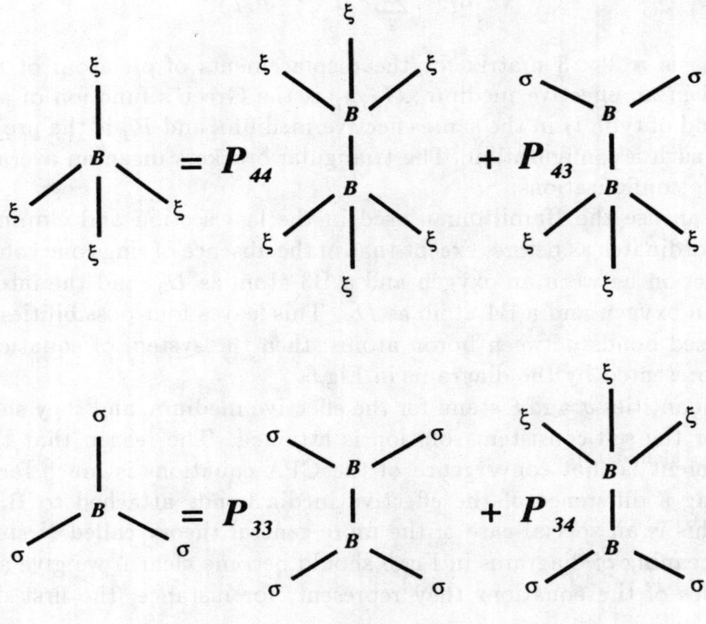

Figure 6: Diagrams representing the CPA equations for an amorphous binary alloy.

The results of the theory when $x = 0$ and $\beta_i = 0$, reproduce the allow bands predicted by a central-force network dynamics model [67]. Furthermore, when $\nu = x/(1+x) = 0.5$ the bands in the DOS follow closely the result of Ref. [64], except that the graphs are smeared out by the disorder. This comparison helped us to fix the value of the only parameter left free in the theory, namely the B3-B4 central force constant α', since better results are obtained in[64] when $\alpha/\alpha' = 0.8$, and α should be near the value 470 N/m when non-central forces are added. The local DOS obtained are rather complicated and show that

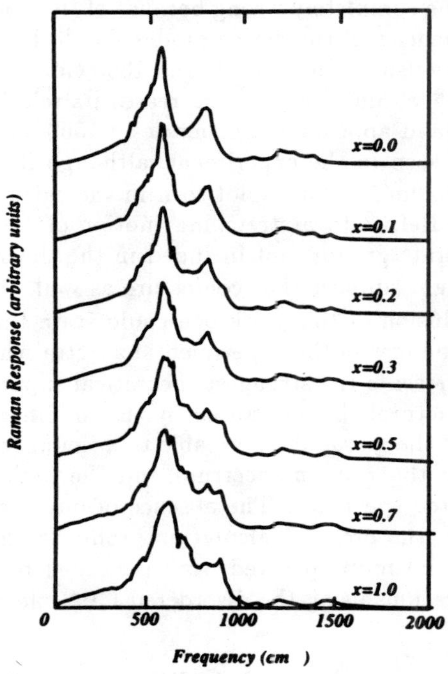

Figure 7: Calculated polarized Raman spectrum for alloys of B3 and B4 configurations.

the B4 vibrations are mainly in the high frequency region of the spectrum, with a sharp peak at $\sim$760 cm^{-1} corresponding to the bending modes. The B3 response is modified with x mainly in the middle region of the spectrum. These correspond to the part in which the various O-B bonds induce disorder, as could be corroborated in the local DOS at the three types of oxygen bonds.

The calculated Raman response is presented in Fig.7 in a fashion similar to the experimental plots in Ref. [5]. It shows a wide band centered at 500 cm^{-1}, which corresponds to in-phase oxygen motion, a peak at 808 cm $^{-1}$, which is not as sharp and intense as in the experiment since no boroxol rings were included in this theory, and a high frequency band, which contains the optical modes with a lot of boron motion and stretching of bonds. This figure should be compared with the experimental ones and then one realizes that the main changes of the spectrum with x are fairly well reproduced by the trends in the theoretical graphs.

Analogous results were obtained for the Infrared response [63]. In this case the intensities should be completely wrong because of the point charge approximation, but the frequencies of the active modes should be correct. The pure glass result presents the usual bands at 720 and 1360 cm $^{-1}$, the low-frequency one does not shift with x, and the other decreases its relative intensity when increasing x. A new band appears at frequencies >1500 cm^{-1}, due to the B4 units. All this is also seen in the experiment, although it must be said that there is a band at 1000 cm $^{-1}$ that is not seen in the calculation. This latter band is attributed in Ref. [6] to a stretching motion of B4 atoms belonging to tetraborate or diborate groups, not included in the theory. Therefore, the present calculation might support that conjecture as well.

As the main conclusion of this work one could state that the calculation reproduces the main changes in the experimental spectra when the modifier is added to B_2O_3. It represents the strongest theoretical support to the hypothesis that the main structural change caused by the modifier is to produce B4 units in the network of the glass. We were able to account for the conspicuous band at 720 cm $^{-1}$ in the Raman spectrum and the high frequency modes appearing in the infrared spectrum. The absence of intermediate range order in the form of rings in the present calculation prohibits an exact agreement with the experiment, and more involved work is needed to incorporate more complex local configurations using the disordered CPA effective medium.

5 Structural changes due to Li_2O, in lithium borate glasses.

5.1 Structure and Growth of Borate Glasses

In this Section we shall be interested in investigating the thermodynamically stable local configurations in borate glasses for small values of x. The main structural change in this regime is the conversion of $B3$ atoms into $B4$ ones.

The theory we will present is an application of the concepts first put forward by Kerner [68] to the lithium borate glasses. More details of the general theory are found in the Chapter by Kerner.

Suppose that in the hot liquid, close to the glass transition temperature T_g, $B-O-B$ bridges start forming. There are three types of these bonds, namely, $B3 - B3$, $B3 - B4$, and $B4 - B4$, with binding energies a_1, a_2, and a_3, respectively. As one cools the liquid down , it is reasonable to assume that medium-range structures similar to the ones found in the quenched solid are forming in equilibrium. When $x \sim 0$ there should be many boroxol rings (BR). Therefore, let us assume that the only building blocks are six-fold rings with zero, one, two and three $B4$ atoms, with relative abundances X_0, Y_0, Z_0, and T_0, respectively, as shown in Fig. 8.

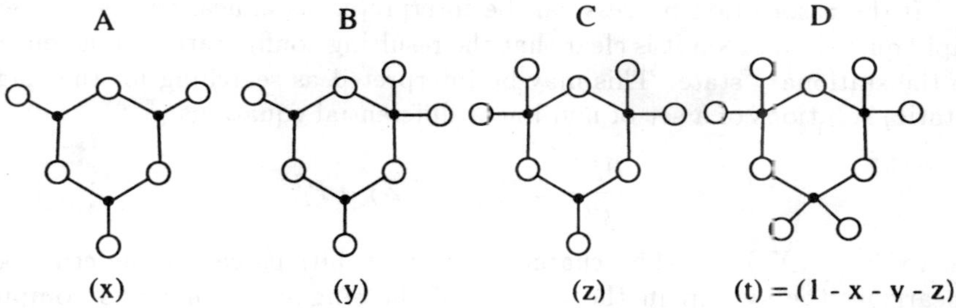

Figure 8: Possible building blocks of six-membered rings with B4 atoms

During the glass transition, at some later stage, these building blocks will cluster together in pairs. If one forbids rings with two and four boron atoms, on the grounds that there are either too costly energetically, or too unstable to survive, there are only 36 different pairs. The probabilities of finding these pairs can be calculated from the binding energies of the newly formed bonds and from the configurational entropy arising from the multiplicity of ways a given pair could be obtained. For instance, a pair formed with two A blocks will appear with probability

$$P_{AA} = \frac{9}{Q} X_0^2 e^{-a_1/kT} \tag{23}$$

and there are three clusters formed with A and C blocks,

$$\begin{aligned} P_{AC3} &= \frac{6}{Q} X_0 Z_0 e^{-a_1/kT} \\ P_{AC4} &= \frac{24}{Q} X_0 Z_0 e^{-a_2/kT} \\ P_{ABC} &= \frac{24}{Q} X_0 Z_0 e^{-2a_2/KT} \end{aligned} \tag{24}$$

where $Q = \sum_{i,j} P_{ij}$ is the proper normalizing factor. In the last case a third ring of type B is formed by creating two new bonds. There are three independent variables, since $T = 1 - X - Y - Z$, and with these probabilities one can evaluate the new probability distribution of elementary blocks

$$P_i^1 = \frac{1}{2}\left(2P_{ii} + \sum_{i \neq j} P_{ij}\right) \tag{25}$$

If the association process can be interpreted as annealing, as opposed to rapid quenching, then it is clear that the resulting configuration is the one close to the stationary state. This may be interpreted as searching for the singular (static) solutions of a set of non-linear differential equations

$$\frac{\partial P_i^1}{\partial t} = P_i^1(X,Y,Z) - P_i^0 \tag{26}$$

where $P_i^0 = X, Y, Z$. The character of the solutions can be determined by linearizing the system in the vicinity of the singular points and computing the eigenvalues. The *saddle point* solutions shall be interpreted as a glassy amorphous state [68]. The system can be solved numerically. The existence of a saddle point solution depends on the temperature and the binding energy values.

To be able to say something about the structure of the real glass, the binding energy parameters a_i have to be estimated by some independent criterion. This is done in Ref. [69] using the dependence of the glass transition temperature T_g with $R = \frac{1}{3}(3 - 3X - 2Y - Z)$, and if one fixes R, as in the experiment, one can translate this curve into the dependence of $T_g = T_g(X,Y,Z)$. Since that curve is nearly linear when $x \simeq 0$, the derivative can be roughly estimated. These data can be compared with the predictions of the model, which can be solved analytically in this limit, giving the relation

$$a_2 - a_1 = kT_0 \ln\left[2 + 4\exp\left(-a_2/kT_0\right)\right] \tag{27}$$

To obtain a_3 one should include non-linear effects. We made use of the NMR results [7] that give the relative abundance of diborate and tetraborate groups, which we could compute. We devised a numerical method to minimize the difference of the departure from the linearity between the experimental and the theoretical curves, and then obtained the best set of parameters

$$\begin{aligned} a_1 &= 0.0783 \\ a_2 &= 0.1226 \\ a_3 &= 0.2129 \end{aligned}$$

in eV. These values are close to the estimations proposed by Uchida *et al.* [70].

We have obtained the behaviour of the glass transition temperature as a function of the effective Li_2O concentrations, which is in good agreement with the experimental curve. One can investigate the range of temperatures in which the system exhibits a saddle point solution, subject to the constriction that R is fixed. In this case the solutions are found as the *projection* of the system (26) on the constriction plane. The result is quite interesting: The constrained system has many saddle point solutions which appear at temperatures within some range around the temperature corresponding to the unconstrained system. The prediction of the model is that this range of annealing temperatures is wider when one increases the Li content. There is some indication that this has been seen experimentally, and because the structure of the glass varies with the annealing temperature, this could account for the difference in the properties of glasses with the same composition, but prepared in different laboratories.

We investigated the R dependence of the density as well. As mentioned before, at low concentrations it increases linearly. This linear increase can be obtained by considering the relative mass and volume changes when one adds Li. However, the slope obtained with this first order approximation is totally wrong. We have included the second order effect [69], consisting in calculating a change when an oxygen bridge is substituted by a $B4$ atom. Our model gives an accurate description of the experimental curve [71]. This result, combined with our annealing model, allows one to make the prediction that for a fixed concentration, the lower the annealing temperature, the lower the resulting density. This is so because the relative abundances Y and Z grow with temperature. This calculation of the density, as trivial as it should appear, gives strong support to the existence of boroxol rings in large numbers, because this second order effect is only possible when one has a bridge between two rings.

5.2 *Existence of Boroxol Rings*

In this section we shall discuss a matter of paramount importance, the very existence of regular low-membered rings, as boroxol rings, in glasses. There is currently a strong feeling against the existence of these regular units, not only in vitreous silica, but even in B_2O_3, where the prevalence of boroxol rings builders of the structure of the glass seemed to be universally accepted and established since long ago.

This feeling comes mainly from people who perform molecular dynamics simulations in large systems [72] simply because in these simulations no regular rings are found in the expected proportion. They claim that the potentials and methods used are good enough as to mimic closely what happens really in Nature. Their results in boron oxide simulations were unable to obtain boroxol rings in the quantities that were expected (only $< 10\%$ of boron atoms in rings) [73].

This situation is in contradiction with recent Inelastic Neutron Scattering experiments [74], that can be coherently interpreted only if there are about 80% of boron atoms in boroxol rings. The strongest objection to our work on the growth and structure of borate glasses [69] is that the theory is based on the assumption that the basic structural units of pure boron oxide are precisely boroxol rings. Therefore, it is essential to investigate the structure of pure boron oxide, following the methods used for the borate glasses.

One has to withdraw the assumption of boroxol rings as starting units. Thus, one starts with simple tripods, that is, a single trigonal boron atom bound to three oxygen atoms with one unsatisfied bond each. Then, one follows the agglomeration method of Ref. [69] , first forming pairs, triplets, quartets, and so on, of these units.

In Fig. 9 this process is sketched up to step 4. In step 2 there is only one structure that could be formed in $3\times3=9$ ways, since one assumes that 2-fold rings are forbidden. In step 3 one can obtain a chain in $4\times3=12$ ways, a boroxol ring in $2\times3=6$ ways, and a branching network in $1\times3=3$ ways. In step 4, a chain is obtained in 4×3 ways, a boroxol ring in $6\times3=18$ ways, a branching network in $3\times3=9$ ways, and a 4-ring in $4\times3\times2=24$. Already here it is evident that the formation of long chains is not favoured, because there are only 4 points in it where a new unit could be attached. On the other hand, the formation of rings and of branching objects is very much favoured by the statistical factors associated with them, which ever grow. It should be noted that in any object at any step there are many ways of obtaining new boroxol rings and one can predict, without any further calculations that in a macroscopic sample there should be many rings.

R. Kerner and coworkers [75] have performed an extremely involved computer calculation up to step 16, and they have obtained good convergence for the existence of a saddle point representing the glass structure. They have counted the number of rings in this situation and they actually found that $\sim 80\%$ of the units *belong to boroxol rings.* Notice that there are rings with more boron atoms and presently they are counting the ring statistics of the amorphous structure, which is expected to be in very good agreement with other estimations. In this work they have found a completely general expression that predicts the sufficient conditions to obtain a glass structure in terms of the glass transition temperature and the energies of formation of few units.

Our interest for rings has lead us to investigate the structure of SiO_2, which

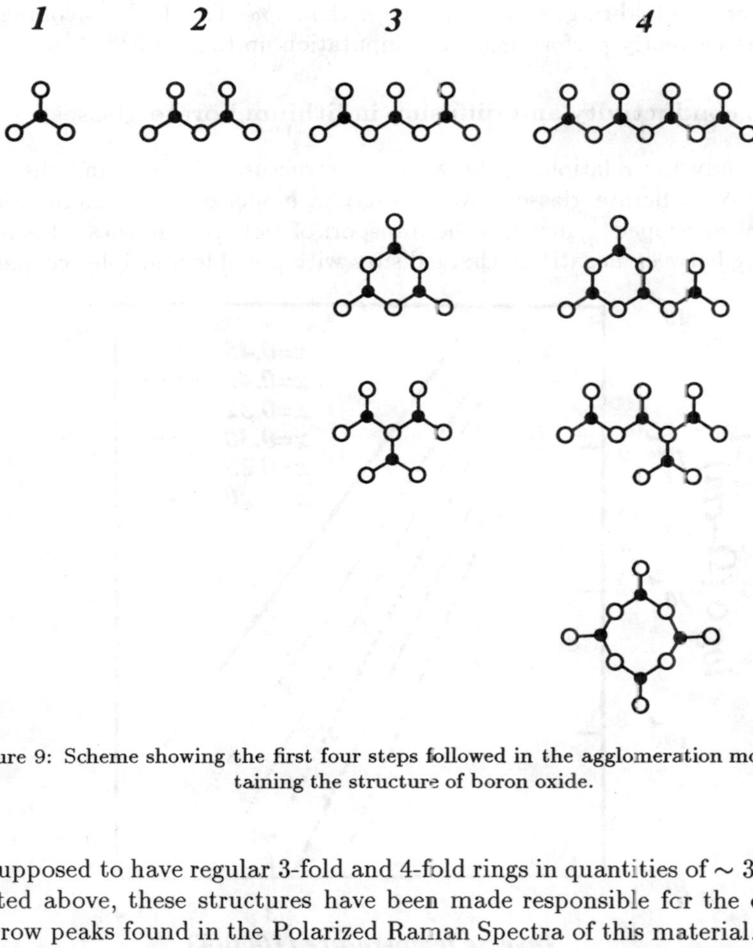

Figure 9: Scheme showing the first four steps followed in the agglomeration model for obtaining the structure of boron oxide.

is supposed to have regular 3-fold and 4-fold rings in quantities of $\sim 3-4\%$. As stated above, these structures have been made responsible for the extremely narrow peaks found in the Polarized Raman Spectra of this material. We have applied the ideas of the agglomeration model to this material, assuming that the basic units are SiO_4 tetrahedra. The building of complex units is in this case much more complicated than for the case of boron oxide, due to the 4-fold coordination of the basic units. The number of different structures at a given step grows enormously, nevertheless, one can neglect some of them on energetic grounds, for instance the formation of structures with edge-sharing tetrahedra, or diborate-like units, should be forbidden. On the other hand it is also true that the formation of very long chains is not favoured by the statistical factors. If one draws all the possibilities, as in Fig. 9 up to step 6, one realizes that the number of 3-fold rings cannot be larger than 7%. This looks encouraging, and we are currently performing the computation up to step 9[76].

6 Ionic conductivity and diffusion in lithium borate glasses

Here we study the relationship between the structural changes and the ionic conductivity in borate glasses. Accordingly, a model of ionic conduction in glasses was developed[77], in which the transport of *classical* particles takes place by hopping between negatively charged sites with possible multiple occupancy.

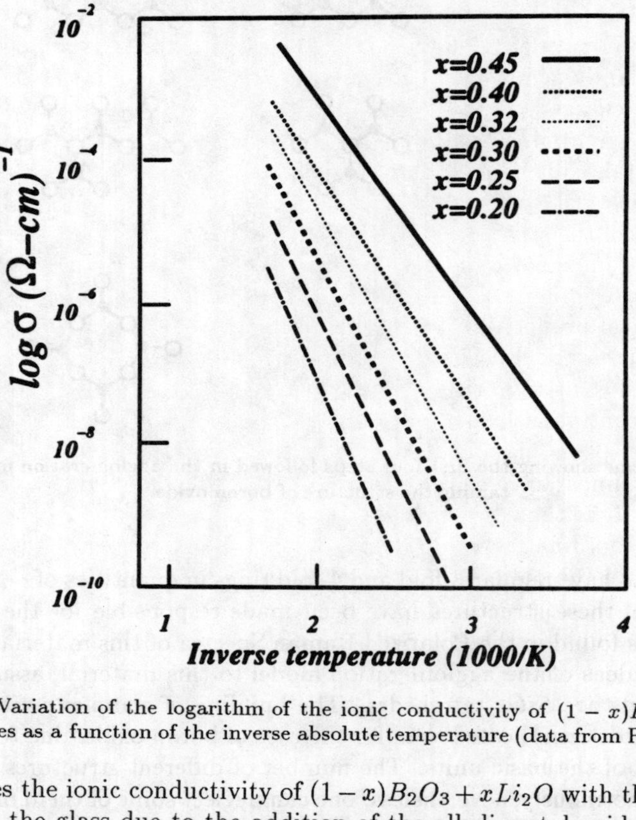

Figure 10: Variation of the logarithm of the ionic conductivity of $(1-x)B_2O_3 + xLi_2O$ glasses as a function of the inverse absolute temperature (data from Ref. [78]).

This relates the ionic conductivity of $(1-x)B_2O_3 + xLi_2O$ with the structural changes in the glass due to the addition of the alkali metal oxide, which give the nature of the negative traps as a function of x. A mean field calculation based on these ideas correctly predicts the behaviour of the activation energy and explains the surprisingly simple Arrhenius behaviour of the experimentally measured conductivity [78] (see Fig.10).

This model is strongly based on the existence of structural changes in the glass when the modifier (Li_2O) is added.

6.1 Ionic Conductivity in $(1-x)B_2O_3 + xLi_2O$

The ionic transport mechanisms in glasses are not well understood, because the global structure and the local environment of the ions are not fixed and generally not known. Furthermore, it is difficult to identify the existence of vacant sites where the ions can hop, since there are usually as many ions as there are available sites in the network, with huge activation energies. Therefore, the theoretical problem of explaining ionic conductivity in glasses is two fold, on one hand one must find a physically plausible mechanism for ions hopping between charged traps, and on the other hand, one should describe such charged local environments as a function of the structural changes induced by the alkali metal oxide.

In the case of $(1-x)B_2O_3 + xLi_2O$, local probes as NMR give a fairly accurate idea of the structural changes in the glass, and also far infrared spectroscopy gives information about the thermal vibrations of the mobile ions in their network residing positions. There are also experimental measurements of various physical properties as a function of x, which are enough to suggest a relatively simple model to describe the ionic transport.

As one adds Li_2O in small concentrations, the physical properties of the glass change continuously, not only the ionic conductivity appears and increases with x, but also the glass transition temperature, the density and the elastic moduli scale almost linearly with $R = x/(1-x)$, which gives the molar fraction of Li atoms. It is thought that for $R \leq 1/3$ (i.e. $x \leq 0.25$) the boron atoms change their coordination from three ($B3$) to four ($B4$) in order to incorporate more oxygen in the network. Glasses with $R \geq 0.38$ are *lithium borates*, and one is likely to encounter local environments for the boron atoms similar to the ones found in the crystalline lithium borates.

NMR studies [7] have been used to make structural models for the glasses by comparing the $B3$ and $B4$ signals to the ones obtained in the crystalline compounds. In Fig. 11 there is a sketch of the structural borate groups found in the glasses, so the concentrations of $B3$ and $B4$ atoms found in similar surroundings are revealed by NMR. The dependence of these concentrations on R is markedly different in three regions: In region I (low alkali content) there is a one-to-one conversion of $B3$ to $B4$, and therefore the concentration of boroxol rings (BR) decreases to zero, as the fraction of $B3$ atoms not in tetraborates or diborates does. Region II is characterized by the increasing presence of metaborates and loose BO_4 (L) units, and the absence of BR.

Also, the total number of $B4$ atoms is constant in this region. Therefore, the way oxygen is incorporated into the lattice is by creating non-bridging oxygens (NBO). In region III (high alkali content) the metaborates and L units are destroyed and isolated molecules like pyroborates and orthoborates increase in number. This explains the depolymerization of the glass network.

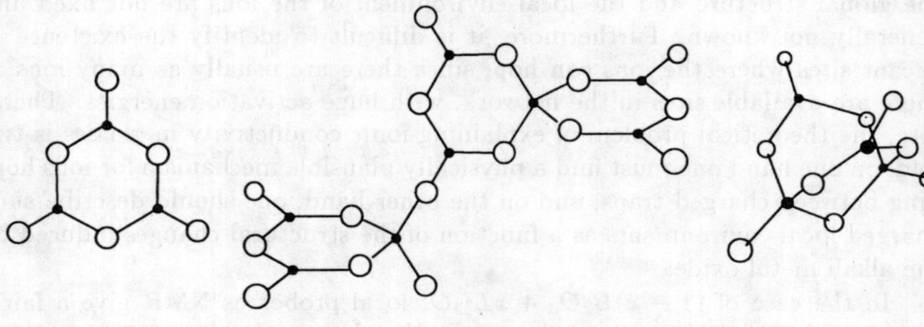

Figure 11: Structural local groups found in alkali borate compounds.

The structural changes in the glass as a function of x have been corroborated by Infrared and Raman studies [6], where the particular signatures of $B4$ atoms vibrating in complex local groups can be identified.

Valuable information about the vibrations of the Li^+ ions around their stable positions in the glass can be extracted from far infrared studies. These have been reported in a series of papers by Kamitsos *et al.* [79,80,81]. They find a wide asymmetric band extending from ~ 100 cm^{-1} to 500 cm^1, which can be deconvoluted in two fairly wide bands centered at $\nu_L = 240$ cm^1, and $\nu_H = 400$ cm^1, respectively, for $x = 0.4$. The interesting properties of these bands are that both are always present, even at very small x, that their weights are very similar in all the range of x, and that the ratio of the central frequencies is nearly constant.

From the NMR studies we know that there are at least two kinds of local negative traps, namely the $B4^-$ and the NBO^- anionic sites, which cannot be identified with the two different environments for Li^+ cations detected by infrared, since there is no correspondence in their behaviour with x. Therefore we assume that the two environments detected by far infrared spectroscopy, in all three regions of Li concentration correspond to singly occupied (s1) and doubly occupied (s2) negative sites. These anionic sites could be simply isolated $B4^-$

or NBO^- or more complicated arrays of negative charges and dipoles, whose nature and number change with x, following the structural changes of the network. Although not detected by Infrared, in thermal equilibrium there must be unoccupied negatively charged sites ($s0$) in equal number as the positively charged doubly occupied ones, so charge balance is accomplished. Hopefully, due to the screening of the lattice this happens in a small volume, just as in a normal ionic solid. This assumption is sound because both anionic sites are really complex configurations that certainly have several equally stable positions around them in which the Li atoms could sit. If one uses a Born- Mayer-type potential to estimate the frequencies of vibration around doubly (ω_2) and singly (ω_1) occupied sites, one obtains that the ratio of the frequencies is

$$\frac{\omega_2}{\omega_1} = \sqrt{\frac{q'_A}{q_A}\left(2 - \frac{1}{2\alpha q'_A}\right)} \tag{28}$$

where q'_A and q_A are the anionic effective charges for the two types of sites, respectively, and $\alpha = 1.64$ is a pseudo-Madelung constant. The effective charges for the anion sites are difficult to derive from a simple physical argument and therefore one must consider them as parameters. For $x = 0.4$ the values given in Ref. [80] are $q_A = q^L_{eff} = 0.37$, and $q'_A = q^H_{eff} = 0.67$, which when used in Eq. (28) give

$$\frac{\omega_2}{\omega_1} = 1.67 \tag{29}$$

very similar to the experimental value of 400/240. The fact that both q_{eff} do not change much with x explains the experimental fact that this ratio is nearly constant.

In this situation one should calculate the d.c. conductivity with

$$\sigma(x,T) = \frac{q_C^2 n(x,T) D(x,T)}{kT} \tag{30}$$

where k is the Boltzmann constant, $q_C = e$ for singly charged mobile cations, n is the number of mobile cations per unit volume, and the diffusion coefficient D is

$$D(x,T) = \frac{l^2(x)}{6} J_{eff}(x,T) \tag{31}$$

where $l(x)$ is the mean hopping length and $J_{eff}(x.T)$ is some average of the hopping rates associated to each process $J_i = \nu_i \exp(-\Delta E_i/kT)$.

For the sake of simplicity, let us assume that there is only one average activation energy and hopping length at fixed x. If there is only one type of anion site (say α, as $B4^-$ in region I), there are only four possible cation jumps

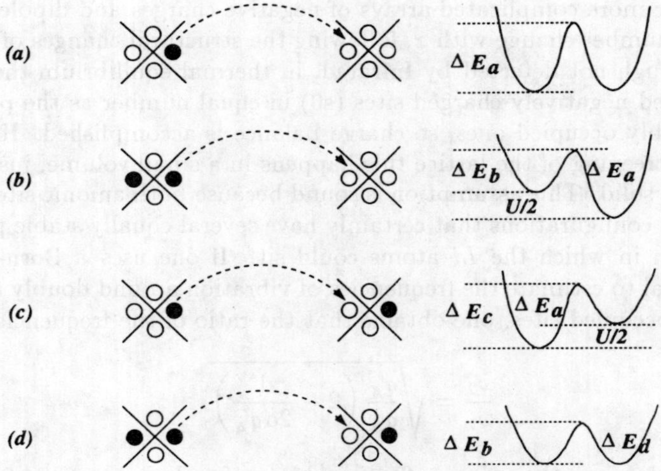

Figure 12: Possible diffusion jumps. In this sketch the cations could occupy four positions around a negative site and the occupied positions are full circles while the vacant ones are white circles. The direction of the jump is from left to right and the possible energy diagrams are sketched as well. The energy cost of having a doubly occupied site is $U/2$.

between neighbouring negative sites, as shown in Fig.12. This is so because one assumes that multiple occupancy, beyond double, is strictly forbidden, and one introduces an on-site Coulomb repulsion U. Then, if one naively assumes that the rattling frequencies measured in the infrared spectra ($\nu_L = \omega_1/2\pi$ and $\nu_H = \omega_2/2\pi$) are the same as the attempt frequencies for jumping, one has

$$\begin{aligned} J_a &= \nu_1 e^{-\Delta E_a/kT} \\ J_b &= \nu_2 e^{-(\Delta E_a - U/2)/kT} = \frac{\nu_2}{\nu_1} J_a e^{U/2kT} \\ J_c &= \nu_1 e^{-(\Delta E_a + U/2)/kT} = J_a e^{-U/2kT} \\ J_d &= \nu_2 e^{-(\Delta E_a - U/2)/kT} = \frac{\nu_2}{\nu_1} J_a e^{U/2kT} = J_b \end{aligned} \quad (32)$$

Then, one can consider a tracer particle and calculate a mean field effective hopping rate between sites of type α. The result is [77]

$$J_\alpha = \frac{J_a}{1+a}\left[\xi + \frac{a}{1+2a}\right] = f_B^\alpha J_a \quad (33)$$

where f_B^α is the correct blocking factor due to dynamical interactions of the tracer with background particles and one defines

$$\frac{J_c}{J_b} = \frac{\nu_1}{\nu_2} e^{-U/kT} = \xi^2 e^{-U/kT} = a^2 \tag{34}$$

Now, if there is another type of negative site then a similar equation

$$J_\beta = f_B^\beta J'_a \tag{35}$$

will be applicable. Thus, if one considers a system with a concentration r of defect sites β at random in the α lattice, one obtains [77]

$$J_{eff} = \frac{1}{\frac{1-r}{J_\alpha} + \frac{r}{J_\beta}} \tag{36}$$

which is the mean field result obtained by Haus and Kher [82].

To evaluate Eq. (30) one needs to give l, f_B^i, r, ν_i, n and ΔE_i $(i = \alpha, \beta)$ as functions of the concentration x. The functions f_B^i and ν_i can be evaluated from the far infrared data and are monotonic in the whole range of x. The remaining ones behave differently in the three regions of x. Therefore, they are related to the structural changes in the network.

If one uses the so called "strong electrolyte", or Anderson-Stuart model [83], the activation energy could be written as

$$\Delta E_a = E_c + E_s \tag{37}$$

where the strain term can be thought of as the energy required to enlarge the radius of a cylindrical hole, of length $l(x)$, by an amount δr, therefore

$$E_s = \frac{\pi G(x)(\delta r)^2}{2} l(x) \tag{38}$$

where $G(x)$ is the shear modulus of the lattice and its functional form can be obtained from experiment. The Coulomb term, for singly charged carriers and sites is

$$E_c = \frac{\alpha e^2}{4\pi\varepsilon r_0} - \frac{2e^2}{4\pi\varepsilon(l(x)/2)} = A\left(1 - \frac{B}{l(x)}\right) \tag{39}$$

where ε is the dielectric constant of the host in units of ε_0. Taking a constant value $\varepsilon = 6.46$ from experiment [84] and the values of α and r_0 given before, the values of the constants in Eq. (39) are $A = 1.83$ eV and $B = 4.88$ Å.

At this stage it turns out that the only important quantity for the model is the hopping length $l(x)$. Let us take

$$l(x) = [n(x)]^{-1/3} \tag{40}$$

and the site concentration $n(x)$ should be related to the structural changes, since one could write

$$n(x) = \frac{1.2R(x)\rho(x)}{70+30R(x)} \times 10^{24} cm^{-3} \tag{41}$$

According to experimental measurements [71], the density $\rho(x)$ can be adjusted by

$$\rho_I(x) = 1.83[1+0.65R(x)](g/cm^3) \tag{42}$$

In region II its behaviour is not monotonic, but it does not vary by more than 7%, and for the sake of simplicity we take it as constant ($\rho_{II} = 2.2 g/cm^3$). Let the defect site concentration be n_I in region I,and n_{II} in region II. These can be calculated with Eq. (41) using the values ρ_I and ρ_{II} respectively for the density. If, for the sake of simplicity, one assumes that there are only two types of anion sites, namely the $B4^-$ (α sites), and the NBO^- (β sites), their concentrations can be written

$$n_\alpha(x) = n_I(x)\theta(x_0 - x) + n_I(x_0)\theta(x - x_0)\,,$$

and

$$n_\beta(x) = [n_{II}(x) - n_I(x_0)]\,\theta(x - x_0) \tag{43}$$

where θ is the step function. The total site concentration is $n_T = n_\alpha + n_\beta$. This picture should be valid up to $x = 1/2$.

The elastic part of the activation energy can be written as $E_s(x) = g(x)n_T(x)^{-1/3}$, where $g(x)$, as the density, could be fitted to experimental elastic measurements giving $g(x) = 0.105[1+3.3R(x)]$eV/Å, in the range $0.15 \leq x \leq 0.25$, and then becomes constant for larger x. Then

$$E_s(x) = n_T^{-1/3}[g(x)\theta(x_0 - x) + g(x_0)\theta(x - x_0)] \tag{44}$$

The activation energy associated to a $B4$ site is

$$\Delta E_\alpha(x) = E_s(x) + E_c(x)\theta(x_0 - x) + E_c(x_0)\theta(x - x_0) \tag{45}$$

The other type of defect site starts to appear at $x = x_0$ and should be envisaged as a NBO in between two $B4$ anions, therefore there is a dipolar contribution to the Coulomb term and

$$\Delta E_\beta(x) = \Delta E_\alpha(x) - \frac{4Ar_0^3}{\alpha} n_\beta(x) \tag{46}$$

This picture should be valid as far as the concentration of NBO is low and the dipolar interaction at large distances between defects holds. When this distance decreases, the attractive term should go as $\sim l(x)^{-1}$. This occurs when $x = 0.46$. Therefore, one should not apply the present model to region III. The defect site concentration r in Eq. (36) can be identified with n_β. The ratio of the frequencies ξ^2 and the frequency $\nu_1(x)$ can be obtained directly from the infrared data, and the quantity a is related to the fraction q_H of cations residing in high energy sites through the relation

$$2a = q_H/(1 - q_H).$$

Figure 11 of Ref. [79] gives the compositional dependence of the fraction $q_H(x)$. There are no adjustable parameters left, since all the quantities involved in the model can either be taken from experiments or simply derived from the

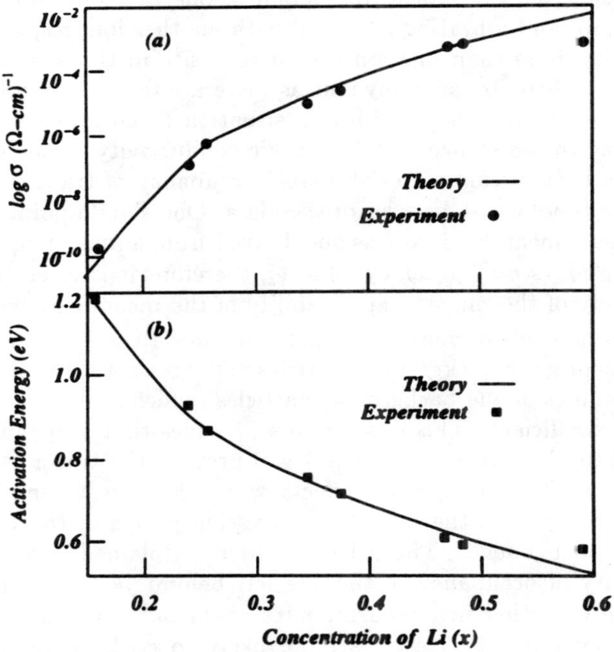

Figure 13: Calculated ionic conductivity of $B_2O_3 + xLi_2O$ as a function of x at 200°C. The experimental points are taken from Ref. [85] (a). Calculated activation energy for the effective medium (b).

model. The results from Eq. (30 are shown in Fig. 13(a) for $T = 473$°K, and compared with the experimental measurements of Ref. [85]. By taking its logarithm one calculates the activation energy in the effective medium, which is shown in Fig. 13(b).

6.2 Model of diffusion of particles allowing double occupancy of sites

The model above was suggested by two broad bands found by spectroscopic studies in the far infrared [80], which could be explained as the vibrational signatures of alkali metal ions loosely bound to two distinct environments, or negatively charged traps. These two traps cannot be four-coordinated boron atoms ($B4^-$) and structures related to non-bridging oxygen (NBO^-) because the two bands are present at any concentration x of ions, and other studies [6] definitely discard the presence of NBO^- at low concentrations. Therefore, it is proposed that the two bands correspond to one *Li* atom vibrating alone in a negative trap, and vibrating in a trap with another ion, respectively. Thus, there could be more than one ion bound to a site in the lattice, as is to be expected, since there are as many ions as there are traps, or available hopping sites in the lattice and the equilibrium situation (each ion bound to a trap) would prevent the existence of a high ionic conductivity. The success of this model means that in this material double occupancy of traps is more important than the topological disorder of the glass. One should point out that this application uses mean-field expressions derived from a master equation for the rates of hopping (see Chapter by Elliott), therefore it is essential to make a thorough study of the range of applicability of the mean-field approximation.

This is important because mean-field assumes a tracer particle moving in an average medium of background particles and does not take into account the detailed dynamics of the background particles, which will definitely diminish the diffusion coefficient. This is so because, besides that a *dynamical blocking factor* (from the background particles) will prevent the tracer to jump freely to any available site, many-body effects will reflect on a *correlation factor,* that takes into account the fact that successive jumps of the tracer *are not* statistically independent. This effect could be explained if one thinks that the probability of occupancy of the site left behind *just after a tracer jump* is zero, while the other neighbouring sites could be occupied by background particles. Therefore, the tracer is more likely to go back than to continue diffusing through the lattice, and the tracer diffusion coefficient should be smaller than calculated by mean-field. It is convenient to remind here that the tracer diffusion coefficient differs from that defined by ionic conductivity if correlation effects are important. However, it has been shown that this difference is small in borate glasses with low alkali ion content.

This complex problem can be approximately handled only in the limits of low (single particle) or high (single vacancy) concentration of particles. In the general situation one should use a reasonable interpolation scheme, and compare them with numerical Monte Carlo simulations to have an idea of the validity of the approximations used. In a recent paper[86] we have done precisely that for a double occupancy model of diffusion, in an utterly simplified version, which nevertheless suffices to investigate the basic assumption of the model, namely, the effects of allowing double occupancy of sites.

Mean-Field Treatment

Consider a perfect hypercubic lattice in d dimensions. The motion of a classical particle that performs hops of length l is characterized by a diffusion coefficient

$$D_0 = \frac{l^2 Z}{2d} J, \tag{47}$$

where Z is the coordination of the lattice and j is the hopping rate, which in thermal equilibrium can be written as

$$J = \nu e^{-(\Delta E/k_B T)}, \tag{48}$$

where ν is the attempt frequency, or a measure of the mean residence time on a site. When there is a concentration c of particles, the motion of background particles will cause a lower value of the tracer diffusion coefficient, and one could write

$$D = f_b f_c D_0, \tag{49}$$

where both, the blocking factor f_b and the correlation factor f_c are less or equal to one. In mean-field all correlations are neglected, therefore $f_c = 1$. The blocking factor is defined as the ratio between the number of successful jumps and the total number of attempted jumps. Therefore, it can be calculated by considering the geometry of the lattice and a mean-field treatment should give its exact value. If $n_i$is the probability that a site is occupied by i particles, then $n_0 + n_1 + n_2 = c$, that is the concentration of particles c. The tracer could only have a probability P_1 of being alone in a site or P_2 of sharing a site with another particle. Then, the hopping rate for the tracer is

$$J^0 = P_1 J + P_2 J', \tag{50}$$

where J' is the hopping rate for making a jump *away* from a *doubly* occupied site. Since diffusion is dominated by the largest of the hopping rates, one obtains that the blocking factor is

$$f_b = (1 - n_2)J^0/J \quad \text{for} \quad J \geq J',$$
$$f_b = (1 - n_2)J^0/J' \quad \text{for} \quad J < J'. \tag{51}$$

Likewise, the tracer diffusion constant will contain the largest of the hopping rates.

The equilibrium values $< n_i(l) >= n_i$ can be found by deriving a rate equation for $d < n_i(l) > /dt$ which involves averages of products of n_i's, which in mean-field can be decoupled into products of single-site averages to give

$$2n_2n_0J' = n_1^2J, \tag{52}$$

which together with the normalization condition gives

$$n_0 = \frac{2a + (1-c)(2a-1) - \sqrt{2a - (2a-1)(1-c)^2}}{2(2a-1)}, \tag{53}$$

where $a = J/J'$. A similar rate equation for the tracer gives the equilibrium condition

$$2P_1(an_1 + n_2) = P_2(2n_0 + n_1), \tag{54}$$

which can be solved to give

$$P_1 = \frac{2n_0 + n_1}{2(an_1 + n_2) + 2n_0 + n_1}. \tag{55}$$

A more detailed derivation of these mean-field results is given in [86].

Many-Body Effects

As mentioned before, many-body correlations can be treated approximately only in the extreme values of c. In the single-particle limit ($c \approx 0$), one observes that only background particles in doubly occupied sites are able to block the way for the tracer, and therefore, they are the only source of correlations. As a consequence one could use expressions derived for the lattice gas model (only single occupancy) through a random-walk approach [87]. In the single-vacancy limit ($c \approx 1$) the lattice gas picture is not valid, since the vacancy could move through the tracer site without exchanging places with it, and the hopping rate of the vacancy is $2J'$. In this situation one obtains

$$f_c = \frac{1 + \frac{1}{2}\cos\theta_0}{1 - \frac{1}{2}\cos\theta_0}, \tag{56}$$

where in two dimensions $\cos\theta_0 = 2/\pi - 1$ is a quantity that measures the imbalance between forward and backward jumps in a random walk. Eq. [56] differs from the one of a lattice gas by the factors $\frac{1}{2}$ in front of the cosines, due to the higher mobility of the vacancy (see Elliott's Chapter).

An interpolation scheme for the lattice gas has been given by Elliott and Tahir-Kheli [88]. One has good reasons to use their expression

$$f_c = \left[1 - \frac{2J^0 n_2 \cos\theta}{\{J' + J^0(1 - n_2)\}\{1 + \cos\theta\}}\right]^{-1}, \tag{57}$$

except that, instead of $\cos\theta_0$, we used $\cos\theta$ that has to depend on c, to give the correct limits. The simplest assumption for this function is a weighted average

$$\cos\theta = \frac{2 - c}{2}\cos\theta_0 + \frac{c}{2}\left[\frac{P_1 \cos\theta_0 + (P_2/2a)\cos\theta_0}{J^0/J}\right]. \tag{58}$$

As we shall see, these expressions provide an excellent interpolation scheme for the many-body correlations.

Numerical Simulations and Results

Numerical simulations of this model in a square lattice of 100×100 sites with periodical boundary conditions were performed using a standard random-walk Monte Carlo method. It was found that in all the parameter space (a, c) 2000 Monte Carlo steps, consisting of 10^4 events each, gave a sufficiently long time to accurately determine diffusion coefficients and render very small finite-size effects. The parameter space was explored by two sets of simulations, in one three extreme values of c were fixed, and a varied in 20 steps from 1/1000 to 1000, and in the other 12 fixed values of a between 1/9 and 9 were chosen to vary c in20 equally spaced steps. The technical details of the simulations are given in [86].

The results for the occupation probabilities agree extremely well with the mean-field calculations, meaning that the numerical simulation is very well done. The numerical calculations of D agree also with the mean-field results, except for extreme values of the parameters, since then there occur very few jumps, even for a large number of Monte Carlo steps. The effects of correlation are best exhibited plotting a comparison of the numerical value of f_c with the results of Eq.[57], which is done in Fig. 14 .

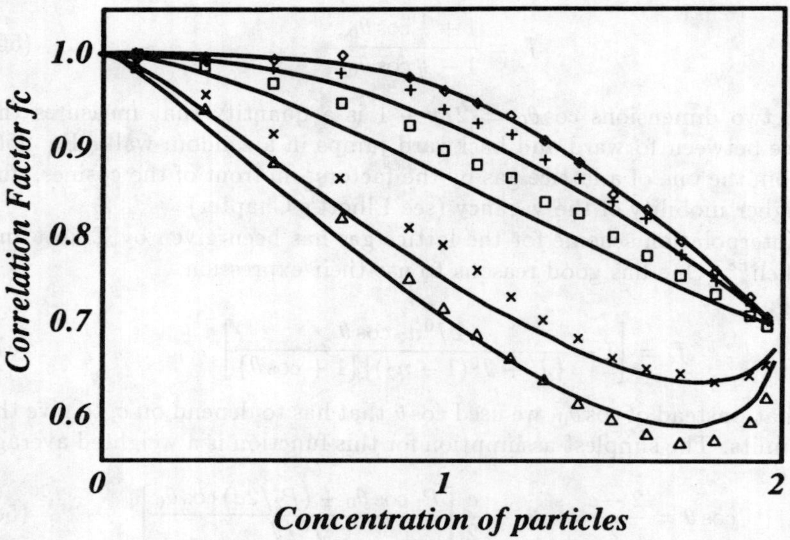

Figure 14: Correlation factor f_c as a function of particle concentration.

Observe that there is a minimum for curves with larger values of a, but all curves tend to the single-vacancy limit of 0.6925 given by Eq.[56]. This behaviour is explained as follows: When a is small, only particles in doubly occupied sites diffuse, and when $c < 1$ there are very few of them, and thus correlation effects are small. On the other hand, when a is large, the doubly occupied sites are nearly static, acting as stationary blocking defects in the lattice and the correlation is large, similar to the one in la lattice gas, where the single-vacancy limit gives $f_c(c \approx 2) = 0.4669$, the value these curves tend to attain. Then, the minimum is a result of the competition between the two terms in Eq.[58].

As a conclusion it should be remarked that this numerical calculation supports the application of a mean-field type theory to the problem of ionic conduction in borate glasses, since although a quantitative agreement is not completely fulfilled for all the range of concentrations and well depths, the discrepancies predicted in the diffusion coefficient are much smaller than any experimental precision.

Acknowledgments

The contents of this Chapter is a brief description of the work during the past four years on lithium borate glasses and related topics done in collaboration with Prof. Sir Roger Elliott, Prof. M. Balkanski, Dr. M. Massot, Prof. R. Kerner, Dr. M. Micoulaut, Dr. J.P. Duruisseau, Dr. G.G. Naumis, Dr. Chumin Wang, Prof. K. Kaski, Dr. L. Perondi, Dr. F.L. Castillo, and our beloved friend F.L. Galeener, deceased in June 1993. I want to thank all of them for the good times spent together and for the illuminating discussions and suggestions they have made to improve the present work.

References

1. A. Levasseur *et al.*, Solid State Ion. **1**. 177 (1980).
2. M.Massot *et al.*, Mater. Sci. Eng. **B 3**, 57 (1989).
3. J. Krogh-Moe, J. Non-Cryst. Solids **1**, 269 (1969).
4. W.L. Konijnendijk, Phillips Res. Rep., Suppl. 1 (1975).
5. M. Massot and M. Balkanski, in *Disorder in Condensed Matter Physics*, edited by J. Blackman and J. Tagüeña (Clarendon, Oxford, 1991), p.74.
6. M. Massot, M. Balkanski, and A.Levasseur, in *Microionics, Solid State Integrable Batteries*, edited by M. Balkanski (North Holland, Amsterdam, 1991), p.139.
7. P.J. Bray, J. Non-Cryst. Solids, **95&96**, 45 (1987).
8. F.L. Galeener, Diffusion and Defect Data **54-55**,305 (1988).
9. F.L. Galeener, Phil. Mag. Letters **B 51**, L1 (1985).
10. R.L. Mozzi and B.E. Warren, J. of Appl. Crystallography **2**, 164 (1969).
11. F.L. Galeener, in *The Physics and Technology of Amorphous* SiO_2, edited by R.A.B. Devine (Plenum, New York, 1988), p. 1.
12. R. Shucker, and R.W. Gammon, Phys. Rev. Lett. **25**, 222 (1970).
13. F.L. Galeener, and P.N. Sen, Phys. Rev. **B 17**, 1928 (1978).
14. R.A. Barrio, F.L. Galeener, and E.Martínez, Phys. Rev. Lett. **52**, 1786 (1984).
15. F.L. Galeener, Phys. Rev. **B 19**, 4292 (1979).
16. M.F. Thorpe,and F.L. Galeener, Phys. Rev. **B 22**, 3078 (1980).
17. F.L.Galeener, A.J. Leadbetter, and M.W. Sringfellow, Phys. Rev. **B 27**,1052(1983).
18. F.L. Galeener, and G. Luckovsky, Phys. Rev. Lett.**37**, 1474 (1976).
19. R.A. Barrio, F.L. Galeener, and E. Martínez, Phys.Rev. **B 31**, 7779 (1985).

20. R.H. Stolen, J.T. Krause,and C.R. Kurkjian, Disc. Faraday Soc. **50**, 103 (1970).
21. F.L. Galeener, J. Non-Cryst. Solids **40**, 527 (1980).
22. J.C. Mikkelsen, Jr., and F.L. Galeener, J. Non-Cryst.Solids **37**, 71 (1980).
23. F.L. Galeener, Kinam **4**,(C), 83 (1982).
24. A.E. Geissberger, and F.L. Galeener, Phys. Rev. **B 28**, 3266 (1983).
25. F.L. Galeener, and R.H. Geils, in *The Structure of Non-Crystalline Materials*, edited by P.H. Gaskell, (Taylor and Francis, London, 1977), p. 223.
26. J.C. Mikkelsen, Jr., and F.L. Galeener, Appl. Phys. Lett. **37**, 712 (1980); F.L. Galeener, and J.C. Mikkelsen, Jr., Solid State Commun. **37**, 719 (1981).
27. F.L. Galeener, and J.C.Mikkelsen, Jr., Appl. Phys. Lett. **38**, 336 (1981); J.C. Mikkelsen, Jr., and F.L. Galeener, J. Mater. **10**, 631 (1981).
28. F.L.Galeener, and A.E. Geissberger, Phys Rev. **B 27**, 6199 (1983); F.L.Galeener, and A.E. Geissberger, J. Phys. (Paris) **44** (C9), 343 (1983).
29. F.L. Galeener and J.C. Mikkelsen, Jr., Phys Rev. **B 23**, 5527 (1981).
30. G.E. Walrafen, P.N. Krishnan, and S.W. Freiman, J. Appl. Phys. **52**, 2832 (1981).
31. W. Primak, *"The Compacted States of Vitreous Silica"*, (Gordon and Breach, NewYork, 1975).
32. R.H. Stolen, and G.E. Walrafen, J. Chem. Phys.**2623** (1976).
33. G. Lucovsky, Phil Mag. **39**, 515 (1979).
34. A.R. Silin, and P.J. Bray, Bull. Am. Phys. Soc. **26**, 218 (1981).
35. F.L. Galeener, J. Non-Cryst. Solids **49**, 53 (1982).
36. C.A. Murray, and T.J. Greytak, J. Chem. Phys. **71**, 3555 (1979).
37. A. Bertoluzza, C.Fagnano, M.A. Morelli, V. Gotardi, and M. Guglielmi, J. Non- Cryst. Solids **48**, 117 (1982).
38. F.L. Galeener, in *The Structure of Non-Crystalline Materials*, edited by P.H. Gaskell, J.N. Parker, and E.A. Davis (Taylor and Francis, London, 1983), p. 337.
39. C.J. Brinker, E.P. Roth, G.W. Scherer, and D.R. Tallant,J. Non-Cryst. Solids **71**, 71 (1985); C.J. Brinker, D.R. Tallant, E.P.Roth, and C.S. Ashley, J. Non-Cryst. Solids **82**, 117 (1986); C.J.Brinker, D.R. Tallant, E.P. Roth, and C.S. Ashley, in *Defects in Glasses*, edited by F.L. Galeener, D.L. Griscom. and M.J. Weber (Vol.61 of Proc. of Mat. Res. Soc., Pittsburgh, 1986), p. 387.
40. F.L.Galeener, J. Phys. (Paris) **42** (C6), 24 (1981).
41. R.M. Martin, and F.L. Galeener, Phys. Rev. **B 23**,3071 (1981).

42. F.L. Galeener, Solid State Commun. **44**, 1037 (1982).
43. M.D. Newton, and G.V. Gibbs, Phys. Chem.Minerals **6** 221 (1980).
44. F.L.Galeener, R.A. Barrio, E.Martínez and ,R.J. Elliott, Phys. Rev. **B 48**, 15672 (1993).
45. F. L. Galeener, in *The Physics of Disordered Materials (a Festschrift for Sir Neville Mott)*,edited by D.Adler, H. Fritzsche, and S.R. Ovshinsky (Plenum, New York,1985), p. 159.
46. M. O'Keeffe, and G.V. Gibbs, J. Chem. Phys.**81**, 876 (1984).
47. F.L.Galeener, R.A. Barrio, E.Martínez, and R.J. Elliott, Phys. Rev. Lett. **53**, 2429 (1984).
48. A.M. Stoneham, V.T.M. Torres, H.R. Schober, and P M. Masri, Phil. Mag. **A 58**, 93 (1988).
49. R.A. Barrio and R. Kerner, in *Lectures on Thermodynamics and Statistical Mechanics*, Proc. XXI Winter Meeting on Statistical Physics, Oaxtepec, México, edited by M. López de Haro and C. Varea, (World Scientific, Singapore, 1992).
50. H.A. Bethe, Proc. Roy. Soc. **A126** (1935) 45; R.E. Peierls, Proc. Camb. Philos. Soc. **32** (1936) 477.
51. C. Domb, Adv. Phys. **9** (1960) 145.
52. M.F. Thorpe, D. Weaire, and R. Alben, Phys. Rev. **B7** (1973) 3777.
53. D. Weaire, and M. F. Thorpe, Phys. Rev. **B4** (1972) 2508.
54. R.A. Barrio, and R.J. Elliott, J. Phys. **C15** (1982) 4493.
55. E. Müller-Hartmann, and J. Zittartz, Phys. Rev. Lett. **33** (1974) 893.
56. R.A. Barrio, Kinam **5** (1983) 337.
57. R.J. Elliott, R.A. Barrio, and M.F. Thorpe, Kinam **4C** (1982) 55.
58. Felix Yndurain, R.A. Barrio, R.J. Elliott, and M.F. Thorpe, Phys. Rev. **B28** (1983) 3576.
59. R.A. Barrio, G.G. Naumis, and Chumin Wang, J. Non-Cryst. Solids, **182**, 22 (1995).
60. R.A. Barrio and Chumin Wang, *Some physical inferences from the quasicrystalline topology of Penrose lattices*, Proc. 4th. Int. Conf. on Quasicrystals, ed. K.F. Kelton, P.C. Gibbons,and A.I. Goldman,(North Holland, Amsterdam, 1992).p. 375.
61. R.A. Barrio and F.L. Galeener, in *Anais do Simposio Latino- Americano de Fisica dos Sistemas Amorfos*, edited by E. Anda (Centro Latino-Americano de Fisica, Rio de Janeiro, 1985), Vol. II. p. 381.
62. R.A. Barrio, F.L. Castillo-Alvarado, and F.L. Galeener, Phys. Rev. **B 44**, 7313 (1991)..
63. R.A. Barrio, and F.L. Castillo-Alvarado, Phys. Rev. **B 46**, 3779(1992).
64. J. Deppe, M. Balkanski, and R.F. Wallis, Phys. Rev. **B 41**, 7767 (1990).

65. F.L. Castillo-Alvarado, G.S. Contreras-Puente, and R.A. Barrio, J. Phys. C **21**, 1887 (1988).
66. J.Avendaño, F.L. Castillo-Alvarado. and R.A. Barrio, J. Non-Cryst. Solids, **137&138**, 299 (1991).
67. F.L. Galeener and M.F. Thorpe, Phys. Rev. **B 28**,5802 (1983).
68. R. Kerner, and Dina Maria L.F. dos Santos, Phys Rev. **B 37**, 3881 (1988).
69. R.A. Barrio, Jean-Paul Duruisseau, and R. Kerner, *"Structural properties of alkali-borate glass* $(1-\nu)B_2O_3 - \nu Li_2O$ *derived from a theoretical model"* (to be published in Phil. Mag.).
70. N. Uchida, T. Maekawa, and T. Yokokawa, J. Non-Cryst. Solids **74**, 25 (1985); *idem* **85**, 290 (1986).
71. D.P. Button, Ph.D. Thesis, Cornell University (1978, unpublished).
72. J.P. Rino, I. Ebbsjö, R.K. Kalia, A. Nakano, and P. Vashishta,Phys. Rev. **B 47**, 3053, (1993).; M. Tetter (*unpublished*).
73. P. Vashishta, (*private communication*).
74. A.C. Wright, and A. Hannon, in *Amorphous Insulators*, Symposium in honor of F.L. Galeener, Winter Park Co., August 1994.
75. R. Kerner, in *Amorphous Insulators*, Symposium in honor of F.L. Galeener, edited by M.F. Thorpe and A.C. Wright (North Holland, Amsterdam, 1995) p. 9.
76. M. Micoulaut, R. Kerner, and R.A. Barrio, (*unpublished*).
77. R.J. Elliott, L.F. Perondi, and R.A. Barrio, J. Non-Cryst. Solids, **168**, 167 (1994).
78. M. Kbala, Thesis (D. d'Etat Sciences), Université de Bordeaux (1984).
79. E.I. Kamitsos, M.A. Karakassides, and G.D. Chryssikos, J. Phys. Chem. **91**, 5807 (1987).
80. E.I. Kamitsos, A.P. Patsis, M.A. Karakassides, and G.D. Chryssikos, J. Non-Cryst. Solids **126**, 52 (1990).
81. E.I. Kamitsos, G.D. Chryssikos, A.P. Patsis, and M.A. Karakassides, J. Non-Cryst. Solids **131-133**, 1092 (1991).
82. J.W. Haus and K.W. Kher, Phys. Rep. **150**, 263 (1987).
83. O.L. Anderson, and D.A. Stuart, J. Am. Ceram. Soc. **37**, 573 (1954).
84. C. Julien, and M. Massot, in *Microionics, Solid State Integrable Batteries*, edited by M. Balkanski (North Holland, Amsterdam, 1991), p.192
85. H.L. Tuller, and D.P. Button, in *'The Role of Structure in Fast Ion Conducting Glasses'*, Proc. Int. Conf. on Transport-Structure Relations in Fast Ion and Mixed Conductors, Ed. F.W. Poulsen *et al.*, Riso Nat. Lab., Denmark (1985) p. 119

86. L.F. Perondi, R.J. Elliott, R.A. Barrio and K. Kaski, Phys. Rev. **B 50**, 9868 (1994).
87. L.F. Perondi, D. Phil. Thesis, Oxford University, 1993.
88. A. Tahir-Kheli and R.J. Elliott, Phys Rev. **B 27**, 844, (1983).

Chapter 4

Properties of Borate Glasses Structure, Vibrational Properties and Transport Experimental Approach

M. Massot

1 Introduction

Since many years, important studies have been made on synthetised fast ionic conductors which can be used as solid electrolytes in various electrochemical devices like solid state batteries, smart windows or sensors[1]. The increasing demand of high electrical power systems able to work in hostile conditions as is founded in polar or space applications have generated the need for electrochemical devices which contain no crystallizing or vaporizing materials to replace the usual liquid electrolyte devices.

Among these materials, fast ionic conducting glasses (doped oxide, sulfur and halide glasses) appear very promising for applications although they do not exhibit markedly higher ionic conductivity than their corresponding crystalline forms. Many types of ions can move in glasses, but cation motion is the most common. Singly charged cations tend to diffuse the most easily in doped oxide glasses, with the smallest ions, e.g. Li^+, or the most deformable, e.g Ag^+, having the highest conductivity of all [2,3].

However, they have the attractive advantages peculiar to the glassy states. They are isotropic materials, without grain boundaries, easy to form in complex shapes and of low cost price. They provide the opportunity to optimise the transport parameters because the large flexibility of their chemical composition which is not restricted to the narrow compositional limits characteristic of crystalline materials.

The development of competitive solid state electrochemical devices from these compounds requires an understanding of the ionic transport mechanism. In the generalised definition of the electrical conductivity σ, given by:

$$\sigma = \sum_i \sigma_i = \sum_i n_i \, Z_i e \, \mu_i \tag{1}$$

each partial conductivity reflects the convolution of three parameters, the mobile carrier concentration n_i, the electrical charge $Z_i e$ and the carrier mobility μ_i. However, in the glass literature, many controversies remain about the relative contribution of carrier density and mobility to the magnitude, composition, and temperature dependence of the ionic conductivity. This is probably the result of the total exclusion of the structural considerations in the characterisation of the ionic conduction in glasses.

In spite of these differences, many results may be learned on fast ionic conductors by the comparison between crystalline and amorphous conductors. Because some structural characteristics are common between in them. They consist of a disordered network of equivalent sites which can be filled by the mobile ions within a highly ordered structural array which is organised to form tunnels, layers or three dimensional arrays derived from a network or sublattice of other atoms. So the knowledge of the structural arrangement in a solid is essential to a detailled understanding of the ionic conductivity or other physical properties.

The main characteristic of glass structure is the absence of long range order in the material and that seems to contradict the second structural condition. However these solids are not completely disordered systems and it exists strong evidence of Short Range Order (SRO) and Medium Range Order (MRO).

Consequently, in the short length scale less than about 3 Å, amorphous substance are characterised by a certain area of short range order (SRO). It describes the atomic arrangement in the neighbourhood of a precise atom and can be well defined in terms of coordination polyhedra. Often, these atomic arrangements correspond to structural units found in crystalline states of the compound. SRO is determined by the basic structural units which constitute the glass network, for exemple the coordination polyhedra BO_3 and BO_4 in the borate glasses or SiO_4 in the silicate glasses. Some of these units have an electrical charge and can receive mobile ions of opposite sign. SRO can be determined by many experimental techniques such as neutrons and X-rays diffraction, Nuclear Magnetic Resonance (NMR), Raman and infra-red spectroscopies (section 3).

Medium Range Order is less easy to deduce from experiment. Its significance and its range are the subject of much discussions. It can be seen as ordering structures with a spatial extension which enlarge the first neighbour units (SRO). In practice, it is usual to separate short-range MRO (3-5 Å), that we call simply Medium Range Order and which concerns the connection between coordination polyhedra (corner-, edge- or face-sharing in chalcogenides glasses), from Long-range MRO which exists at larger length scale (at about 5-10 Å) and can be associated to correlation in the covalently bonded network. In some particular glass systems, Short-range MRO is characterised by the presence of very well defined superstructural units consisting of several basic SRO units (rings or clusters of atoms of particular shapes and size) and, such superstructural units can be regarded as forming the essential building blocks of the amorphous structure (section 4).

Long-range MRO can be regarded as being associated with the local dimensionality of the covalently bonded glassy network. If in an isotropic material the corresponding local dimensionally is three, this can be reduced by the type of connection between the coordination polyhedra, like in chalcogenides glasses, where 1D chain-like structures are formed by edge-shared tetrahedra. This can also result from the depolymerization due to the addition of a network modifier responsible of the formation of non-bridging oxygen atoms (NBO) at the expense of the bridging ones. Here, the local dimensionally is defined as the dimension of the percolation paths connecting neighbouring NBO sites, and the result of the NBO formation is to make this less than three. In alkali silicates, 2D sheet-like internal surfaces bordered by negative NBO sites are formed and they can acts as conducting paths for ionic transport. Long-range MRO is the origin of "abnormal" behaviour observed in dynamical properties such as specific heat, thermal conductivity, acoustic attenuation, low frequency Raman scattering, infra-red absorption, inelastic neutron scattering. Long-range medium range order in borate glasses will be discussed in detail in section 5 (Long Range Order in borate glasses).

In the fast ionic conducting glasses, the ionic conductivity results from the motion of monovalent ions such as alkali (essentially lithium or sodium) or silver cations. For temperatures below the glass-transition temperature T_g, the ionic conductivity of the vast majority of glassy conductors σ verifies the Arrhenius law:

$$\sigma = \sigma_o / T \exp\left(-E_{\sigma T} / kT\right) \tag{2}$$

where σ_o, the prefactor is related to the number of mobile ions and $E_{\sigma T}$ the activation energy for the ionic transport (section 6).

A common characteristic of ionic transport in borate glasses and also in most of the fast ionic conducting glasses is a relatively low and concentration independent value of the prefactor σ_o which does not agree with the prediction of a simple model including the diffusion of an isolated ion, while the activation energies can be understood by structural changes induced by modifications in the glass composition. This will be discussed in section 7 as a function of the different degrees of disorder (Short, Medium and Long Range Order).

2 Alkali borate glass formation

2.1 *Introduction*

Glass formation is not a very frequent phenomena but it arises for certain families of compounds. Most of the elements or minerals, give rise on melting to liquids with small viscosity (10^{-2} poises). When cooling such liquids crystallise rapidly when the temperature reaches the melting point temperature and does to, for a large cooling rate domain. However there exist some compounds for which the viscosity of the melt is about 10^5 to 10^7 poises. For such liquids as the temperature decreases this becomes so high than the compound has all the characteristics of a solid. This process is call vitrification and the material is obtained in the vitreous state: It is a glass [4].

Some examples of compounds which can vitrify easily are found in different families such as:

– elements of the group V and VI of the periodic table (phosphorus, sulphur, selenium).

– oxides. Oxides like SiO_2, GeO_2, B_2O_3, P_2O_5, As_2O_3, Sb_2O_3, give glasses alone or in association with other oxides (alkali or metallic oxides).

– chalcogenides. Glasses are obtained in the binary systems *As-S* (or *Se*), *Ge-S* (or *Se*), *Si-S* (or *Se*), or by mixing sulphides or selenides with *Ga, In, Sn, Te, Bi, Pb, Sb, Tl*, etc.

– halides. BeF_2 and $ZnCl_2$

These compounds are called glass formers. They form cross-linked chains of covalently bonded units and serve to constitute the skeleton of the glass network. A great variety of glass-forming systems can be obtained from the many possible combination of the different formers.

The range of the possible glasses is enormously extended by including some of the numerous other oxides or chalcogenides during the melting process. These additives are used as glass modifiers and they introduce ionic bonds into a structure which was previously covalently bonded. This leads to structural changes in the glass which are reflected in physical properties (melting point, glass transition temperature, density, etc.). A further type of major constituents are the doping salts, which, unlike the modifier enter interstitially into the glass with generally, little or no direct effect on the network.

2.2 *Preparation of fast ionic conducting alkali borate glasses*

Boron oxide (B_2O_3) is an excellent glass former because it cannot be obtained easily in its crystalline form. Its melting point is about 475 °C. Monocrystals can be obtained [5] by the slow cooling of the melt at a pressure between 1 GPa and 2 GPa. B_2O_3 can be prepared from melts obtained by dehydration of H_3BO_3 at progressively increasing temperature and it then forms glass even with slow cooling.

Alkali borate glasses such as $(1\text{-}x_m)B_2O_3\text{-}x_mLi_2O$ are prepared from the melt of boron oxide with a glass modifier like Li_2O or other alkali oxide or carbonate in appropriate amount. They are commonly call binary alkali borate glasses. The binary alkali borate glass formation domain are shown in figure 1.

In the alkali borate systems, the liquidus temperature increases rapidly with the addition of alkali. Melting times of about 10 to 30 mn at a temperature in the range of 800 °C - 1000 °C are necessary to obtain clear and bubble free melts. A platinum crucible is used for the glass elaboration. Then, because they are much easier to crystallise than B_2O_3 itself, glass samples are obtained by rapid quenching of the melt between copper blocks. In figure 1, the extensions of the vitreous domain obtained by the use of twin-roller quenching are also reported [6].

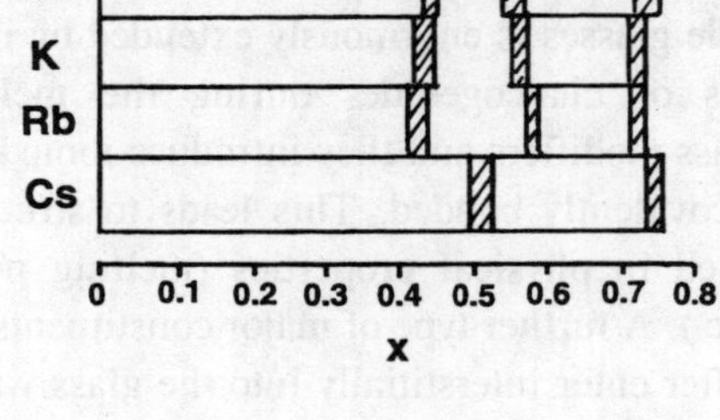

Figure 1. Glass forming regions of alkali borate glasses. The shaded areas indicate vitrification of small batches by splat or roller quenching (from reference 6).

The network modifier, as its name implies, serves to induce structural changes in the glass network of the boron oxide. It introduces ionic bonds into the structure initially covalently bonded of B_2O_3. Alkali borate glasses become fast ionic conductors of the lithium cations and the ionic conductivity increases with the increase of the modifier content.

After the addition of a modifier alkali oxide, the glass former B_2O_3, is able to dissolve a third constituent and considerably more lithium can be added in the binary glasses without causing devitrification by incorporating it in the form of lithium salts such as lithium halide, sulphate or phosphate [7]. In this way, the ionic conductivity obtained is significantly higher than for glasses containing alkali oxide alone. Figure 2 shows the vitreous domain obtained for lithium chloro- and sulfo-borate glasses.

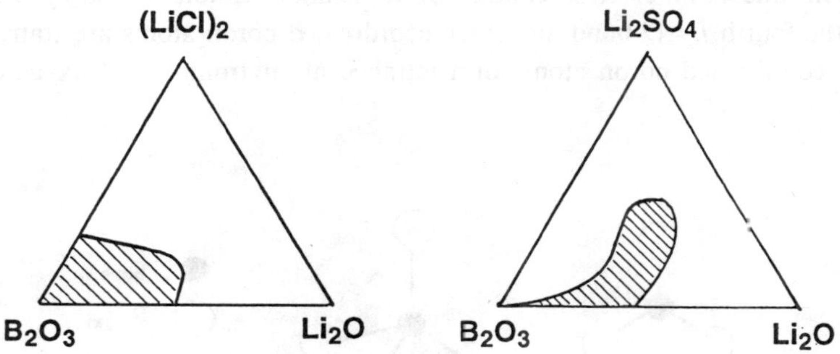

Figure 2. Vitreous domain of the two "doped" glass systems B_2O_3-xLi_2O-$yLiCl$ and B_2O_3-xLi_2O-yLi_2SO_4. (from reference 7 and 8).

The ternary glasses are prepared with the same experimental procedure as for the binaries. Some authors have noticed that considerable vapour losses (in halide glasses) occur during melting [8,9]. The amount of halide retained is believed to depend upon the temperature and duration of the melting. and they have tried to minimise such loss by melting during short times and by covering the crucible.

3 Short range order (SRO) in alkali borate glasses

In 1932, Zachariasen [10] have proposed what is now considered as the basic model of the atomic structure of vitreous B_2O_3. Boron oxide is viewed like a three-dimensional "continuous random network" (CRN) of corner-shared BO_3 structural units. In earlier NMR work, Silver and al. established that all of the BO_3 are essentially planar [11]. Each of them form an equilateral triangle with an oxygen at each corner (figure 3). This model was confirmed by Mozzi et al. by means of X-Ray diffraction [12]. BO_3 triangles are also evidenced in infrared spectroscopy where they can be characterised by the stretching frequency of the B–O bonds observed at 1400 cm^{-1} [13,14].

When alkali or metal oxides M_2O are incorporated into the B_2O_3 to form B_2O_3-xM_2O binary glasses, the oxygen atoms are incorporated into the boron-oxygen network. It is assumed, at least for small x, that for each alkali or metal atoms added, an electron coming from the alkali cation is used by the boron to form the fourth B—O bond and three-coordinated boron atoms are transformed into four-coordinated boron atoms in a tetrahedral environment of oxygen atoms (figure 3).

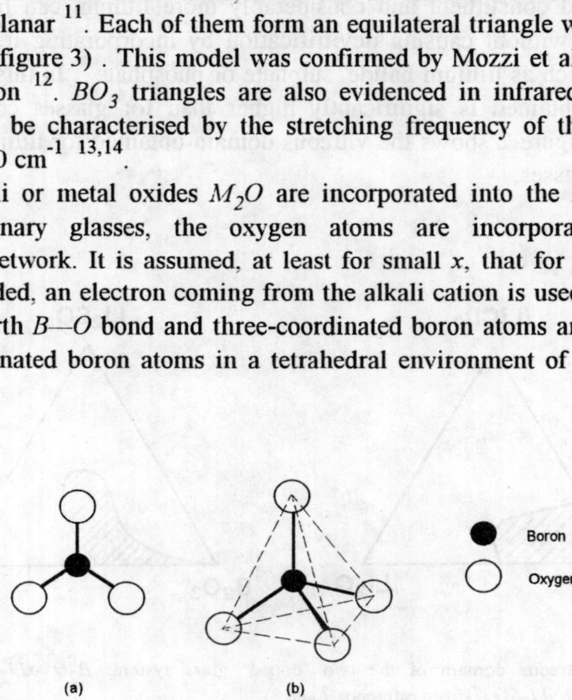

Figure 3. The three- and four-coordinated boron atoms in their oxygen environment: (a) planar BO_3 triangles; (b) BO_4^- tetrahedron.

This reaction can be described by the following equation:

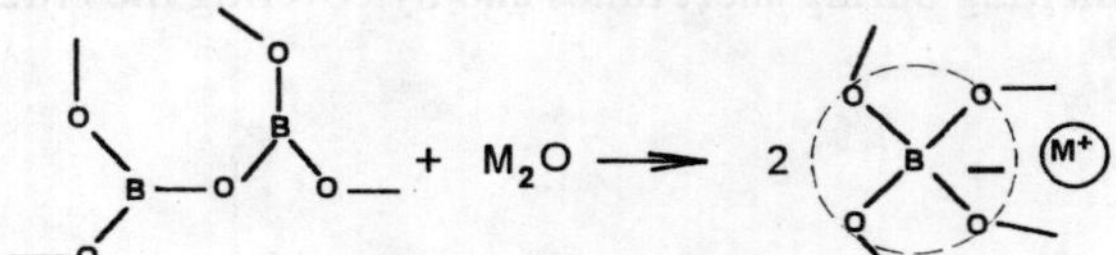

(3)

In the covalent network, the BO_4 units are negative sites while the M^+ ions move around to provide local charge neutrality [13]. This assumption has been verified by ^{11}B NMR [14], where a correspondence is obtained between the number of alkali atoms and the number of four-coordinated boron. Bray et al. [15] have shown that, N_4, the fraction of four-coordinated boron atoms in alkali borate glasses of composition $x_m R_2O\text{-}(1\text{-}x_m)B_2O_3$ ($R = Li, Na, K, Rb, Cs$) follows the relation:

$$N_4 \propto x_m / (1\text{-}x_m) \tag{4}$$

up to approximately $x_m = 0.3$ mol%, where it rises a maximum value (figure 4).

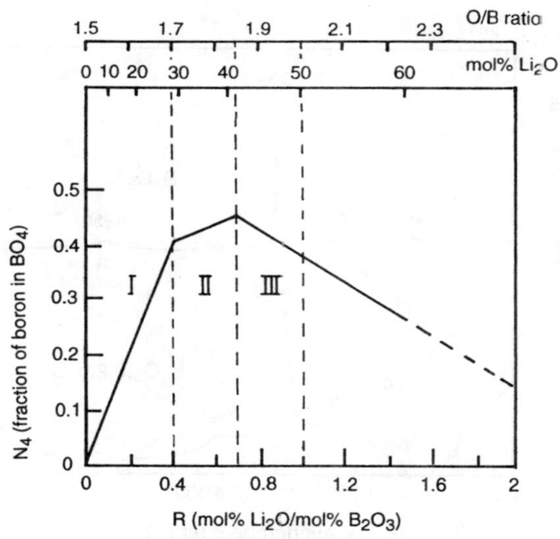

Figure 4. Plot of the fraction of four coodinated boron atoms N_4, versus composition in lithium borate glasses (from reference 16).

Systematic neutron diffraction experiments performed on different metal oxide modified borate glasses $x_m M_2O\text{-}2B_2O_3$ ($M = Ag, Li, Na$) have shown that the short range order of the boron-oxygen network, involving the first B-O, B-B, and O-O interatomic distances, is almost identical for the silver, lithium, sodium borates of same modifier concentration. A detailed analysis of the first peak in the atomic pair correlation function reveals two different *B-O* nearest-neighbour distances, 1.37 and 1.47 Å, which are attributed respectively to BO_3 and BO_4 groups [17].

These structural changes are also detected by vibrational spectroscopies (Raman or infrared). In Raman scattering, the formation of four-coordinated boron atoms with the modifier addition is evidenced by the appearance of new bands at about 780 and 960 cm^{-1} and by a change in the shape of the broad band between 1300 cm^{-1} and 1400 cm^{-1} [18] (figure 5). They correspond to internal vibrations of well defined complex structural units containing only BO_3 (806 cm^{-1}) or both BO_3 and BO_4 groups (780 cm^{-1}) These groups constitute a second step of structural organisation in the boron-oxygen network and this latter point will be discussed in the section on Medium Range Order .

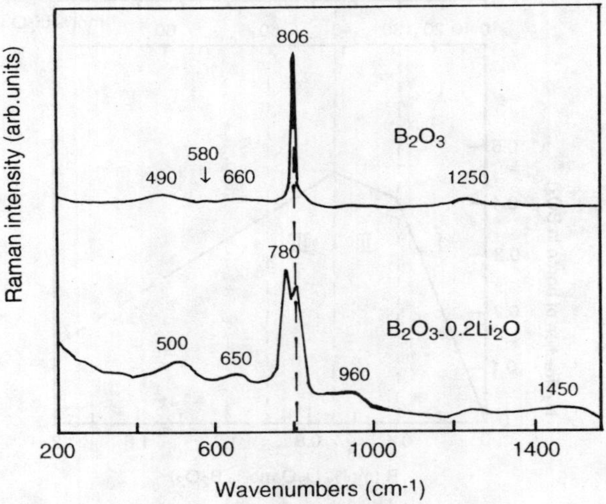

Figure 5. Modifications of the Raman spectra of the borate glasses under the modifier addition: (a) glassy B_2O_3; (b) B_2O_3-$0.2Li_2O$.

In lithium borate glasses B_2O_3-xLi_2O, the vibrational signature of the BO_4 tetrahedral units is clearly observed in infrared spectra where a new band appears at about 1000 cm^{-1}. It corresponds to the stretching vibration of the *B—O* bonds in BO_4 tetrahedra while those at 1300 cm^{-1} is attributed to the same motion but in planar BO_3 triangles (figure 6) [19]. The variation of the intensity of this band is similar to those reported for the number of BO_4 units measured by NMR. The band at about 700 cm^{-1} is characteristic of the vibration of bending of the *B–O–B* bridges constituting the covalent network of the alkali borate glasses. As has been observed by Raman scattering, another band, at 1260 cm^{-1}, is also the signature of vibrational motion in structural units forming the Medium Range Order of the network.

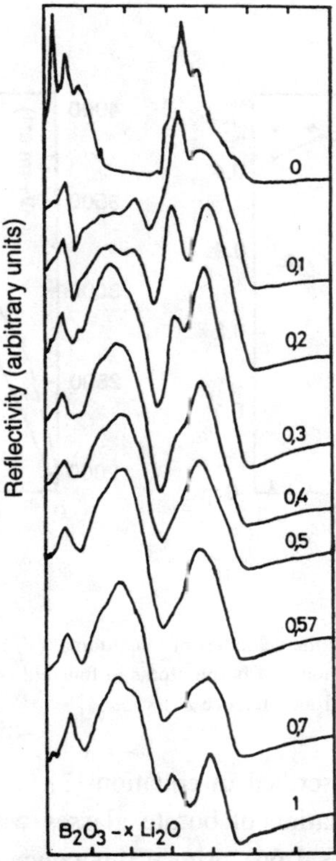

Figure 6. Infrared reflectivity of binary glasses B_2O_3-xLi_2O with $0 < x < 1$.

The transformation $B_{III} \rightarrow B_{IV}$ greatly increases the rigidity of the glass network. This has been revealed by the changes of the acoustic properties of binary borate glasses B_2O_3-xLi_2O reported by Lorösch et al. [20]. The variations of longitudinal and transverse velocities with the lithium oxide concentration are given in figure 7. As x increases, the small Li^+ cations are inserted into interstices in the network. At the same time, the rigidity increases with the Li_2O concentration. Comparison is made in the figure with the number of four-coordinated boron atoms obtained from NMR results [15].

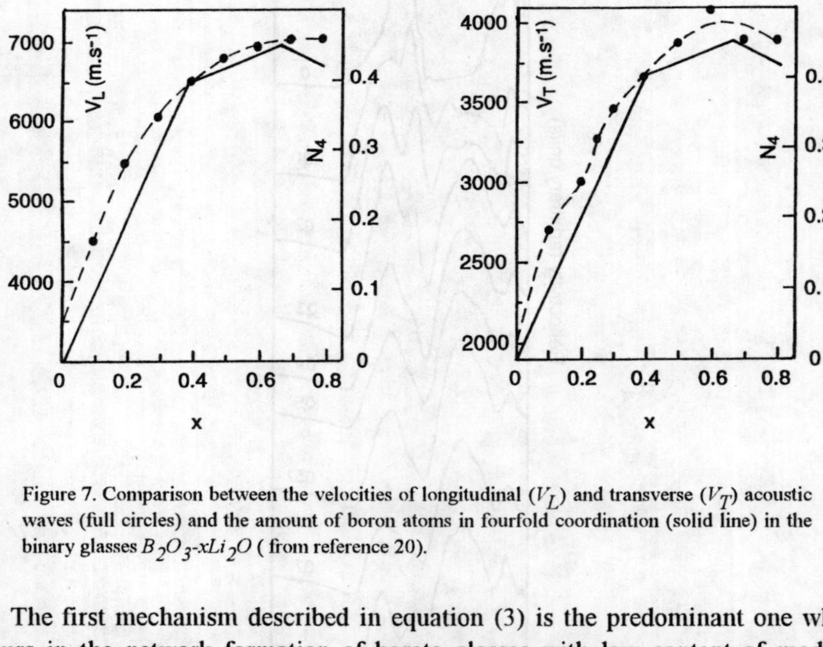

Figure 7. Comparison between the velocities of longitudinal (V_L) and transverse (V_T) acoustic waves (full circles) and the amount of boron atoms in fourfold coordination (solid line) in the binary glasses B_2O_3-xLi_2O (from reference 20).

The first mechanism described in equation (3) is the predominant one which occurs in the network formation of borate glasses with low content of modifier oxide ($x \leq 30$ mol% or $O/B \leq 1.66$). Above this concentration, the introduction of network modifiers is responsible for the creation of negatively charged non-bridging oxygen (NBO) centres at the expense of the bridging oxygens, thereby causing depolymerisation of the network. This new process is described by the equation (5):

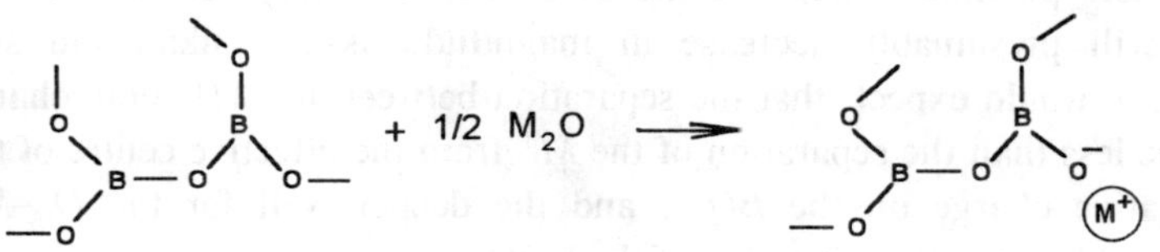

(5)

The coordination of the boron atoms remains three and triangular BO_3 units with non-bridging oxygen atoms are formed into the network.. The frequency of the stretching vibration of the non-bridging $B—O^-$ bonds in BO_3 with NBO is found near 1250 cm^{-1} in the infrared spectra (figure 5) [13]. Simultaneously, the observation by NMR, of a decrease of the fraction of four-coordinated boron atoms suggest that another mechanism is also responsible of the formation of these new type of triangular BO_3 units [13,21]. In the proposed new mechanism, borate groups containing BO_4 tetrahedral units are broken and thus constitute short elements of chain which contain one BO_3 with NBO connected with one other BO_3 triangle.This other way would be described by the equation (6):

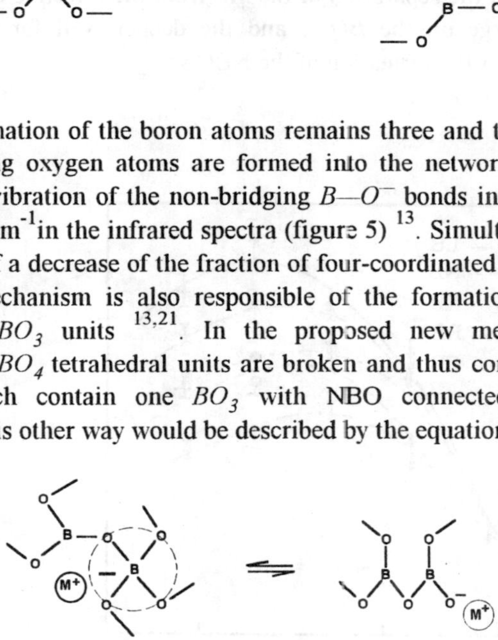

(6)

In their NMR experiments, Bray et al. [22] have reported that the size of the alkali cation influences the relative abundance of three and four coordinated borons when the alkali atoms are replaced by an other one while maintaining the alkali content constant. The figure 8 shows that the number of four-coordinated boron atoms for a network with a given oxygen-boron ratio decreases as one goes from lithium to caesium borate glasses.

This result also appears in the composition dependence of the sound velocities [20]. They follow directly the composition dependence of the number of boron atoms with coordination four. As the size of the cation increases, the decrease of N_4 reduces the network rigidity and the two sound velocities decrease.

The dependence of N_4 on the specific alkali ion arises presumably from a competition between two processes that utilise the alkali ion in the glass. One of these processes is the formation of four-coordinated borons in tetrahedral BO_4^- units. The other process is the formation of a negative NBO on BO_3 units. The negative coulombic potential energy of the BO_4^-–M^+ configuration (M^+ is the alkali cation) will presumably decrease in magnitude as the alkali ion size increases. But one would expect that the separation between the M^+ ionic charge and the O^- to be less than the separation of the M^+ from the effective centre of the distributed negative charge on the BO_4^-, and the deeper well for the O^-–M^+ configuration would favour the formation of the NBO's.

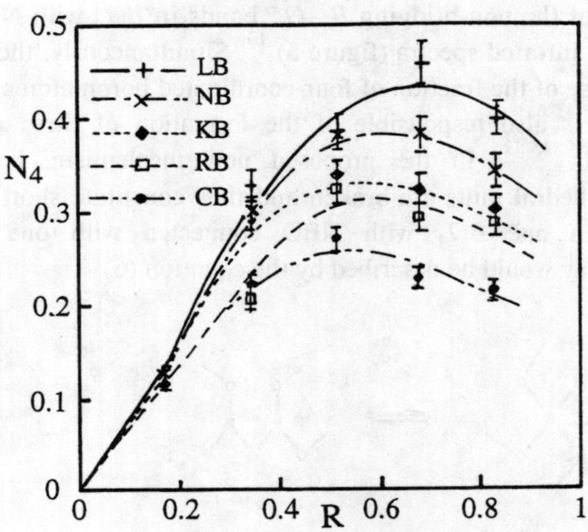

Figure 8. N_4 as a function of R for alkali borate glasses. LB = lithium borate, NB = sodium borate, etc. Here R = mol% alkali oxide/mol% B_2O_3 (reference 22).

Other experimental results like glass transition temperature T_g or density ρ measurements can be understood in terms of an initial strengthening of the network by the systematic replacement of the planar BO_3 units by tetrahedrally bonded BO_4 units with the increase of the O/B ratio, followed by a weakening at larger alkali oxides concentrations due to the increase of the number of non-bridging oxygen and the decrease of those of the four-coodinated boron. In each case, a sharp increase is reported in the dependence of T_g or in ρ with increasing the O/B ratio for $O/B \leq 1.66$, then, in a concentration range $1.66 < O/B < 1.7$, a maximum, followed by a monotonic decrease, is observed with increasing O/B. (figure 9) [23].

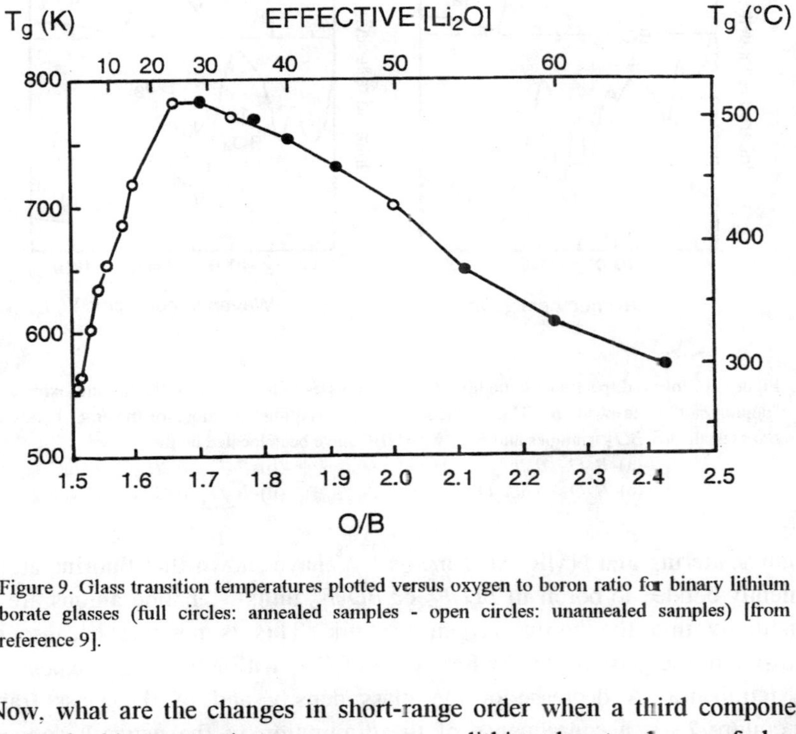

Figure 9. Glass transition temperatures plotted versus oxygen to boron ratio for binary lithium borate glasses (full circles: annealed samples - open circles: unannealed samples) [from reference 9].

Now, what are the changes in short-range order when a third component is added the the binary matrix to prepare ternary lithium borate glasses of chemical formulae B_2O_3–xLi_2O–yLi_nX^n ("doped"glasses) The answers come from many different experiments like NMR [24], Raman or infrared spectroscopies [25], T_g and density measurements [26]. It is clear that the same set of structural basic units are found in the ternary glass network. Figure 10 shows the infrared reflectivity of some lithium borate glasses with different composition in modifier and in doping salt. At varying composition, the vibrational features assigned to the three basic units of the matrix have different intensities determined in a first approximation by the values of the O/B ratio. However, changes in the relative amount of the three basic units (BO_3, BO_4, or BO_3 with NBO) are reported when doping salts are added in the boron-oxygen network [25]. Except for fluoroborate glasses, in which

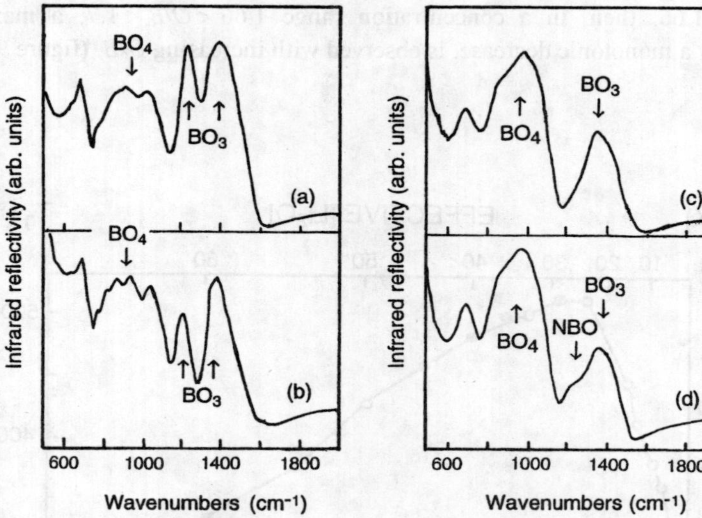

Figure 10. Infrared spectra of some lithium borate glasses as a function of the lithium oxide and "doping salt" concentrations. The frequency of the stretching vibrations of the *B–O* bonds in BO_4 tetrahedra, BO_3 triangles and BO_3 with NBO have been located on the picture.

-(a)- B_2O_3–$0.2Li_2O$ -(b)- B_2O_3–$0.2Li_2O$–$0.1LiCl$
-(c)- B_2O_3–$0.6Li_2O$ -(d)- B_2O_3–$0.6Li_2O$–$0.24LiCl$

Raman scattering and NMR experiments [27,28] have shown that fluorine atoms are covalently bonded to boron in BO_3F_3 or BO_2F_2 units [29], halide anions are added interstitially into the boron oxygen network. This is possible because of the expansion of the glass matrix by formation of BO_3 with NBO at the expense of the BO_4 tetrahedra. A decrease of the glass density and of the glass transition temperature T_g is a consequence of the diminution in the network connectivity (figure 11) [23]. It is interesting to note that for a given R_2O/B_2O_3 ratio, these effects are amplified when large size cations are added [30].

Non-borate groups are not excluded from the network when the doping salt contains oxyanions like SO_4^{-2} or PO_4^{-3} which can participate in the network structure (glasses doped with lithium sulphate or lithium phosphate) [5]. In lithium borophosphate glasses B_2O_3-xLi_2O-yLi_3PO_3, no isolated phosphate ions have been observed inside the glass matrix by Raman scattering [31]. It is reasonable to suppose that phosphate ions react with the boron-oxygen network and that they transform

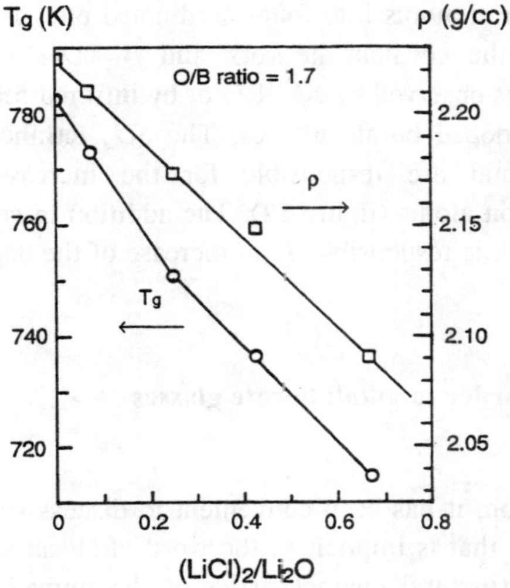

Figure 11. Glass transition temperature T_g and density ρ of lithium chloroborate are plotted against $(LiCl)_2/Li_2O$ at a fixed O/B ratio of 1.70 (from reference 23).

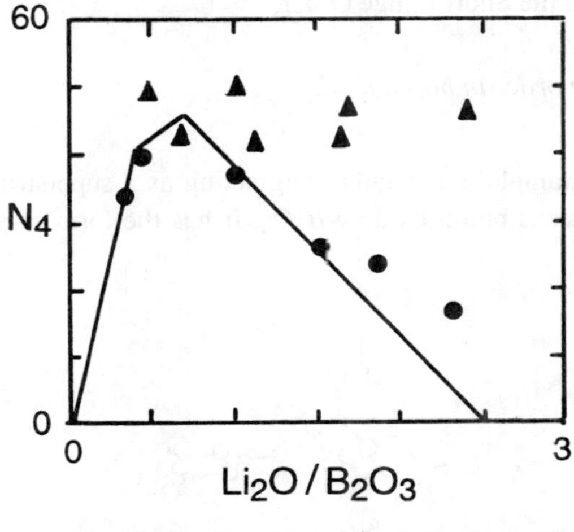

Figure 12. Fraction of the four-coordinated borons in alkaline borate glasses (solid line and circles) and in sulphoborates (triangles) versus the Li_2O/B_2O_3 molar fraction (from reference 32).

three-coordinated boron atoms into four-coordinated one. Tetrahedral PO_4 groups are integrated into the covalent network and B_{IV}–O–P–O–$B_{III(IV)}$ are formed. Similar behaviour was observed by NMR [32] or by infrared reflectivity [33] in the case of lithium sulphate doped borate glasses. The SO_4 tetrahedra participate to the covalent network. and are responsible for the increase of the number of four-coordinated boron atoms (figure 12). The addition of such oxyanions into the boron-oxygen network is responsible of an increase of the degree of connectivity in the covalent network.

4 Medium range order in alkali borate glasses

In the previous section, it has been convenient to discuss structural parameters in terms of length scale that is implicit in the word of "local structure". This can be taken to mean the structural characteristics of the immediate environment of a given specie out to second neigbours. At the other extreme of the the length scale, glasses are clearly disordered. Between the limits for local and long range structure comes the "medium range" structure. The Medium Range Order (MRO) is defined as the type of connection between the coordination polyhedra describing the Short Range Order in the glassy material and it can be considered as the next level of organization beyond the Short Range Order.

4.1 Medium-range order in boron oxide

One classical example of a regular ring acting as a superstructural unit is the boroxol ring in vitreous boron oxide v-B_2O_3. It has the form of a planar hexagon

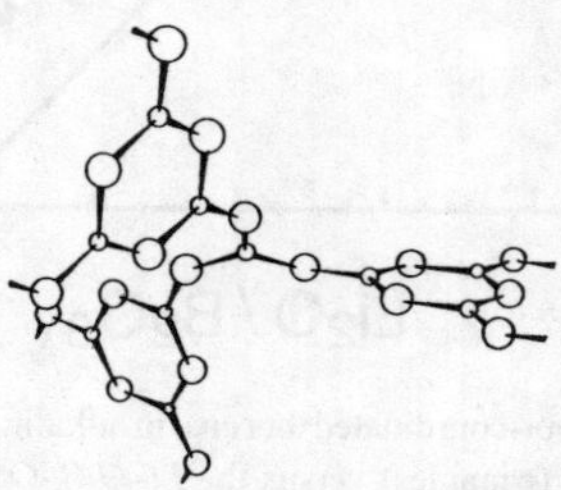

Figure 13. Perspective vue of the boroxol rings used to describe v-B_2O_3.

consisting of three corner-linked planar BO_3 triangles connected together as the shape of an equilateral triangle (figure 13). The presence of boroxol groups was first deduced from the observation of an extremely intense, very narrow for an amorphous material (~10 cm^{-1}) and highly polarised peak at 808 cm^{-1} in the Raman spectrum of v-B_2O_3 [34,35] (figure 14).

By isotopic substitutions of boron or oxygen, it was found that this line involves little or no motion of the boron atoms [36]. It is thus interpreted as due to the breathing mode ν_2 of a boroxol ring [37] (figure 15). The sharpness of the line is ascribe to the motion being localised to a single ring and to the small spread of the angle in the ring [38]. One other peak, observed at about 500 cm^{-1} is due to the ring breathing mode ν_3 which corresponds to in phase movements between the boron

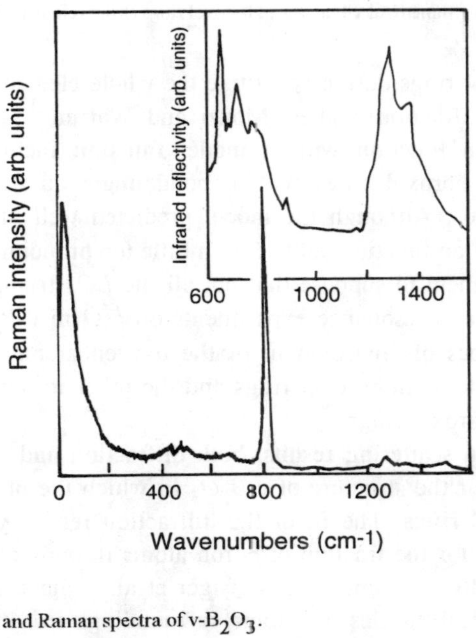

Figure 14. Infrared and Raman spectra of v-B_2O_3.

atoms and the bridging oxygen. The breadth of the line is ascribed to the motion being extended and to the distribution of bridging oxygen *B—O—B* angles outside the rings. Another typical feature attributed to the boroxol rings is observed at

1250 cm^{-1} in the infrared reflectivity spectra (figure 15) and it corresponds to the vibration mode ν_1 which involves symmetric motions of the boron against the bridging oxygen atoms[37].

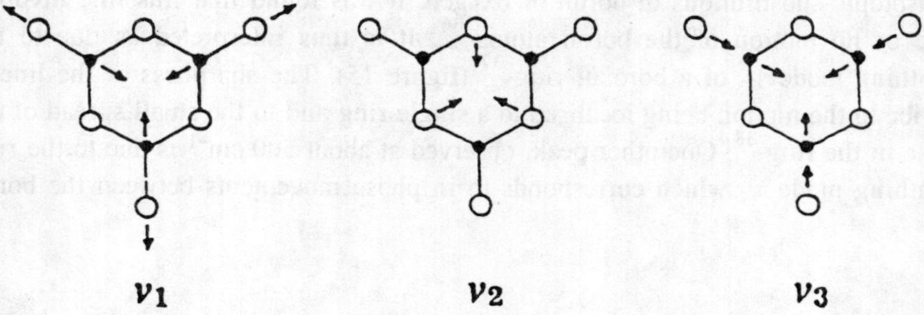

Figure 15. The three symmetrical vibrations of boroxol rings (from reference 37).

However, boroxol rings do not constitute the whole element of the network of $v\text{-}B_2O_3$. In X-ray diffraction studies, Mozzi and Warren[12] have compared the real-space correlation function with a model function including a completely random network of boroxol rings with a bond angle on the bridging oxygen between rings of 130°. Although the model predicted well all the peaks of the experimental correlation function out to 6 Å, a little too pronounced structure in the model curve allows them to suppose that not all the BO_3 triangles are in boroxol rings. Nuclear magnetic resonance experiments on ^{17}O in $v\text{-}B_2O_3$ have provided evidence for two types of environment of the oxygen atoms. One of these was ascribed to oxygen atoms in boroxol rings and the other to oxygen atoms in BO_3 triangles out of the boroxol rings[39].

Recently, neutron scattering results, both diffraction and inelastic scattering, support this model for the structure of $v\text{-}B_2O_3$ in which the majority of the boron atoms are in boroxol rings. The fit of the diffraction results yields an estimated value of 0.80 ± 0.05 for the fraction of boron atoms in boroxol rings[40]. Through two-dimensional NMR experiments, Zwansiger et al.[41] have obtained a detailed picture of the chemical species and their surroundings in vitreous B_2O_3. Their dynamic angle spinning experiments yields sufficient resolution in ^{11}B spectra to detect fine details in the glass structure. In glassy boron oxide, the boron in boroxol rings and in loose BO_3 triangles were resolved and the relative populations of each boron site estimated: boroxol rings sites ~ 65% and non-ring BO_3 sites ~ 35%.

4.2 *Medium Range Order in binary glasses B_2O_3-xLi_2O*

As oxygen is added to B_2O_3, by the addition of alkali oxide modifiers, this additional oxygen is accommodated by conversion of trigonal boron atoms into four-fold or tetraedral coordinated boron atoms. The four oxygens of each BO_4 are

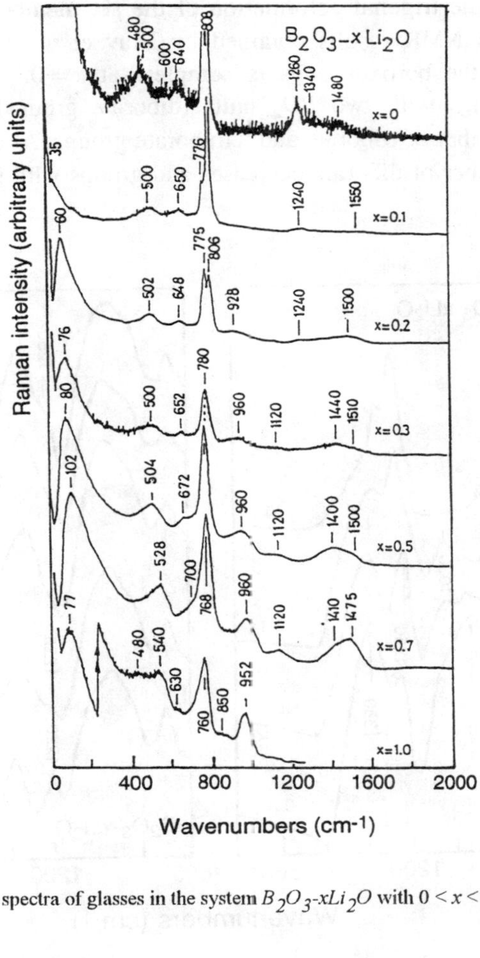

Figure. 16. Raman spectra of glasses in the system B_2O_3-xLi_2O with $0 < x < 1$.

then bridging oxygen and these units increase the cohesion of the glass network. This transformation is clearly seen in Raman scattering experiments in the binary glasses B_2O_3-xLi_2O where a new peak appears at about 780 cm^{-1} (see section 3). The intensity of the breathing mode of the boroxol rings at 805 cm^{-1} decreases and vanishes when x reaches 0.3 (figure 16).

According to Krogh-Moe's structural model [42], for small alkali oxide content, the addition of modifier oxide to borate glasses leads to the formation of rings with one BO_4 unit (triborate or tetraborate groups). It is therefore reasonable to attribute this new band to the trigonal deformation of the six membered borate ring with BO_4 units [43]. The NMR results obtained by Bray et al. confirm also that the disappearance of the boroxol rings is complete at $x = 0.3$ [44]. At higher alkali concentrations, rings with two BO_4 units (diborate groups) are created at the expense of the number of triborate and tetraborate groups. At concentrations above 30 mol%, the number of diborate decreases and groups with non-bridging oxygen

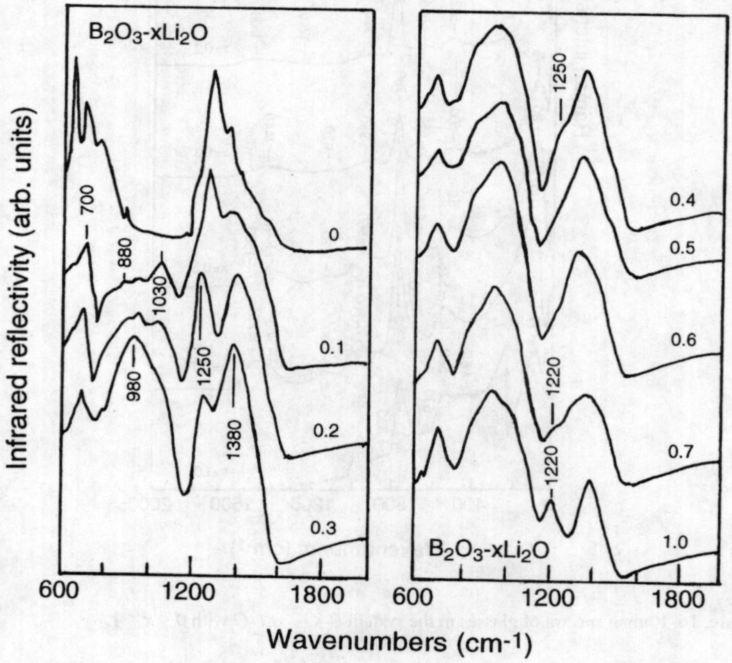

Figure 17. Infrared spectra of the binary borate glasses B_2O_3-xLi_2O with $0 < x < 1$.

atoms (NBO) are formed. Each type of ring has a characteristic Raman signature in the frequency range between 700 and 850 cm^{-1} [33]; the boroxol rings being responsible of the Raman band at 805 cm^{-1} (breathing mode) and the borate rings with one BO_4 unit at 775-780 cm^{-1}. The presence of one four-coordinated boron in the ring results in the weakening of *B-O* bonds in the ring and for these rings the frequency of the breathing mode will be lower than in the boroxol rings.

Borate rings with two BO_4 units show lines in the range 760-770 cm^{-1}. As for borate rings of the previous type, the presence of a second four-coordinated boron in the ring contribute to a more important lowering of the Raman frequency for this specific vibration mode.

In the mid infrared spectra, (see figures 6 and 17), the main observation, under the increase of the oxide content, is the growth of a new band between 850 and 1100 cm^{-1} characterised by a fine structure with three components.By comparison with the spectra of polycrystalline compounds [37], it is possible to relate the central component at 980 cm^{-1} to the *B–O* bond stretching of BO_4 units grouped in pairs to form diborate units while the other two near 890 and 1035 cm^{-1} involve the same vibration but in tetraborate units. Perspective views of these particular borate groups are drawn in figure 18.

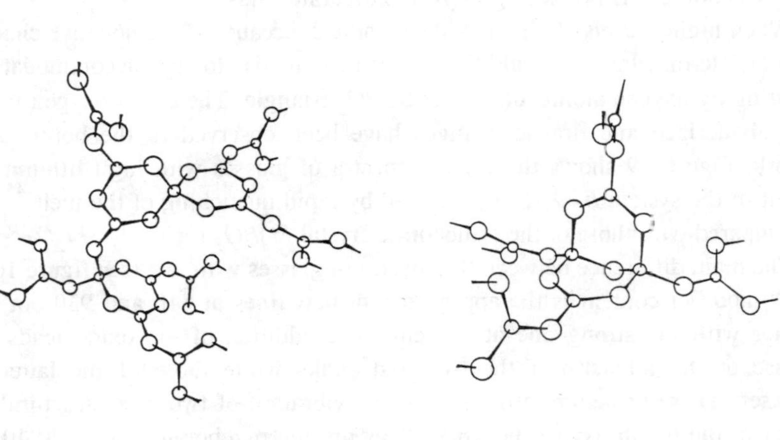

(a) (b)

Figure 18. Perspective views of the tetraborate (a) and diborate (b) groups.

These two units have the same number of four-coordinated boron atoms, but in the case of the diborate, the two BO_4 tetrahedra are connected and have the same atomic surroundings. On the contrary, in tetraborate groups, the two BO_4 are in different atomic environments; one is included in a six-membered ring while the other links two perpendicular six-membered borate rings. The evolution of the shape of this band with the increase of the lithium oxide content, is caused by the increase of the number of BO_4 units in the network and by the change of the relative number of each two different borate groups. The existence of the band at 1250 cm^{-1} in the spectra until $x = 0.5$ (ν_1 vibration of the boroxol rings) indicates that this vibration exists also in borate rings with BO_4 tetrahedra. For this reason, it might be considered as the signature of six-membered rings with one four-coordinated boron atom (triborate units). These units are one of the rings constituting the tetraborate groups which can be considered as the connection of a triborate with a pentaborate group.

These results demonstrate the existence of a second level of organisation inside the network of borate glasses and give a description of the changes in the Medium Range Order with the lithium oxide addition. At low modifier content, BO_3 triangles and BO_4 tetrahedra are grouped to form triborate rings isolated or included into tetraborate rings. Then, with the increase of x, the smoothing of the shape of the BO_4 band and the position of its maximum at 980 cm^{-1} indicates the transformation of the previous groups into diborate rings.

When higher levels of alkali oxide are added, because of the negative charge on the BO_4 tetrahedra, the additional oxygen tends to be accommodated as non-bridging oxygen atoms, one for each BO_3 triangle. The boron-oxygen network is depolymerized and drastic changes have been observed in the boron oxygen network. Figure 19 shows the Raman spectra of glasses with high lithium oxide content in the system B_2O_3-Li_2O obtained by rapid quenching of the melt [45]. They are compared with those of the orthoborate crystal Li_3BO_3 (or B_2O_3-$3Li_2O$).

The main difference between the spectra of glasses with low (see figure 16) and higher modifier content is the appearance of new lines at 840 and 930 cm^{-1} seen together with the strong one at 770 cm^{-1}.The addition of the oxide leads to an increase of the intensity of the two first peaks while those of the latter peak decreases. These peaks are attributed to the vibration of two new structural units present in the boron-oxygen network. They are the pyroborate $B_2O_5^{4-}$ (840 cm^{-1}) and orthoborate BO_3^{3-} (930 cm^{-1}) groups and they constitute the main part of the network as soon as the Li_2O content reaches 50 mol%. By a deconvolution of the Raman spectra into six Gaussian lines, the relative amount of the structural units is obtained for these binary glasses. Borate rings including BO_4 units with bridging oxygen atoms which form the main part of the covalent network, have a maximum of concentration at a composition of 33 mol% (or $x = 0.5$). The addition of the modifier oxide reduces their concentration while the number of pyroborate and orthoborate groups which do not form a covalent network increases. Finally, in the glass of composition B_2O_3-$2.33Li_2O$, the Medium Range Order is not now formed by structural units assembled in a covalent network but by discrete borate anions such as the pyroborate and the orthoborate groups (table 1).

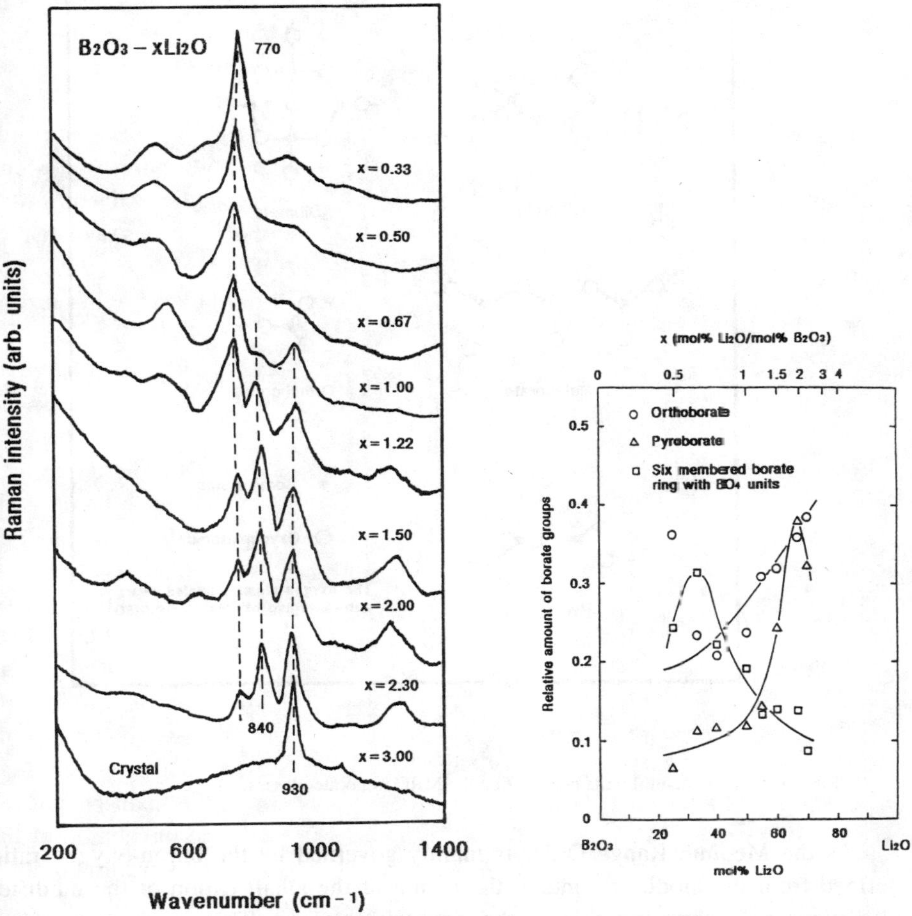

Figure 19. Raman spectra of glasses in the system B_2O_3-xLi_2O with $0.33 < x < 3$ (left) and relative amounts of borate units deduced of the Raman spectra (right) (from reference 45).

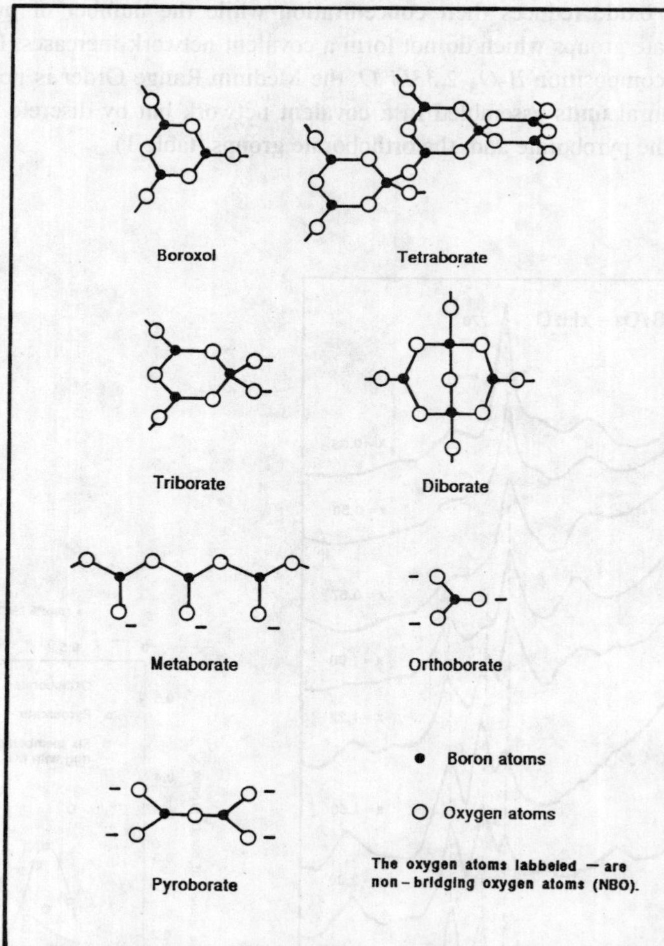

Table 1. Some structural units constituting the MRO in borate glasses.

As the Medium Range Order is mainly governed by the boron-oxygen ratio defined from the modifier content, the nature of the alkali cation of the modifier also plays a determining role in the network obtained. This is observed when lithium atoms are substituted by potassium atoms in diborate glasses B_2O_3-$0.5Li_2O$ and B_2O_3-$0.5K_2O$ [46]. In the Raman spectra (figure 20), there is evidence first of a downward frequency shift of the peak assigned to the symmetrical vibration of the six-membered rings with BO_4 tetrahedra. The shift from 780 to 760 cm^{-1} of the frequency of this peak is indicative of six-membered rings formed with more weakly bound BO_3 triangles and BO_4 tetrahedra.

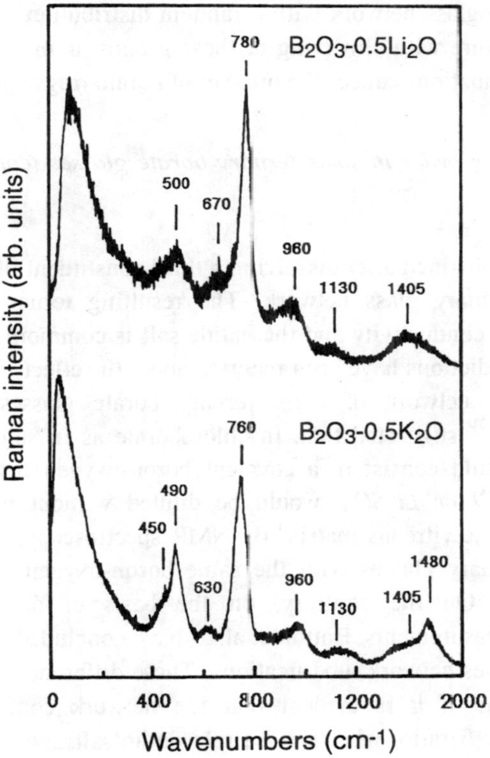

Figure 20. Raman spectra of B_2O_3-$0.5Li_2O$ and B_2O_3-$0.5K_2O$ (from reference.46).

Differences are also reported in the band assigned to the in-phase motion of the *B–O* bridges (near 500 cm^{-1}) and in the range between 1400-1480 cm^{-1} where the spectra is assigned to vibrations of groups containing non-bridging oxygen atoms. The variations of these latter bands reflect differences in the arrangement of the bridging and the non-bridging oxygen atoms included in the structural units defining the Medium Range Order in the boron-oxygen network. This is explained by the disappearence of large borate rings such as tetraborate or diborate and the formation of borate units of smallest size containing NBO's. The metaborate chain is one example of such groups with non-bridging oxygen atoms (table 1).

Similar observations have been made by Soppe et al [47] in caesium borate glasses, Cs_2O-B_2O_3, where the addition of Cs_2O to the borate glass leads to an enhanced creation of NBO's, compared with lithium borate glasses. The size of the Cs^+ ion favours a glass network with a random distribution of Cs^+ ions and BO_4 units than a structure with clustering of these groups as in lithium borate glasses and the NBO's formation reduces the number of borate rings with two BO_4 groups.

4.3 *Medium range order in some ternary borate glasses (commonly call "doped glasses")*

These glasses are obtained after dissolving a third constituent like alkali sulphate or halide into the binary glass network. The resulting ternary glass displays an enhanced cationic conductivity and the halide salt is commonly called the "doping salt". Some contradictions have been reported about the effects of a "doping salt" on the boron-oxygen network of these ternary borate glasses. Irion et al.[48] and Levasseur et al.[49,50] suggested that, in chloroborate as in sulphoborate glass, the glass structure would consist in a covalent boron-oxygen network in which the "doping salt", *LiCl or* Li_2SO_4, would be diluted without producing detectable interactions with the vitreous matrix. By NMR spectroscopy, Geissberger et al.[24], in binary and ternary glasses with the same boron-oxygen ratio, have observed identical spectra. On the contrary, on the basis of density and transition temperature T_g measurements, Button et al.[19] have concluded that *LiCl* addition in borate glasses causes network modifications. These different results give rise to an important question: if it is evident that the network conformation is mainly governed by the *O/B* ratio, what precise role dothe salt anions play in modifying these glasses to increase the ionic conductivity. Some answers were found in structural investigations of the glass network by Raman [51,52] and infrared spectroscopy [13] which display important modifications in the medium range order caused by the anions of the "doping salts".

4.3.1 *Medium range order in halogenoborate glasses*

The medium range order in halogenoborate glasses B_2O_3-xLi_2O-$yLiX$ where X represents the halide anions F^-, Cl^-, Br^-, I^- will be discussed in series which differ by the modifier content x. The first one, in which $x = 0.2$ is interesting because the starting boron-oxygen network contained simultaneously boroxol rings and complex borate groups with four-coordinated boron atoms while the second, where $x = 0.6$, has been chosen because in this lithium oxide concentration the number of four-coordinated boron atoms is maximum.

The Raman spectra of the glasses with the lower lithium oxide content are indicated in figure 21. Two strong peaks at 808 cm^{-1} and 784 cm^{-1} are observed for the B_2O_3-$0.2Li_2O$ binary glass. The first one has been attributed to the boroxol rings, while the second one to the symmetric vibration of six-membered borate rings with BO_4 tetrahedral units. With the addition of lithium halide to the binary($y = 0.1$ for the glass spectra shown in the figure 21a) [52], the relative

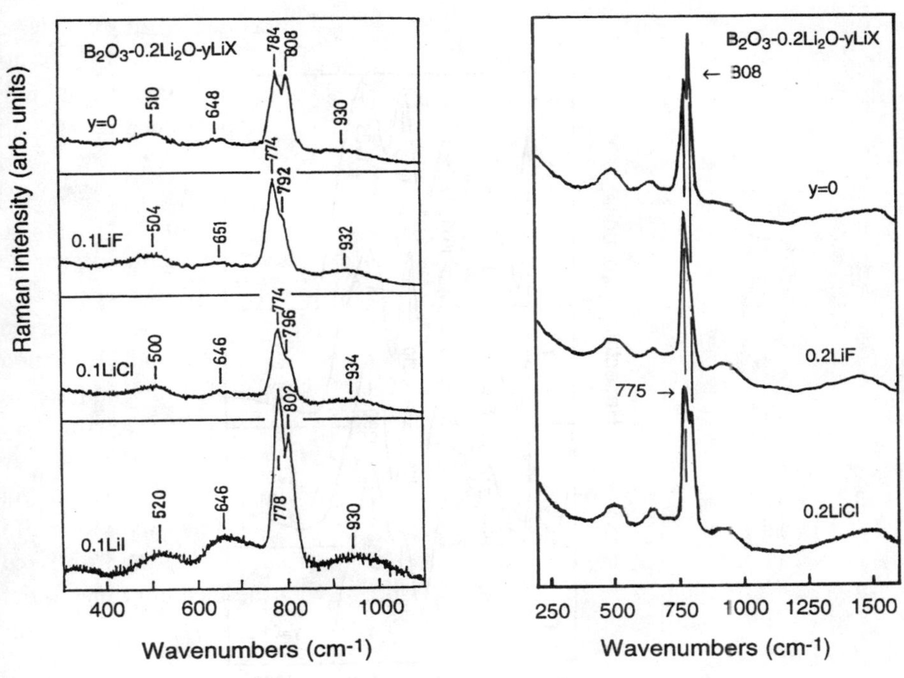

Figure 21. Raman spectra of lithium halogenoborate glasses B_2O_3-$0.2Li_2O$-$yLiX$.
(a) y=0.1 and $X = F$, Cl or I (from reference 52).
(b) y=0.2 and $X = F$ or Cl (from reference 28).

intensity of the peak at 808 cm^{-1} decreases at the expense of the peak at 784 cm^{-1}. The former band practically disappears in the glass with lithium fluoride; it remains for the iodide glass. An intermediate situation is expected in the case of chloride glass. Wang et al.[28] have reported similar observations on glasses with the same Li_2O/B_2O_3 ratio ($x = 0.2$) but in which the lithium halide concentration has been fixed at 0.2 (figure 21b). With this higher concentration in halide salts, the peak at 808 cm^{-1} disappears completely for the *F*-glass, while it remains for the *Cl*-glass. On this basis, it is proposed that the addition of lithium halides such as *LiCl*, *LiBr* or *LiI* in borate glasses with low modifier concentration transforms the boroxol rings into rings with four-coordinated boron atoms but it is also suggests that *LiF* has a singular behaviour in comparison with those of the other lithium halides.

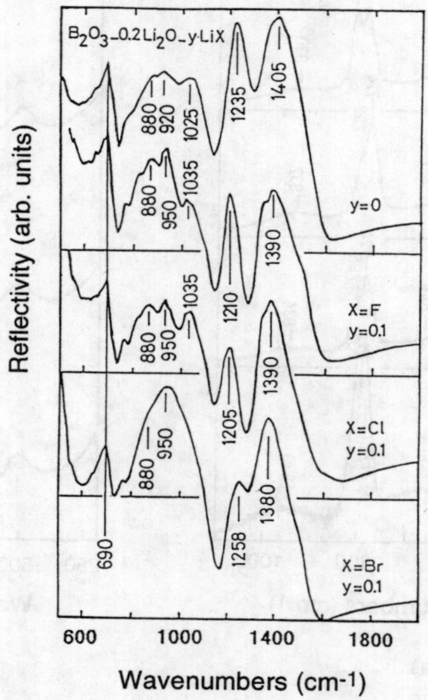

Figure 22. Infrared spectra of halogenoborate glasses B_2O_3-$0.2Li_2O$-$yLiX$ with y=0.1 and $X = F$, *Cl* or *I*.

More interesting results about the medium range order in these glasses are deduced from the infrared reflectivity spectra (figure 22). Four bands at about 700, 850-1100, 1230 and 1350-1400 cm^{-1} are found in the mid-infrared reflectivity spectra of the binary glass. The conversion of B^{III} into B^{IV} appears in the increase of the intensity of the BO_4 stretching vibration band at 850-1100 cm^{-1} when the halide concentration or when the halide anion is changed. Simultaneously the intensity of the narrow band near 1230 cm^{-1} becomes smaller. This latter band is attributed to boroxol rings and shows the decrease of their number under the halide salt addition. As for the binary, the analysis of the shape of the BO_4 band shows that in the fluoride glass, the BO_4 tetrahedra are included preferentially in a tetraborate structure. With the other halide anions, these BO_4 units are redistributed in the network into diborate groups and in the iodide glass, the four-coordinated boron atoms are mostly grouped into these latter structural units.

This behaviour is explained by the fact that halide anions of increasing size are

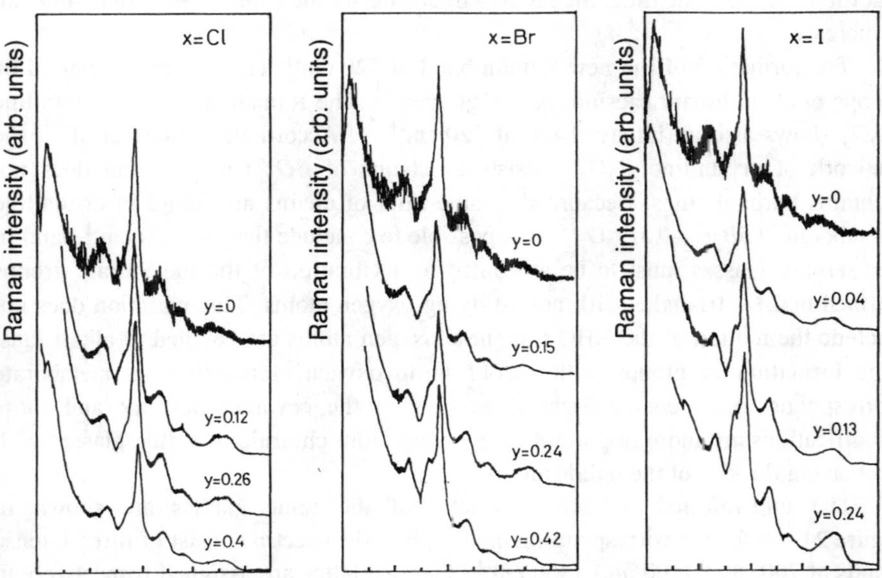

Figure 23. Raman spectra of halogenoborate glasses B_2O_3-$0.6Li_2O$-$yLiX$ with $X = Cl$,Br or I as a function of the lithium halide content y (from reference 51).

located in different positions in the glass network. In fluoroborate glass, tetrahedral units in which the fluoride is covalently bonded with boron are expected while chloride, bromide and iodide anions are in interstitial positions. ESR measurements on chloroborate glasses in which Cl^{-2} radicals are formed upon X-ray irradiation have confirmed that Cl^- ions enter into the glass network interstitially as a free ions [28]. These size effects, which affect the medium range order, arise during the melting process. At this step, to insert the halide anions into the network, *B–O–B* bridges are probably broken with the transforming of tetraborate units into diborate. Chloroborate glass constitutes an intermediate state where tetraborate and diborate coexist in similar numbers.

In halide glasses with higher Li_2O/B_2O_3 ratio ($x = 0.6$), the Raman spectra have a strong peak at 770 cm^{-1}, characteristic of borate rings containing BO_4 units (figure 23). The boron-oxygen network does not contain boroxol rings but modifications of the network are revealed by the appearance of a shoulder at 720 cm^{-1} under the lithium halide addition [51]. The frequency of this new peak shifts toward the lower frequency and it intensity is enhanced upon the halide salt addition. At the same time, the relative intensities of the bands at 960 and 1480 cm^{-1} increase.

The formation of the new Raman band at 720 cm^{-1} has been also reported by Soppe et al. in binary caesium borate glasses [47]. The Raman spectra of crystalline B_2O_3 shows also an intense band at 720 cm^{-1} [53]. According to Gurr et al.[54], the network of crystalline B_2O_3 consists of chains of BO_3 triangles and does not contains boroxol rings. Because the same kind of chains are found in crystalline metaborate $0.5B_2O_3$-$0.5Li_2O$ [55], it is possible to conclude that the 720 cm^{-1} band in the ternary glasses must to be attributed to a vibration of the metaborate groups formed by BO_3 triangles with non-bridging oxygen atoms. This vibration does not include the motion of the NBO's as these oxygen atoms are coupled to alkali ions. The formation of groups with non-bridging oxygen atoms such as metaborate corresponds to a depolymerization process in the covalent network and these modifications are more important as one goes from chloride to iodide glasses with increasing the size of the halide anion.

The mid-infrared reflectivity spectra of the same glasses are shown in figure 24. As for the corresponding binary glass the spectra consist of three intense bands at 700, 800-1100 and 1350 cm^{-1}. The two latter are assigned respectively to stretching vibrations of *B–O* bonds in tetrahedral BO_4 and in BO_3 triangles. The broad maximum at 970 cm^{-1} observed in the BO_4 bands indicates that diborate groups form the main part of the network. When lithium halide is added, a new band appears in the reflectivity spectra at 1245 cm^{-1} in the fluoride glass. By

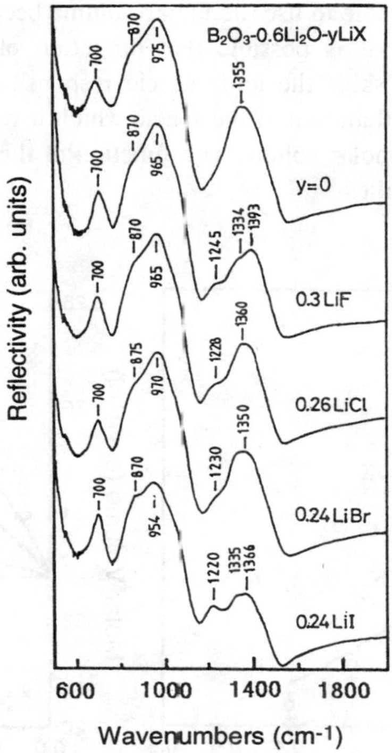

Figure 24. Infrared spectra of halogenoborate glasses B_2O_3-$0.6Li_2O$-$yLiX$ with $X = F$, Cl ,Br or I.

comparison with measurements on the binary glasses, this band is attributed to vibration of BO_3 triangles with one non-bridging oxygen atom as is reported before in Raman scattering experiments. Its frequency decreases with halide anion exchange and reaches 1220 cm^{-1} in the iodide glass. In these glasses, the result of the addition of lithium halide as doping salt, is a decrease of the connectivity of the network.

As it is shown by vibrational spectroscopy, the accommodation of the doping salt into the network is dependant on the degree of connectivity of the covalent network fixed by the ratio between the oxygen and the boron atoms. That explains very well the decrease reported in the glass transition temperature T_g and, in spite of the addition of a third component in the glass, the variations of the density [26] and of the molar volume of the ternary glasses [56]. In the discussion of the structural properties it is preferable to use the molar volume because when data are analysed in terms of density, it is possible that dilatation of the glass network can be undetected if the mass of the ionic specie responsible of the dilatation effect is considerably greater than that of the specie which it replaces. The figure 25 shows the evolution of the molar volume as a function of the concentration in halide salt for the two glass families.

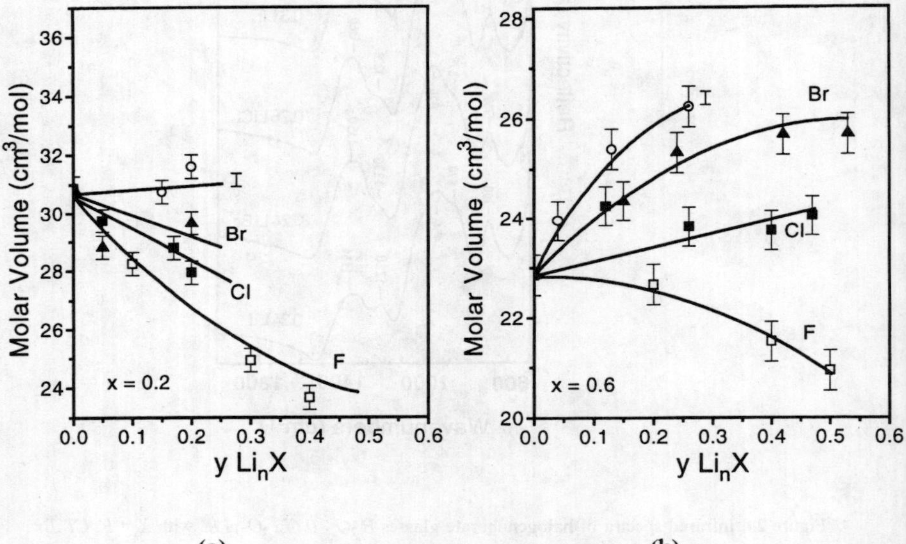

Fig. 25. Molar volume as a function of "doping" salt concentration (y) in ternary glasses B_2O_3-xLi_2O-$yLiX$ with X= F(□), Cl(■), Br(Δ), or I(○):

(a) glasses of series with x = 0.2

(b) glasses of series with x = 0.6

In glasses with low lithium oxide content (x = 0.2), the molar volume remains practically constant or decreases a little under the lithium halide addition. The smallest variations are obtained when large size anions are included in the network. On the contrary, in glass with higher Li_2O concentration, the molar volume increases with the modifier content except for lithium fluoride doped glasses. The effect is more important as the size of the halide is larger. In glasses of the first type, the accommodation of the halide anions can be easily achieved without any distortion because the initial network has a small connectivity. In the opposite case, in glasses with high O/B ratio (x = 0.6) which have an higher degree of connectivity, the insertion of the halide anion is responsible of an expansion of the network with, as a consequence, the formation of non-bridging oxygen atoms.

Different processes depending of the nature of the halide salt, can be proposed to describe the origin of the medium range order in the halide borate glasses. In the particular case of the fluoride glass, the fluoride addition is responsible of the shrinkage of the boron-oxygen network which results from the direct participation of the fluoride anions in the matrix by forming four-coordinated boron atoms. Two mechanisms can be proposed:

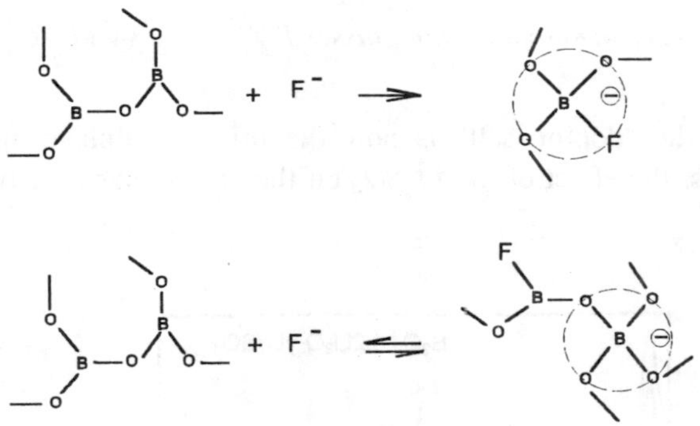

(7)

and because it carries an increase in the network connectivity by forming BO_4 tetrahedra, the second one seems to be the more probable. For the other halide salt, their addition produces the formation and/or the redistribution of the BO_4 units inside of the network, more especially when the size the halide anions is larger. In glasses with low *O/B* ratio, because of the small degree of linkage of the matrix, this is essentially done by the transformation of triborate and/or tetraborate into diborate groups resulting in a small or zero expansion of the network. In glasses with high *O/B* ratio, where the connectivity is very important because the higher number of four-coordinated boron atoms, the redistribution is associated with the breaking of *B–O–B* bridges to form trigonal BO_3 with non-bridging oxygen atoms. This becomes more important as the size of the halide anions is increased. In all the cases, the grouping of the BO_4 by pairs in diborate groups is promoted by the electrostatic interaction between the salt anions and the negatively charged BO_4 tetrahedra

From another viewpoint, the boron-oxygen network can be considered to be formed by diborate units diluted into an expanded network formed by BO_3 triangles with and without non-bridging oxygen atoms in relative concentrations fixed by the boron-oxygen ratio. The observed modifications in the medium range order of the ternary glasses induced by the halide anions produce changes in the degree of connectivity of the boron-oxygen network which become more open and yields more easily the halide addition. This model of glass structure describes also the shift to lower frequency observed for the stretching vibration of BO_3 units with NBOs when the halide anion is exchanged. As anions of large size are introduced interstitially in the network, the network is locally expanded and the length of the bond between boron and non-bridging oxygen is increased. This results in a decrease of the force constant of the bond between these two atoms.

4.3.2 Medium range order in sulphoborate glasses B_2O_3-xLi_2O-$y\ Li_2SO_4$

In these glasses, the "doping salt" is now the lithium sulphate and as for halogenoborate glasses, the effect of the Li_2SO_4 on the boron-oxygen network will

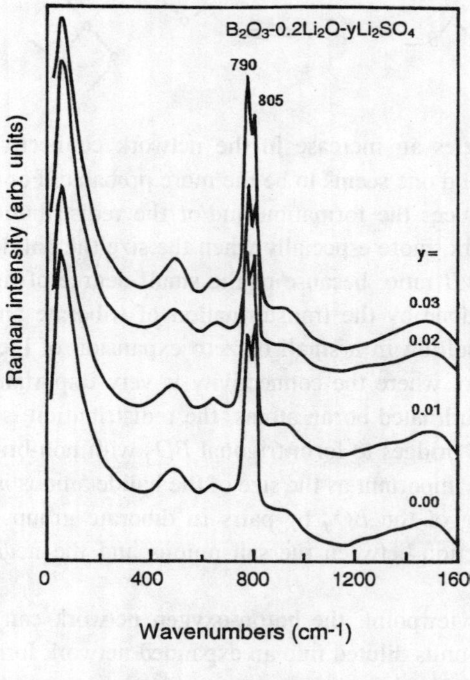

Figure 26. Raman scattering in sulphoborate glasses B_2O_3-$0.2Li_2O$-yLi_2SO_4. (reference 57)

be described in two families of glasses with lithium oxide content within the glass forming domain (x = 0.2 and x = 0.7).

The Raman spectra obtained by Katmitsos et al. [57], on glasses with the lower lithium oxide content are presented in figure 26. The undoped glass spectrum is dominated by two peaks. The sharp one at 805 cm^{-1} is assigned to the breathing mode of the boroxol rings while the other one at about 780 cm^{-1} and a little less intense, is the indication of the presence of borate rings with four-coordinated boron atoms. The effect of the addition of a very small quantity of lithium sulphate is clearly responsible of a continuous increases in the relative intensity of the latter peak. No peaks coming from vibration of the sulphate ions are observed in the ternary compounds because of the very small lithium sulphate content (less than 0.03). This implies an increase of the number of borate rings with one BO_4 tetrahedron at the expense of the boroxol rings.

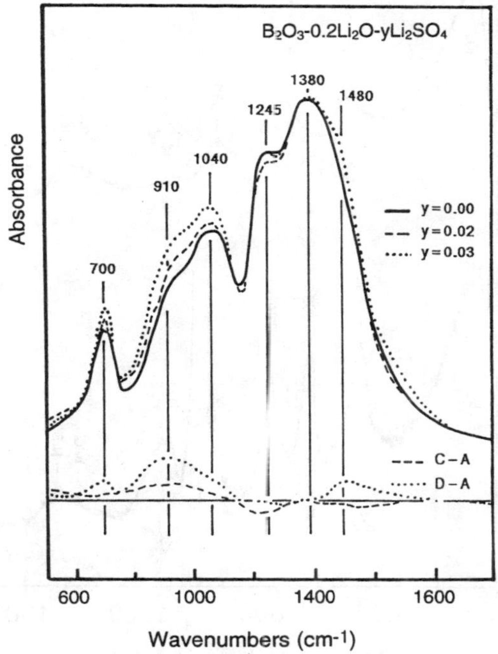

Figure 27. Infrared absorption spectra of ternary glasses B_2O_3-0 2Li_2O-yLi_2SO_4. (from.reference 57)

The infrared spectra of the same glasses are shown in figure 27 where in order to show the changes caused by the Li_2SO_4 addition, the spectra have been normalised to give the same intensity at the BO_3 band. After this scaling, a systematic increase of the intensity of the band attributed to the stretching vibration of the BO_4 units is observed upon increasing the sulphate content. This is confirmed in the difference spectra where a broad band is observed at 910 cm^{-1}. An increase of the intensity of the band at 700 cm^{-1} is also reported. As that latter band corresponds to the bending vibration of the *B–O–B* bridges, this indicates to an increase of the number of such bridges with the lithium sulphate addition. It can be understood as a consequence of the increase of connectivity of the boron-oxygen network by formation of additional BO_4 tetrahedra.

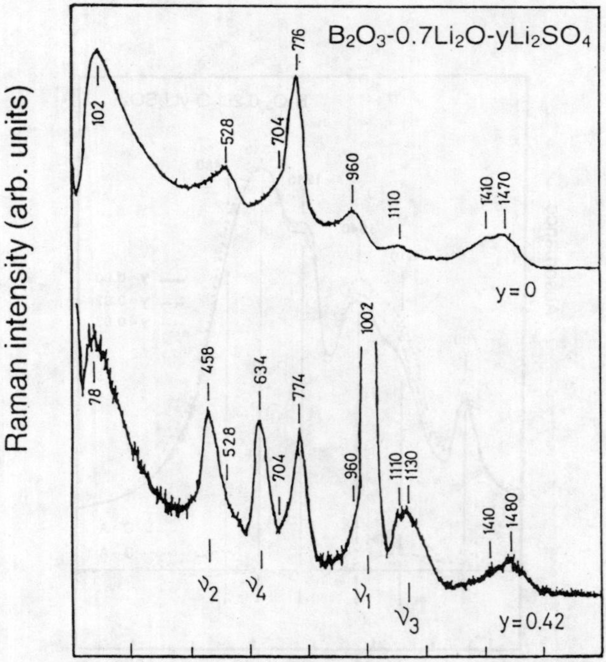

Figure 28. Raman spectra of sulphoborate glasses B_2O_3-$0.7Li_2O$-yLi_2SO_4.

The addition of lithium sulphate has also been studied in glasses with higher lithium oxide content ($x = 0.7$) where the degree of connectivity of the covalent network defined by the number of four-coordinated boron atoms is maximum. The Raman spectra of these ternary glasses are shown in figure 28. By comparison with the spectrum of the starting binary, the spectra of the "doped glasses" can be understood as the superposition of the B_2O_3-$0.7Li_2O$ and the Li_2SO_4 spectra. Four peaks whose the intensity increases upon the lithium sulphate addition, are clearly seen at about 458, 634, 1002, and 1130 cm^{-1}. They are attributed to internal vibration of tetrahedral SO_4^{-2} anions and correspond respectively to the ν_2, ν_4, ν_3 and ν_1 vibrations of these anions [58]. The values of these frequencies indicates that lithium sulphate is completely dissociated during the glass forming process and does not interact with the alkali cations in the ternary glasses. The dissociation of the alkali sulphate has been verified by experiments on diborate glasses B_2O_3-$0.5Li_2O$ respectively doped with the same amount of Li_2SO_4 and K_2SO_4. The two glasses give quite identical Raman spectra (figure 29) and that indicates that the sulphate dissociation is unaffected by the nature of the cation included in the "doping salt" [59].

By thermal annealing of the two glasses, it has been observed that the boron-oxygen network crystallises into a diborate structure characterised in Raman scattering by three intense lines at 725, 780, and 1030 cm^{-1} while the formation of lithium sulphate or potassium sulphate is shown by the observation of the ν_1 vibration of the SO_4 tetrahedra at 1016 and 982 cm^{-1} respectively. These two frequencies are identical to those reported in the crystalline forms of the two alkali sulphates. In the ternaries, the medium range order of the boron-oxygen network is mainly formed by borate rings with four-coordinated boron atoms characterised by their breathing vibration found here at about 774 cm^{-1}. Even for higher lithium sulphate content, the Raman spectra are not significantly modified, except for an increase of the intensity of the sulphate anion peaks with Li_2SO_4 concentration [59].

On the contrary, the analysis of the infrared reflectivity reveals important transformations on boron-oxygen network (figure 30) [60]. In the binary, the shape of the BO_4 band centred at 970 cm^{-1} is characteristic of a boron-oxygen network in which the diborate groups are the main constituent. However, the presence of a small band at 1235 cm^{-1} shows the existence of groups containing BO_3 with non-bridging oxygen atoms. As was reported previously about Raman scattering, new bands are observed in the infrared reflectivity spectra of the ternary "doped glasses". These bands whose intensity increases with the sulphate content are identified as the ν_4 and ν_3 vibration modes of tetrahedral SO_4^{-2} ions. They are

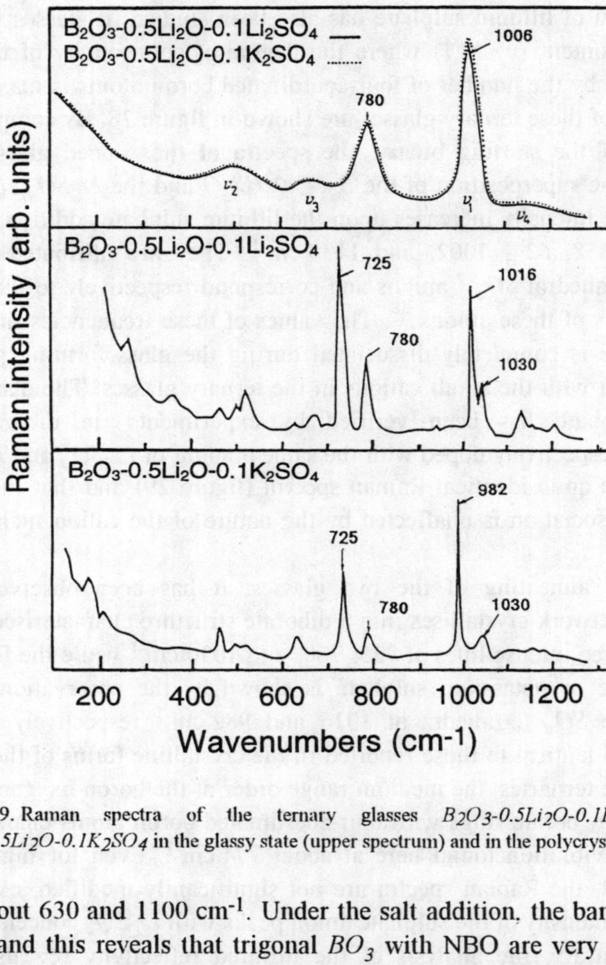

Figure 29. Raman spectra of the ternary glasses B_2O_3-$0.5Li_2O$-$0.1Li_2SO_4$ and B_2O_3-$0.5Li_2O$-$0.1K_2SO_4$ in the glassy state (upper spectrum) and in the polycrystalline state.

found at about 630 and 1100 cm^{-1}. Under the salt addition, the band at 1235 cm^{-1} disappears and this reveals that trigonal BO_3 with NBO are very sensitive to the lithium sulphate addition. They are transformed into BO_4 units as is shown by the increase of the ratio of the intensity between the BO_4 and BO_3 bands (I_{970} / I_{1390}) and which is plotted in the box included in figure 30. The mechanism proposed in equation (6) can be used to describe this structural transformation. The shape of the BO_4 band seems to be unaffected by the previous transformation, so no information can be deduced about the structural grouping which are formed preferentially.

The density of such glasses does not change with the salt addition, and this corresponds to an expansion of the glass matrix associated with an increase of the degree of linkage of the network when sulphate ions are accommodated by the glass. This is confirmed by the increase of the molar volume under the lithium sulphate addition (figure 31) and leads us to expect that the direct participation of the SO_4^{-2} ions into the network is at the origin of the formation of BO_4 tetrahedra distributed mainly into diborate units.

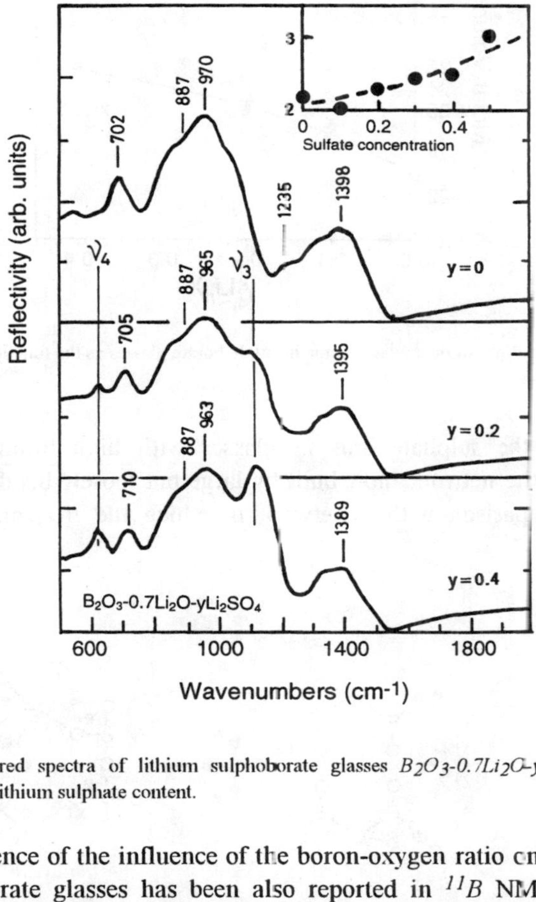

Figure 30. Infrared spectra of lithium sulphoborate glasses B_2O_3-$0.7Li_2O$-yLi_2SO_4 as a function of the lithium sulphate content.

Similar evidence of the influence of the boron-oxygen ratio on the structure of lithium sulphoborate glasses has been also reported in ^{11}B NMR MAS (Magic Angle Spinning) spectroscopy experiments [32]. Figure 12 in section 3 displays N_4 versus the Li_2O/B_2O_3 molar ratio for binary and sulphoborate glasses. Although Li_2SO_4 promotes the conversion of B^{III} atoms into B^{IV} atoms, the salt does not simply act as a modifier oxide. Large compact clusters with BO_3 and BO_4 units are

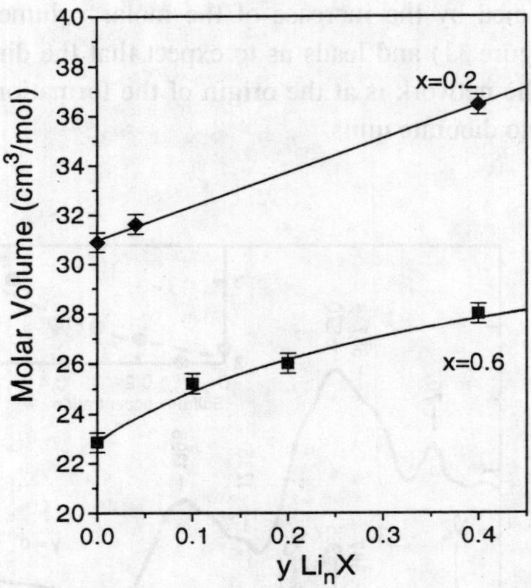

Figure 31. Variations of the molar volume in sulphoborate glasses as the function of the lithium sulphate content.

formed around the sulphate ions in glasses with high lithium oxide content (figure 32) and the network, now built by large but loosely bonded borate units is softened in comparison with a network in which the BO_3 and BO_4 units are randomly

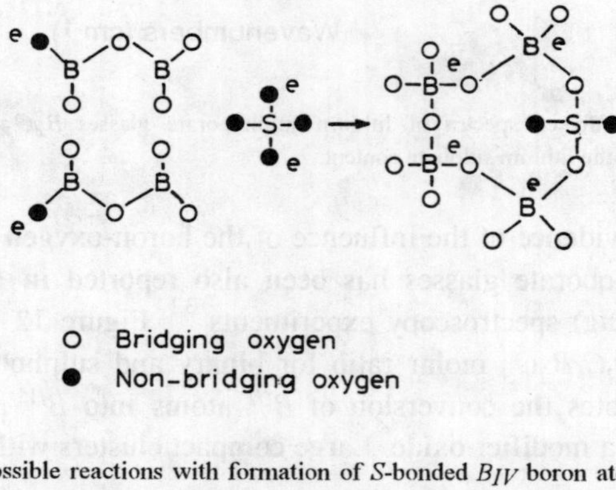

Figure 32. One of the possible reactions with formation of *S*-bonded B_{IV} boron atoms. (from reference 32)

distributed. In sulphoborate glasses, the sulphate ions participate to the reconstruction of the connectivity of the network. They have a "gluing action" contrasting with the bond-breaking effect observed when the borate glasses are "doped" with halide salt.

5. Vibrational properties and Long Range Order in borate glasses

The understanding of the ionic condutivity in vitreous fast ionic conductors demands the undestanding of both local and long-range range aspects of the ionic motion. Information about the local motion can be deduced from the description of Short Range and Medium Range Order structure in the glass network which show evidence the existence of negative sites for the ionic transport (BO_4^- tetrahedra) and the inital stages of structural organisation into well defined borates units. The ionic transport which involves long range motion in the disordered network, requires the existence of correlation between the structural units containing the BO_4^- tetrahedra. Long range correlations in glasses or amorphous materials are related to the anomalous vibrational behaviour exhibited by these non-crystalline materials, one manifestation of which is the "boson peak" observed in the Raman spectrum at low frequency [61].

5.1 Low frequency Raman scattering in glasses

Although a great deal of information has been obtained from the vibrational spectra of glasses, very little is known about the nature of the "boson peak" which is predominant in the low frequency part of the spectra ($\omega < 150$ cm^{-1}). In the figure 33 where the Raman spectra obtained for vitreous and crystalline silica are compared [62], the "boson peak" is the most noticeable feature of the Raman spectra of the glasses which appears as a very broad band, with a maximum in the frequency range 20-100 cm^{-1}. It is evident that one cause for the observed spectra is the disorder of the glass structure which leads to the violation of the wave vector selection rules.

The phenomenon is commonly ascribed to disorder induced first order light scattering from acoustic modes. It was first suggested that it reflects the crossover between long wavelength vibrations propagating through the continuous medium and highly localised acoustic vibrations in ordered domains [63]. Recently, it has been shown that there is an excess in the density of vibrational states in the low

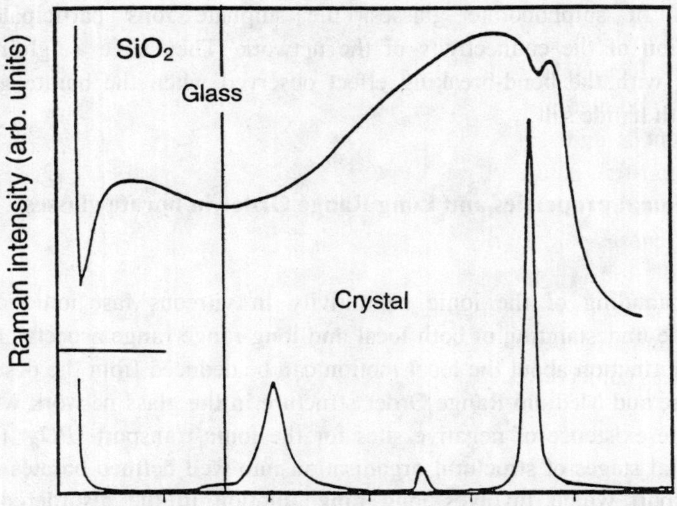

Figure 33. The Raman spectra of vitreous and crystalline silica (from reference 62).

energy range which is responsible of the "boson peak" [64]. A structural correlation length (SCL) $2R_c$ characteristic of the crossover can be deduced from the frequency of the maximum of the boson peak ω_{max} and the sound velocity V according to the relation $2R_c \sim V/\omega_{max}$ [65]. Several attempts to correlate the "boson peak" with the first sharp density peak (FSDP) which appears in the static structure factor $S(q)$ [66,67] have been made but recent observations show that the previous "universal" correlation between the positions of the FSDP and the "boson peak" fails [66-69].

The universality of the boson peak in glasses or amorphous materials [70] gains additional significance with the recognition of the fact that the low-frequency dynamical properties of glasses display a number of anomalies when compared to crystalline materials. These anomalies manifest themselves in a substantial excess of vibrational density of states over the Debye model. This has been directly observed in infrared absorption [72] and inelastic neutron scattering experiments [73]. Such anomalies leads to an excess of the specific heat C_p in glasses as compared to the T^3 law predicted by the Debye model for a crystal in the temperature range 5-10 K [74] and by the "plateau" found in the thermal conductivity in the same temperature range [75].

These anomalies in the low energy vibrational behaviour in glasses and amorphous materials are ascribed to the strong phonon scattering caused by domains of fluctuating density in the amorphous network. Inside these domains, short-range and medium-range order is conserved and outside the material presents an homogenous and isotropic structure. So, phonon localisation occurs when the phonon mean-free path l becomes comparable to the size of the domains ($l \sim 2R_c$) and that is responsible to the excess of vibrational states observed in the low vibrational energy range [76].

From the Ioffe-Regel condition ($q.l \sim 1$), if q is the phonon wavevector and V the sound velocity ($q = \omega/V$), the Structural Correlation Length $2R_c$ define as the size of the domains, can be deduced from the frequency ω_m of the maximum of the boson peak as conjectured above [65]:

$$2R_c \approx V/\omega_m \tag{8}$$

Using the assumption that the correlation lengths of the vibrational modes are small compared to optical wavelengths, Shuker and Gammon [63] have shown that the intensity of the disordered induced one phonon contribution to the Stokes-Raman spectra can be represented by :

$$I_{ij}(\omega,T) = C_{ij}(\omega).g(\omega).[1+n(\omega)].1/\omega \tag{9}$$

where $n(\omega)$ denotes the Bose distribution at temperature T, C_{ij} the Raman coupling coefficient, i and j the polarisation components of the incident and scattered light, and $g(\omega)$ is the vibrational density of states. The anti-Stokes scattering intensity is simply obtained by replacing $1+n(\omega)$ by $n(\omega)$ in equation (9). From this equation, it would be possible to determine the vibrational density of states from the Raman spectra if the frequency dependence of $C(\omega)$ is known.

Different analytical forms have been proposed to describe the spectral dependence of the coupling coefficient. Shuker and Gammon [63] have supposed that $C(\omega)$ = constant, but this behaviour does not reflect the correct frequency dependence of the density of state. Later, using a plane wave approximation for the acoustic vibrations, Martin and Brenig [77] have shown that $C(\omega)$ may have a more complicated frequency dependence which is :

$$C(\omega) \propto \omega^2 \exp -(2\omega R_c / V)^2 \tag{10}$$

where V is the sound velocity and $2R_c$ the correlation length of the amorphous structure. Then, $C(\omega)$ has a maximum at the frequency $\omega_{max} = V / 2R_c$.

However, it was later shown, by comparison between the Raman spectra of amorphous materials and inelastic neutron scattering data which yield directly the vibrational density of states $g(\omega)$, that some of the predictions or assumptions in this model are wrong: the vibrational density of states does not have Debye behaviour and the spectral dependence of the light-phonon coupling coefficient has no maximum in the frequency range of the "boson peak"[78].

Comparison of low frequency Raman scattering spectra of various glasses with heat capacity data have shown that in a broad frequency range, the Raman intensity is proportional to the number of vibrational excitations [vibrational density of states $g(\omega)$ multiplied by the Bose factor $n(\omega)$][79]. In a large frequency range (from 5-8 cm^{-1} up to about 100 cm^{-1}) $C(\omega)$ is well approximated by a linear frequency dependence. This linear dependence of $C(\omega)$ have been recently confirmed by comparison of low frequency Raman scattering results to vibrational density of states obtained from inelastic neutron scattering. In inorganic glasses such as SiO_2 or B_2O_3, it was found that $C(\omega) \propto \omega^2$ in the low frequency domain ($\omega \leq 20$ cm^{-1}) and that $C(\omega) \propto \omega$ at slightly higher frequencies ($\omega \geq 20$ cm^{-1})[80].

5.2. *Low frequency Raman scattering and long range order in binary alkali borate glasses*

Lorösch et al.[20] have first reported Raman scattering measurements obtained for B_2O_3 and alkali borate glasses B_2O_3-xM_2O ($M = Li, Na, K, Rb, Cs$ and Tl). The "boson peak is interpreted on the basis of the Martin and Brenig model[77] described in the previous section. In the absence of experimental data concerning the vibrational density of states $g(\omega)$, in the limit of long-wavelength acoustical phonons, these authors have supposed that $g(\omega)$ follows a Debye law $g(\omega) \propto \omega^2$ as was used previously by Nemanich for chalcogenides glasses[81]. With these assumptions, from the observed Stokes Raman intensity $I_{obs}(\omega)$, they defined the "reduced" Stokes Raman intensity $I_R(\omega)$ as:

$$I_R(\omega) = I_{obs}(\omega) / \omega \, [n(\omega)+1] \propto C(\omega) \tag{11}$$

This reduced Raman intensity has been fitted using equations (10) and (11) and its typical shape is shown in figure 34.

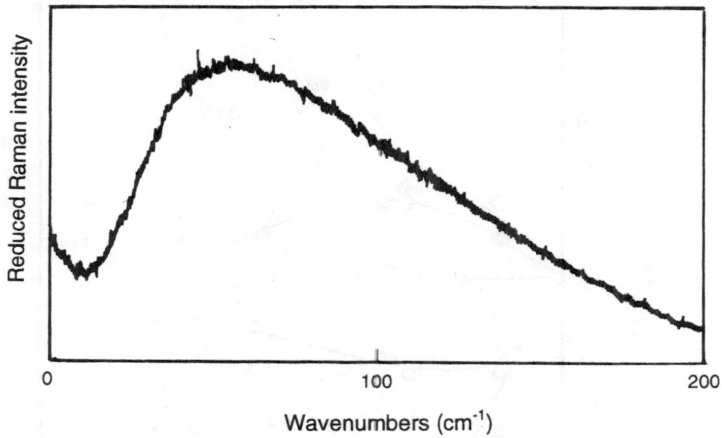

Figure 34. Typical shape of the reduced Raman intensity calculated by equation (11).(from reference 56)

The structural correlation lengths (SCL) deduced from the maximum of the curve $I_R(\omega)$ are plotted in figure 35 as a function of the modifier concentration for the different studied systems B_2O_3-xM_2O. Hence for B_2O_3, the SCL is about 8 Å, while the size of a boroxol ring is approximately 5.6 Å. This seems to indicate that medium range order in boron oxide involves more the one boroxol ring. With the addition of Li_2O to B_2O_3, a marked decrease is observed to 5.5 Å at x = 0.3 and to 5 Å at x = 0.8. For these authors, the medium range order in lithium borate glasses does not exceed the dimension of one six-membered ring.

In figure 35 Lorösch et al. have also reported that for glasses with a given value of x, the SCL increases as the size of the M^+ cations increases. For the biggest cations such as Rb^+ and Cs^+, the correlation length becomes greater than those of pure B_2O_3 (8-9 Å at $x = 0.3$). This corresponds to the incorporation of structural edifices with large size into the covalent network and must be correlated with the decrease of the number of four-coordinated boron atoms observed in NMR [22]. The accommodation of large size cations into the network favours the formation of packed domains with BO_4 units connected by metaborate chains with BO_3 triangles with bridging or non-bridging oxygen atoms. The number of these increases as the size of the modifier cation is bigger.

Another approach of the low frequency Raman scattering have been made by Souto [56]. Using now the assumption that the "boson peak" is not the signature of a maximum of the coupling coefficient $C(\omega)$ but of an excess of the density of

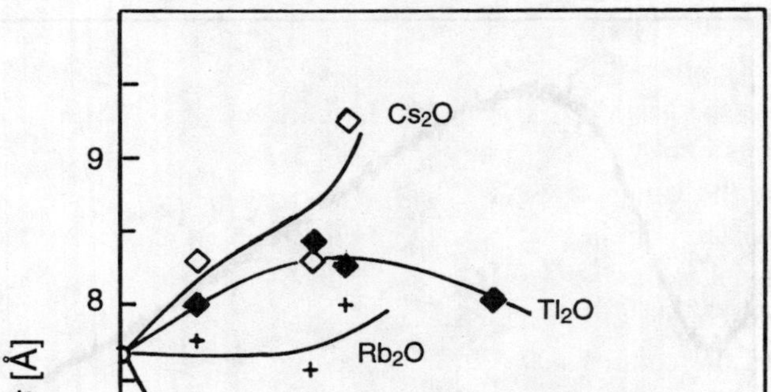

Figure 35. Structural Correlation Lengths for various B_2O_3-xM_2O glasses (from reference 20).

vibrational states [72]. For these reasons taking $C(\omega) \propto \omega$ [74], it is possible to obtain the vibrational density of states $g(\omega)$ from the observed Raman spectra $I_{obs}(\omega)$ by the relation:

$$g(\omega) = I_{obs}(\omega) / [n(\omega)+1] \tag{12}$$

For vitreous boron oxide, the shape of $g(\omega)$ calculated by this method is in good agrement with those obtained by inelastic neutron scattering (figure 36) [82]. The structural correlation length calculated from the maximum of the density of states deduced of the relation (12) are shown in figure 37 for binary lithium borate glasses B_2O_3-xM_2O. For boron oxide the SCL is about of 10 Å and is comparable with the size of two boroxol rings. With the addition of lithium oxide, the SCL decreases rapidly up to $x = 0.3$ and reaches a practically constant value of around 5 Å between $x = 0.3$ to $x = 1$. The first domain corresponds to the transformation of the boroxol rings into structural units with BO_4 tetrahedra under the Li_2O addition and at large x, the value of the SCL is comparable with the size of one diborate units (4.5 Å).The structural correlation length obtained from the phonon localisation model of the "boson peak" agrees very well with the results obtained in infrared spectroscopy [19] and discussed in section 4.2. The addition of the lithium oxide increases the connectivity of the network by the transformation $B_{III} \rightarrow B_{IV}$ giving the formation of a random network where the main units are diborate rings connected by BO_3 triangles. At higher lithium oxide content, the SCL is not affected by the depolymerization process due to the formation of non-bridging oxygen atoms which are formed by breaking of $B–O–B$ bridges on BO_3 triangles.

5.3 *Long range order in ternary borate glasses*

The phonon localisation model of the "boson peak" is an interesting tool to obtained informations about the long range order in glasses and in ternary glasses, Souto et al.[56,83,84] have studied the addition of lithium salts in two series of binary lithium borate glasses with different lithium oxide content (B_2O_3-$0.2Li_2O$-yLi_nX and B_2O_3-$0.6Li_2O$-yLi_nX where if $n = 1$, $X = F, Cl, Br, I$ or if $n = 2$, $X = SO_4$). In these two systems with different degrees of connectivity for the starting network, they have shown that the "doping salt" produces modifications of the long range order and that they depend of the concentration in lithium salt and of the size of the anions of the "doping salt". For the two glass systems the frequency ω_{max} of the maximum of the "boson peak" are shown in figures 38-(a) and 38-(b) as a function of the concentration of the different "doping salts" The frequencies are obtained by fitting of the low frequency Raman spectra with a log-normal law which describes very well the universal shape of the "boson peak" [56,85,86].

With the increase of the doping salt concentration, in glasses of the series $x = 0.2$ (figure 38-a), the frequency ω_{max} increases with the same behaviour for glasses doped with LiF, $LiCl$, $LiBr$ while it decreases in glasses doped with LiI. In the sulphoborate glasses, the frequency of the "boson peak" seem to be practically constant with the lithium sulphate addition. In the glasses of the series $x = 0.6$ (figure 38-b), excepts for the glasses doped with LiF where the frequency ω_{max} remains constant, ω_{max} decreases with the salt addition. For the lithium halide

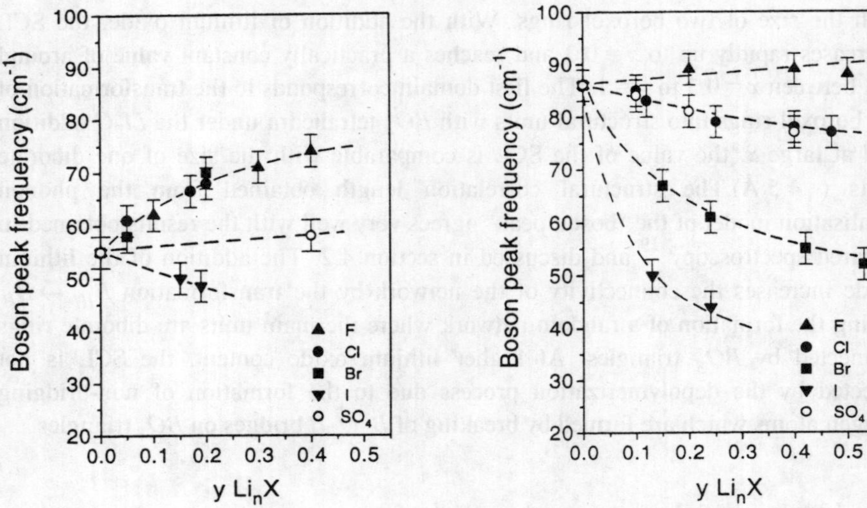

(a) (b)

Figure 38. Variation of the frequency of the "boson peak" ω_{max} with the "doping salt" content in: -(a)- B_2O_3-$0.2Li_2O$-yLi_nX.and -(b)- B_2O_3-$0.6Li_2O$-yLi_nX with $X = F, Cl, Br, I$ or SO_4 (from reference 56)

doped glasses the shift to lower frequency is the more important as the size of the halide anion of the "doping salt" is greater. In the sulphoborate glasses, the frequency shift is quite similar to those observed in *LiCl* doped glasses though the ionic radius of the SO_4^{-2} anion would be similar to those of the iodide anion. In the latter series, the behaviour of ω_{max} found for the chloroborate glasses is similar to those reported by Soppe et al.[87] or by Börjesson et al.[88].

As for the binary borate glasses, with the theoretical approach of the low frequency Raman scattering proposed by Souto [56], it is possible to deduce of the vibrational density of states $g(\omega)$ from the intensity of the observed spectra I_{obs} by the way of the equation (6). The vibrational densities of states (VDOS) obtained for the two series of ternary glasses (B_2O_3-xLi_2O-yLi_nX) are shown in figures 39 ($x = 0.6$) and 40 ($x = 0.2$). They are characterised by an important contribution of the vibrational states between 0 and 80-100 cm^{-1} to the low frequency part of the VDOS which depends of the nature of the "doping salt". In glasses of the serie corresponding to $x = 0.6$, by comparison with the VDOS of the undoped material, an excess of vibrational states is systematically observed in all the ternary glasses in the low frequency part of the VDOS. For the halogenoborate glasses, the increase of the

VDOS becomes more important as the size of the halide anions of the "doping salt" increases and it appears as a peak in *LiI* doped glasses.

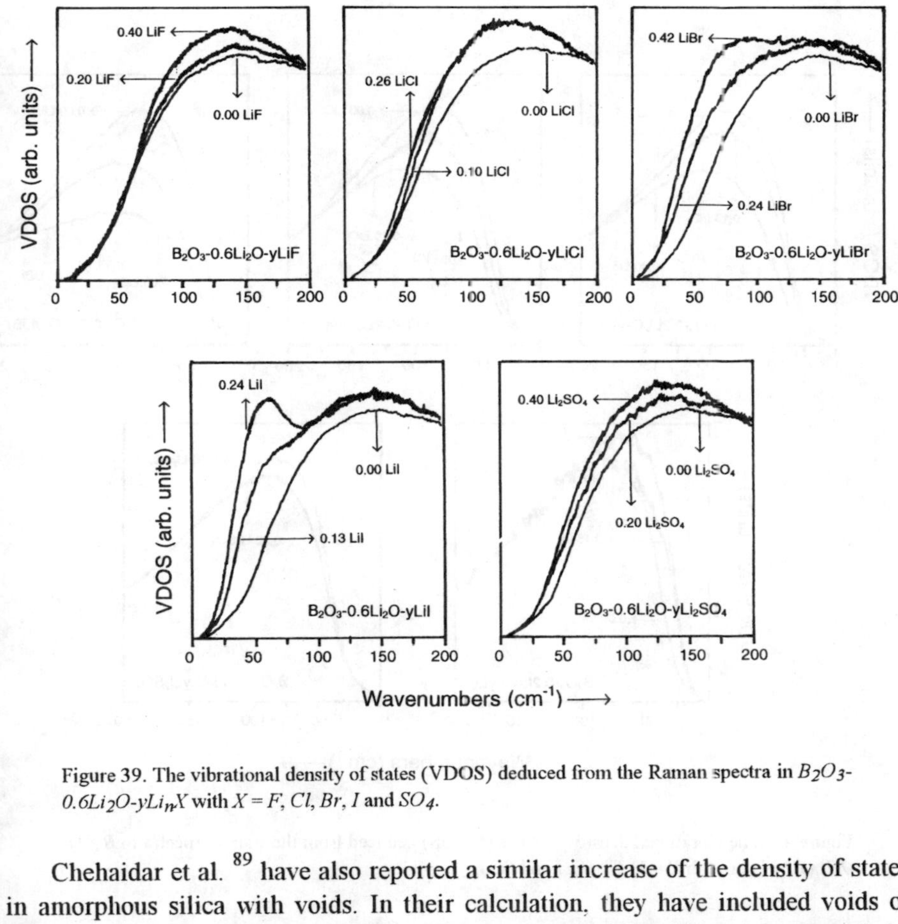

Figure 39. The vibrational density of states (VDOS) deduced from the Raman spectra in B_2O_3-$0.6Li_2O$-yLi_nX with $X = F, Cl, Br, I$ and SO_4.

Chehaidar et al. [89] have also reported a similar increase of the density of states in amorphous silica with voids. In their calculation, they have included voids of different sizes and various concentrations and they observed an increase of the VDOS which comes from local vibration modes of low frequency in comparison with those of the pure amorphous silica and that these become more important as the number and/or the size of the voids increase.

A different and more complex behaviour has been observed by Souto [56], in borate glasses with low modifier content ($x = 0.2$) where the VDOS decreases in glasses doped with *LiF*, *LiCl* and *LiBr*, while it increases in those doped with *LiI*.

As it has been reported in the description of the acoustic properties of the borate glasses, these two regimes are related to changes in the network connectivity induced when salt anions of increased size are accommodated into the interstices of the network.

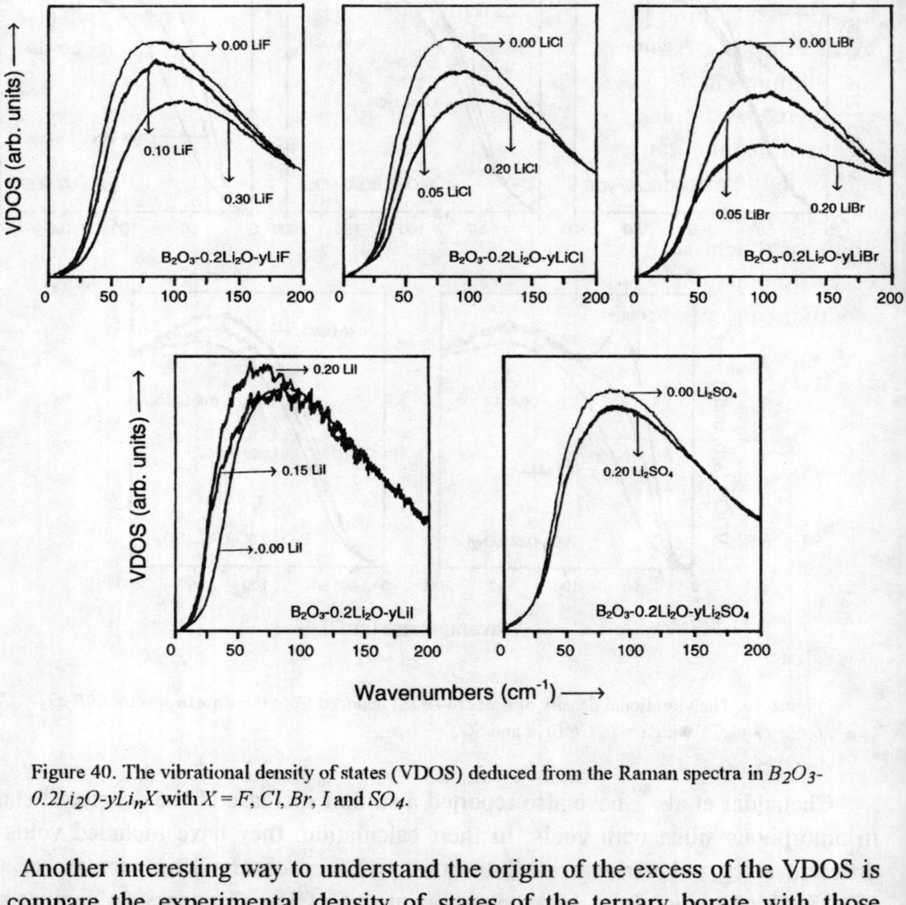

Figure 40. The vibrational density of states (VDOS) deduced from the Raman spectra in B_2O_3-$0.2Li_2O$-yLi_nX with $X = F, Cl, Br, I$ and SO_4.

Another interesting way to understand the origin of the excess of the VDOS is to compare the experimental density of states of the ternary borate with those calculated for the binary system B_2O_3-xLi_2O by Barrio et al.[90]. In this work, the vitreous boron-oxygen network was simulated by a Bethe lattice in which three fundamental features are incorporated:(1) the presence of the tetrahedral boron sites, (2) the disorder, both topological and substitutional, and (3) the activity of the different modes in the experiments. However, this calculation does not take account of the presence of complicated structural units such as boroxol rings, tetraborate or diborate groups. So the agreement with the experimental data is not complete. But, it appears that the VDOS is affected by the changes in the connectivity of the covalent network. In this work, in the low frequency range, the VDOS is mainly due to the vibration of the bridging oxygens between two three-coordinated boron atoms and this contribution increases with the increase of the concentration in lithium oxide probably because the appearance of non-bridging oxygen atoms located on BO_3 triangles. The increase in the VDOS observed with the addition of the lithium salt Li_nX probably has the same origin. The glass network can be considered as formed by domains with a high concentration of BO_4 tetrahedra localised mainly in diborate rings and connected by BO_3 triangles. Some involving non-bridging oxygen atoms.

The structural correlation length (SCL) obtained from the maximum of the vibrational density of states are plotted on figure 41 - (a) and (b) - as a function of each lithium salt concentration. Two different behaviours depending of the modifier concentration are observed. In the first series of glasses ($x = 0.2$), the SCL

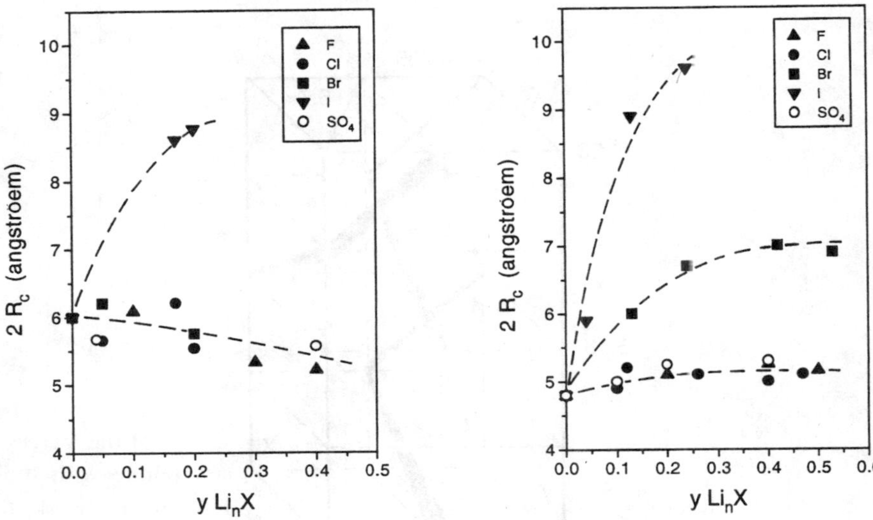

(a) (b)

Figure 41. Dependence of the Structural Correlation Length (SCL) with the "doping salt" content in B_2O_3-$0.2Li_2O$-yLi_nX (a) and B_2O_3-$0.6Li_2O$-yLi_nX (b) glasses.

decreases slowly with the concentration of salt in glasses doped with *LiF*, *LiCl* and *LiBr* (from 6 Å to 5.5 Å) while a marked increase is observed in *LiI* doped glasses (from 6 Å to 8.7 Å). In the second family of glasses ($x = 0.6$), the SCL seems practically constant in glasses doped with *LiF* and *LiCl* (~ 5 Å) but it increases in glasses doped with *LiBr* and *LiI* and reaches respectively ~ 7 Å and ~ 9.5 Å. In the two series, the enhancement is the more important as the size of the halide anion is greater (figures 41 & 43).

In section 4 about the Medium Range Order, it has been said that in the halogenoborate glasses, the BO_4 tetrahedra are mainly grouped in diborate rings. So the comparison of the structural correlation length $2R_c$ with some characteristic lengths measured in crystalline form of the diborate [91,92] shows that the SCL is of the same order of magnitude as typical distances between diborate cages (figure 42): - 5.38 Å - the distance between cages in one diborate chain, - 6.7 Å - between corresponding cages having the same z-coordinate but in different chains and- 8.44 Å - between diborate cages belonging to different chains in the diagonal direction. If the first distance gives an estimation of the size of a single diborate ring, clusters with size $2R_c$ contain respectively about 3 and 8 diborate groups when $2R_c$ rises to 7 and 10 Å.

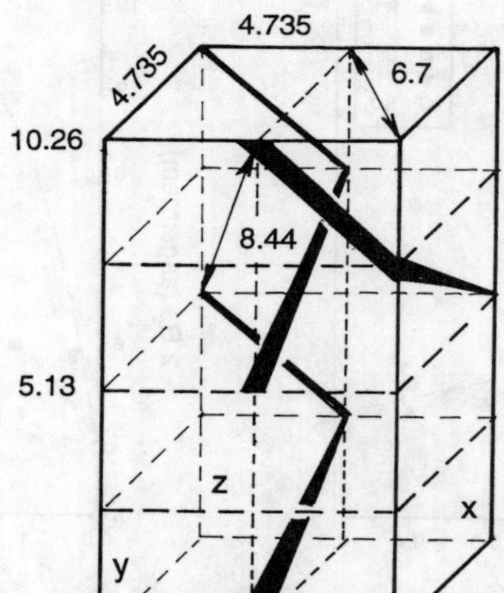

Figure 42. Schematic picture of the crystalline lithium diborate structure. Two system of chains are visualised (from references 91 & 92).

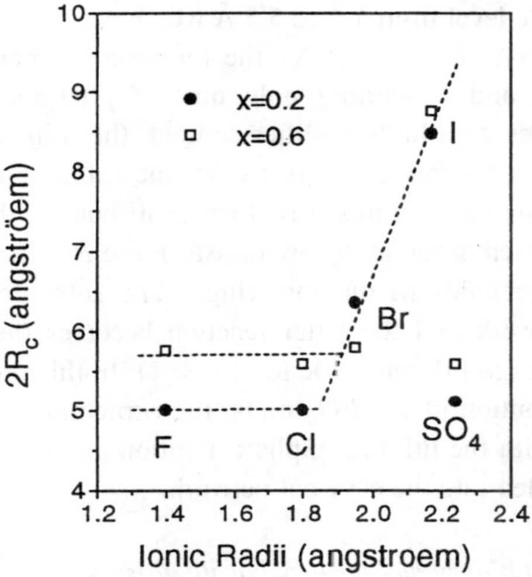

Figure 43. Dependence of the Structural Correlation Length with the size of the anion of the "doping salt" in B_2O_3-$0.2Li_2O$-yLi_nX (●) and B_2O_3-$0.6Li_2O$-yLi_nX(□) glasses.

It is now possible to describe the structural modifications induced by the addition of alkali halogenides as doping salts into the glassy network. Except for the fluoride doped glasses, the anions of the doping salt are accommodated in interstices inside the boron oxygen network and as the size of these interstices increases with the size of the anions. this induces the formation of regions with a higher concentration of BO_3 triangles around the X^-anions, and the formation of BO_4 tetrahedra which are redistributed into domains formed by diborate rings. The size of such domains of higher connectivity is more important when the size of anion is large (figures 41 & 43). This idea is similar to those already proposed by Button et al. in reference 23 for alkali chlorodiborate glasses.

For the two series of glasses, it is interesting to note that for sulphate doped glasses, the enhancement of the low frequency VDOS is quite similar to those produced by the same concentration of $LiCl$ in the binary network though the size of the SO_4^{-2} ions (2.23 Å) is of same order than those of iodide anions (2.16 Å). In glasses with high Li_2O content ($x = 0.6$) the VDOS is enhanced while in the second series of glasses ($x = 0.2$), it decreases with the "doping salt" addition. This indicates that sulphate ions play a different role than those of the halide anions and we proposed that sulphate ions acts as a modifier because it induced the formation of BO_4 units in glass with low modifier content. The small decrease of the SCL with the lithium sulphate level from 6 Å to 5.5 Å which is comparable with the size of the boroxol ring (5.6 Å) is evidence for the formation of borate rings without long range correlation and containing only one BO_4 tetrahedra, such as the triborate rings. In glasses with high modifier content, the lithium sulphate can be considered as a second glass former. In this case, the sulphate ions form bridges between metaborate chains and in this way, they contribute to the increase of the number of four-coordinated boron atoms by transforming BO_3 triangles with NBO into BO_4 which are included in diborate rings. The sulphate ions restore the connectivity of the network and the latter reaction becomes the more important process in glasses with high lithium oxide level ($x > 1$). In this process, which does not require the redistribution of the BO_4 units, the Structural Correlation Length $2R_c$ remains constant with the lithium sulphate addition and the diborate rings are homogeneously distributed into the covalent network.

5.4 *Structural model of binary and ternary borate glasses*

Before discussing the correlation between the network structure and the ionic conductivity in the borate glasses, it is necessary to review and schematise the different steps of the structural organisation observed into the boron-oxygen network of the borate glasses.

The addition of network modifier to vitreous B_2O_3 results either in the conversion of the triangular BO_3 structural units to BO_4 tetrahedra or to the formation of non-bridging oxygen atoms. These units constitute the Short Range Order (SRO) and their relative concentrations depends of the modifier concentration and also of the nature of the modifier cations.

Unlike other glass systems, borate glasses contain superstructural units comprising a well defined arrangement of BO_3 triangles and BO_4 tetrahedra. They are characteristic of the presence of Medium Range Order (MRO) and correspond to a second degree of structural organisation inside the boron-oxygen network. The nature of the superstructural units depends on the type and on the concentration of the modifier oxide. To accommodate a great number of lithium ions and/or modifier cations of large size, the BO_4 units are also redistributed into the network so that domains with high concentration of such units occur connected between them by domains where the BO_3 triangles are the predominant units. If the Structural Correlation Length (SCL) defining the Long Range Order (LRO) of the network gives the order of magnitude of the size of the BO_4 rich domains, the decrease of the SCL with the lithium or sodium oxide concentration while the number of boron atoms in four-coordination increases, indicates that the MRO units consists mainly of diborate rings with two BO_4 in neighbouring positions. On the other hand, the substitution of the Li^+ cations by cations with larger size is favorable to the formation of large domains containing borate groups with one BO_4 tetrahedra such as triborate and pentaborate, because of the decrease of N_4 with the increase of the size of the cations.

In the ternary glasses, the addition of a third constituent in the glass such as alkali halides *LiX*, induced changes in the MRO and in LRO. By electrostatic interactions between the negatively charge BO_4 tetrahedra and the halide anions, the redistribution process the BO_4 units is amplified. They are now grouped in diborate rings rather than in groups with one BO_4 such as the triborate or pentaborate groups and the size of the domains containing these BO_4 units increases as the size of the anion increases in order to produces interstices able to accept the salt *LiX* (figure 44). These interstices are included in the BO_3 rich

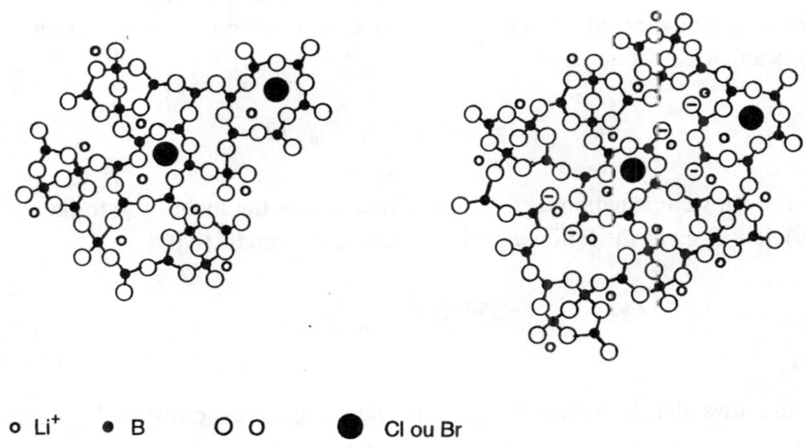

Figure 44. Two-dimensionnal schematic structures proposed in the system of the alkali halogenoborate glasses. In the chloroborate (left) the BO_4 are clustered in diborate ring while in bromoborate (right) they formed large clusters with about 3 diborate rings. Large interstices with additional NBO are also present in the latter network.

regions whose the size is increased with the size of the X^- ions. This is the origin of the expansion observed in the alkali halide borate glasses except for those doped with lithium fluoride.This mechanism which occurs during of the glass formation, is absent when the salt (i.e. lithium sulphate) acts as a second glass former. In this case, the size of the BO_4 rich domains and consequently the Long Range Order are practically unaffected by the "doping salt".

6 Ionic transport in alkali borate glasses

6.1 *Introduction to ionic transport*

Under the assumption that the ionic motion can be described with a simple diffusion model characterised by isolated jumps of the mobile ions with a random walk and that the ionic conductivity occurs only by the movement of one ionic species, a detailed expression can be obtained for the ionic mobility introduce in equation (1). In this model ions are considered as vibrating species with a characteristic frequency ν_o inside a potential well of depth E_m. Assuming Boltzmann statistics, the diffusion constant D depends on the jump distance d and on a geometrical factor α, as:

$$D = \alpha d^2 \nu_0 \, exp(-E_m / kT) \tag{13}$$

The Einstein equation $\mu kT = ZeD$ which relates the mobility μ to the diffusion coefficient D gives the following expression for the conductivity:

$$\sigma = \frac{n(Ze)^2 \alpha d^2 \nu_0}{kT} exp(-E_m / kT) \tag{14}$$

which shows the Arrhenius temperature dependence experimentally observed in fast ionic conductors.

In classical ionic conductors, the concentration of electrical carriers depends of the formation of defects by a Schottky process (cation-anion vacancy pairs) or by a Frenkel process (cation or anion vacancy-interstitial pairs) and in either case, the defect concentration is given by:

$$n = n_0 \, exp(-E_d / kT) \tag{15}$$

where E_d is the half of the defect-pair formation energy. The fraction of mobile ions n/n_0 is often designated as the symbol β. Since E_d is generally high

($E_d \sim$ 1eV in NaCl), the fraction of defect sites is usually quite small. The conductivity is given now by :

$$\sigma = \frac{n_0 (Ze)^2 \alpha d^2 v_0}{kT} exp - (E_m + E_d)/kT \qquad (16)$$

$$\sigma = \frac{n_0 (Ze)^2 \alpha \beta d^2 v_0}{kT} exp - (E_m/kT) \qquad (17)$$

Due to the characteristic disorder of the mobile ion sublattice in fast ionic conductors, a high concentration of defects is expected without the need for thermal generation. In this case for FIC's, β may be considered as independent of temperature and much larger than in other normal ionic conductors. For these reasons, the activation energy for fast ionic conductivity should represent that of migration alone.

Because of the structural disorder of glasses, many terms of equation (17) are not known and are difficult to specify. That explains the general form commonly used to describe the temperature dependence of the ionic conductivity. This simplified form of the Arrhenius law is given by:

$$\sigma = \sigma_0 / T \, exp(-E/kT) \qquad (18)$$

6.2 *Ionic conductivity in borate glasses*

Exceptionally higher values of the lithium ion conductivity in lithium borate glass system were first measured by Otto in 1966 [93]. The bulk conductivity values were extracted from the data of complex impedance spectroscopy [94,95]. The figure 45 shows the results of the frequency dependence of the complex impedance for the glass of composition $0.75B_2O_3$-$0.25Li_2O$ at the temperature of 253 °C [96]. The bulk impedance is modelled by a parallel *R-C* electrical network and corresponds to the low impedance semicircle while the high impedance semicircle is attributed to the impedance of the electrodes. From this diagram, the bulk impedance is determined by the point on the real axis at which the semicircle meet. To investigate the dependence of the conductivity σ on the glass composition, *Log*σ values, derived from the impedance diagrams, are plotted as a function of composition at fixed temperature.

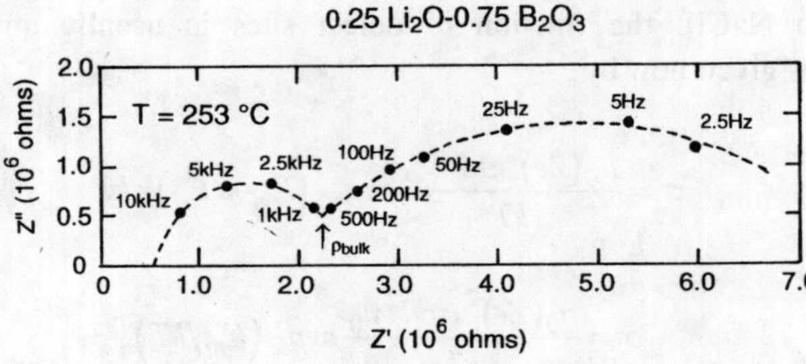

Figure 45. The real and imaginary part of the complex impedance of the glass $0.75B_2O_3$-$0.25Li_2O$ as a function of the frequency (from reference 96).

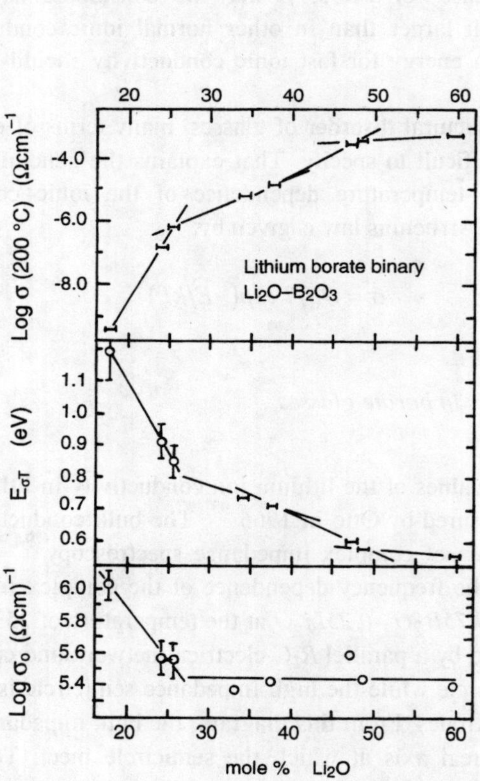

Figure 46. Conductivity data in the binary lithium borate versus Li_2O concentration (reference 97)

For B_2O_3-Li_2O glasses up to ~ 25 mol% Li_2O, from Button et al.[97], the increase in lithium oxide gives an important increase of the conductivity σ and a practically linear decrease of the activation energy called by the authors $E_{\sigma T}$ (figure 46). Above 25 mol%. σ and $E_{\sigma T}$ are considerably less dependant on the modifier content while the logarithmic plot of the pre-exponential factor, *Log* σ_0, takes a nearly constant value of about 5.4. In the first concentration range, *Log* σ_0 , is a decreasing function of Li_2O content then it become constant when the modifier content is higher than ~ 25 mol%. From the figure, it is also clear that the increase of the conductivity is essentially due to a corresponding decrease of the activation energy.

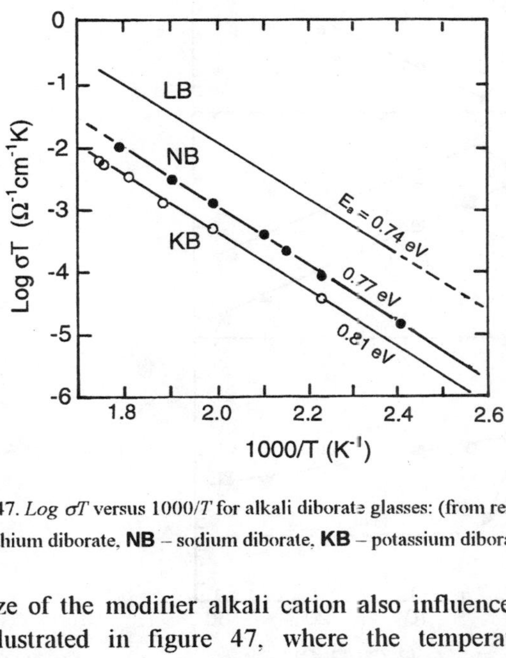

Figure 47. *Log* σT versus $1000/T$ for alkali diborate glasses: (from references 26 & 98)
LB – lithium diborate, **NB** – sodium diborate, **KB** – potassium diborate

The size of the modifier alkali cation also influences the ionic conductivity. This is illustrated in figure 47, where the temperature dependence of the conductivity of lithium, sodium and potassium diborate glasses are compared and one finds an inverse relationship between the ionic conductivity and the size of the alkali cations [98].

Large increases in the ionic conductivity of glasses can be obtained in glasses having a fixed total alkali content by substituting partially the alkali oxide by, for example, an alkali halide. The comparison of the conductivity parameters obtained for lithium and sodium dichloroborate glasses (figure 48), indicate that the increase in the conductivity observed in lithium and sodium glasses is directly tied to a corresponding decrease of the activation energy $E_{\sigma}T$ while the pre-factor σ_o is

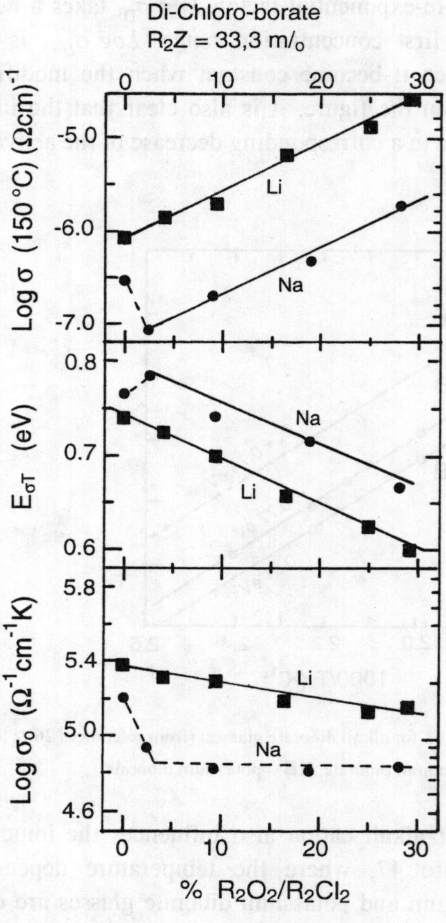

Figure 48. Conductivity parameters for a series of *Li* and *Na* di–chloroborate glasses as a function of the fraction of alkali chloride (from reference 97).

insensitive to the halide substitution [97]. In contrast to the results obtained for glasses with smaller alkali cations, chloride substitution appears to have little effect on the conductivity enhancement in potassium glasses. The activation energy for the motion of the K^+ ions is weakly dependent on the chloride content while those for the Li^+ and Na^+ motion are strongly reduced [98].

Levasseur et al. [99] have studied the ionic conductivity of the chloroborate glasses (B_2O_3-xLi_2O-$yLiCl$) inside the vitreous domain limited by $0 \leq x \leq 0.85$ and $0 \leq y \leq 0.85$. When y is small, the conductivity increases rapidly with x while for higher values of y, it seems that the conductivity reaches a saturation regime (figure 49).

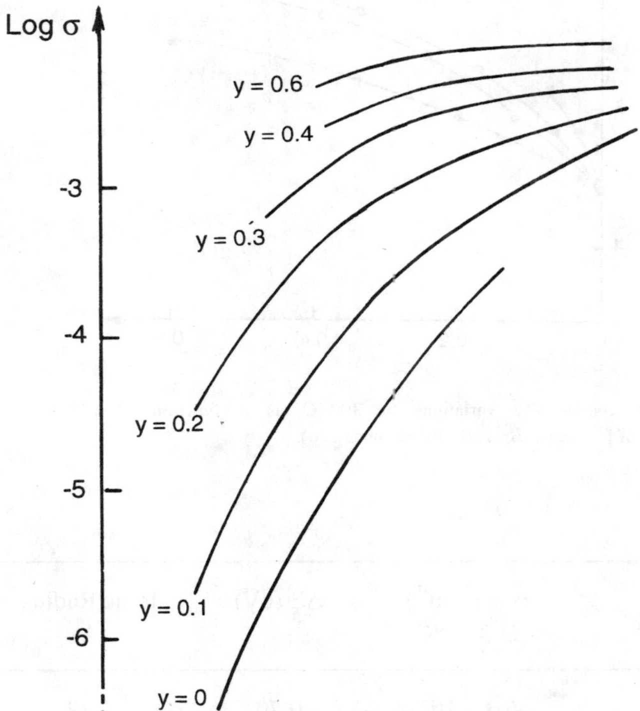

Figure 49. Ionic conductivity variations at 300 °C as a function of Li_2O and LiCl rates for B_2O_3-xLi_2O-$yLiCl$ glasses(from reference 99).

The size of the alkali anions is also another parameter which changes the ionic conductivity. Levasseur et al.[100,101] have shown that, the ionic conductivity increases with the size of the anion X (figure 50) in glasses with fixed boron-oxygen ratio (B_2O_3-$0.6Li_2O$-$yLiX$ with $X = F$, Cl, Br or I). They have also

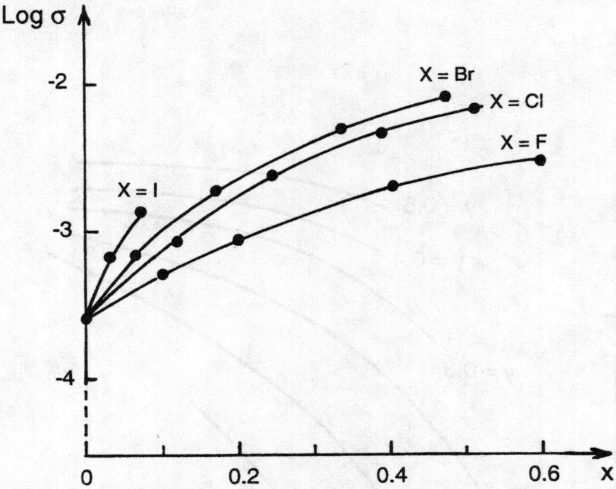

Figure 50. Ionic conductivity variations at 300 °C as a function of LiX rate for B_2O_3-xLi_2O-$yLiX$ ($X= F, Cl, Br, I$) (from reference 100).

X	σ ($\Omega^{-1}cm^{-1}$) at 300 °C	ΔE (eV)	Ionic Radius (Å)
F	6.3×10^{-4}	0.70	1.29
Cl	1×10^{-3}	0.60	1.81
Br	1.3×10^{-3}	0.58	1.96
I	3.2×10^{-3}	0.52	2.20

Table 2. Conductivity parameters of B_2O_3-$0.56Li_2O$-$0.08LiX$ glasses doped with $X = F$, Cl, Br, I.

reported that the activation energy call ΔE in reference 100 decreases simultaneously when, in glasses with same composition, the halogen anions are varied from fluorine to iodine (table 2).

To improve the conductivity of these ternary vitreous materials, Levasseur et al. [99,101] have replaced the lithium halide by lithium sulphate. These materials exhibit an excellent electrical conductivity when they have high concentrations in lithium oxide and lithium sulphate ($\sigma = 2.10^{-2}\ \Omega^{-1}cm^{-1}$ at 300 °C and an activation energy around 0.58 eV).

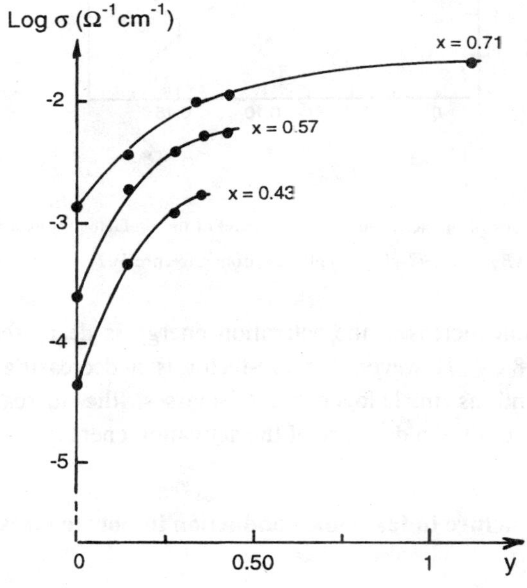

Figure 51. Variation of the ionic conductivity at T = 300 °C as a function of Li_2O and Li_2SO_4 in B_2O_3-xLi_2O-yLi_2SO_4 glasses (from reference 101).

The figure 51 shows the dependence of the ionic conductivity at a temperature of 300 °C as a function of the Li_2SO_4 content in glasses with increasing modifier content. For glasses with a fixed *O/B* ratio, the ionic conductivity increases with the lithium sulphate concentration. The figure 52 shows the variation of the activation energy and those of the prefactor σ_o in the case of diborate glasses doped with lithium sulphate (B_2O_3-$0.5Li_2O$-yLi_2SO_4 with $y < 0.15$) [102]. As the

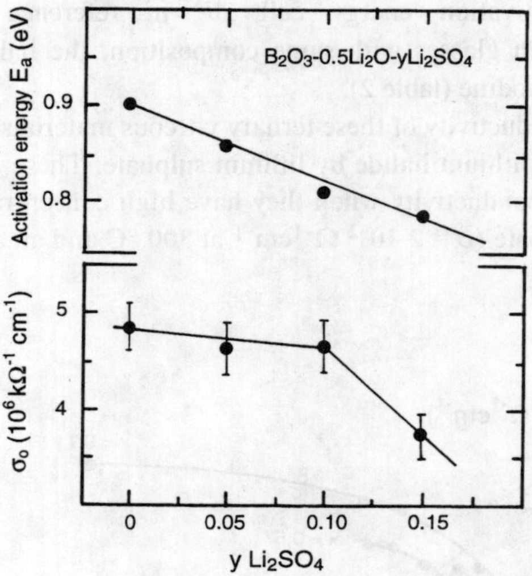

Figure 52. Variations of the activation energy E_a and of the prefactor σ_o as a function of the Li_2SO_4 content in B_2O_3-xLi_2O-yLi_2SO_4 glasses (from reference.102)

dopant concentration increases, the activation energy is decreased by about 10% from 0.9 eV to 0.8 eV. However, the pre-factor is a decreasing function of the Li_2SO_4 content and as in halogenoborate glasses, the increase of the ionic conductivity is the result of a decrease of the activation energy

7 The role of structure in fast ionic conduction in borate glasses

In the previous sections, data were presented which established that both structure and transport properties are influenced by the composition of the borate glasses. Modifications in the Short and Medium Range Order (SRO and MRO) as well in Long Range Order (LRO) were described. The aim of this section is to demonstrate that these changes in structure and ionic conductivity are related.

We examine first how the structural changes observed with composition in the binary alkali borate glasses are correlated with transport properties. In pure glassy B_2O_3, the ionic conductivity is negligible because of the absence of mobile ionic species in the pure covalent network. With the addition of a modifier oxide M_2O, the SRO is modified by the transformation of BO_3 triangle into tetrahedral BO_4

units. By the accommodation of the oxygen, the network become partially ionic and the ionic conductivity appears. Because of the excess of negative charge on the boron atoms, it is reasonable to suppose that the M^+ is located in the vicinity of the BO_4 units and the activation energy for the ionic transport is mainly fixed by the depth of the potential well of the M^+–BO_4 configuration.

In lithium borate glasses, with the increase of the modifier content x, the BO_4 units have a tendency to form domains with triborate and pentaborate units at low x, then with diborate units containing two adjacent BO_4 tetrahedra when x becomes higher. During these structural modifications, the distance between two negative sites is reduced and this leads, by increasing the overlap between the potential wells, to the lowering of the energy barriers which drive the ionic motion and contributes to the enhancement of the ionic conductivity of the binary glasses. The fact that the Structural Correlation Length (SCL) decreases and does not exceed the size of one diborate ring confirms the existence of great number of such units in glasses with higher conductivity.

When the alkali level in the glass reaches about 30 mol% (x = 0.5), borate groups containing planar BO_3 triangles with non-bridging oxygen atoms (NBOs) appears (see section 3). On these groups the potential wells are deeper than those on BO_4 tetrahedra and they play the role of traps for the mobile species. The enhancement of the ionic conductivity with the alkali content reflects the competition between two processes: - the increase of the number of boron atoms in coordination four which lowered the activation energy and the formation of NBO atoms which are responsible of a decrease of the number of mobile ions.

With the change of the modifier cation, the activation energy increases as the size of the cation increases and the ionic conductivity decreases. This can be explained with the help of the ideas developed previously: the decrease of the number of four-coordinated boron atoms which is responsible of a reduction of the overlap between the potential wells on each BO_4 sites and the increase of the number of groups with NBO coming from the local expansion necessary to accommodate more larger cations. The size of the domain containing BO_4 units is larger than a diborate ring and these domains are probably composed of triborate and pentaborate units with only one BO_4 site per unit.

In the lithium borate with a very high lithium oxide content ($x > 1$), the structural groups gradually transform into smaller ones containing non-bridging oxygen atoms such as orthoborate and pyroborate units. Some studies of evaporated thin films with $1 < x < 3$ show an increase of the ionic conductivity with x which reaches a maximum value for B_2O_3-$3Li_2O$ glasses [103]. However, the previous discussion cannot be retained to discuss of the electrical properties of these materials. In fact, in these materials, the covalent network is more and more destroyed by the depolymerisation process due to the formation of non-bridging oxygen atoms[45,104] and eventually a pure ionic amorphous structure is expected.

The previous discussion has permitted us to correlate the decrease of the activation energy responsible of the enhancement of the ionic conductivity with structural changes in the covalent network in binary alkali borate glasses. When a third compound such as an alkali halide is added into the glass to increase the conductivity, the main structural change is a redistribution of the BO_4 tetrahedra into the network with the formation of diborate rings. Two mechanisms are responsible for the enhancement of the conductivity by lowering of the activation energy. The first one which predominates in glasses doped with LiF and $LiCl$ is the redistribution of the BO_4 with formation of diborate rings containing two of these negative sites at the expense of groups with only one BO_4 (triborate and/or pentaborate). Then with the increase of the size of the halide anion of the doping salt and/or with the increase of the salt concentration, the diborate rings form clusters of increasing size. In such clusters, the overlap of the potential wells associated with each negative site is enhanced and this induces the lowering of the activation energy for the ionic motion observed when the type and/or the concentration of the salt is changed.

In doped glasses with high modifier content (above 30 mol% Li_2O) which correspond to glasses having a network with high degree of connectivity, this process gives rise to the formation of BO_3 triangles with non-bridging oxygen atoms. In such a group, the negative charge is localised on the NBO which acts as a trap for the mobile cation. The relative fraction of mobile cations becomes less important under the salt addition and the ionic conductivity has a tendency to saturate in glasses with high doping salt content. This can be understood as the result of the competition between the decrease of the activation energy due to the increase of the size of the clusters and the decrease of the number of mobile cations by trapping of some of them in $M^+–BO_3^-$ bonds.

The correlation between ion transport in glasses and the structure of the network appears clearly in figure 53. In halogenoborate glasses with same composition, where the Structural Correlation Length (SCL) which describes the long range structure of the network can be changed by the exchange of the halide anion, the figure shows that the ionic conductivity increases linearly with the increase of the SCL obtained by the anion exchange.

In the case where the "doping salt" (i.e. lithium sulphate) behaves as a glass modifier, it participates directly to the covalent network. In the sulphoborate glass system, whatever the composition, the alkali sulphate are incorporated into the

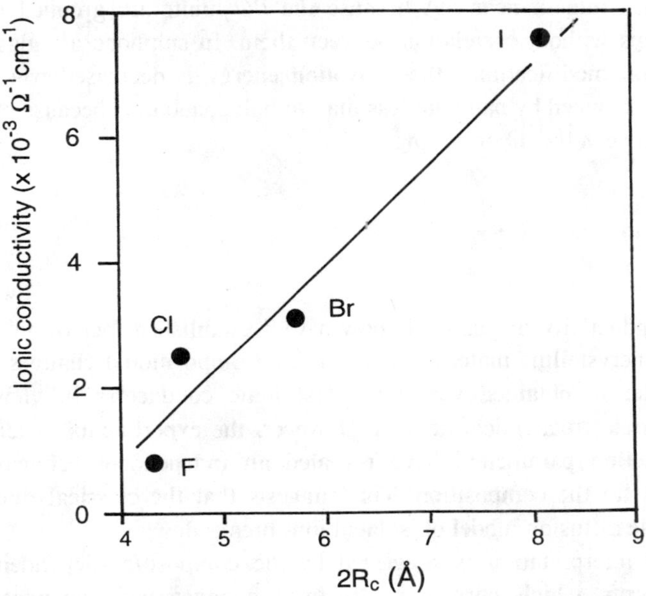

Figure 53. Correlation between ionic conductivity and Structural Correlation Length in the ternary borate glasses B_2O_3-$0.6Li_2O$-$0.24LiX$ (from reference 84).

network by forming B_{IV}–O–SO_2^{-2}–O–B_{IV} covalent bonds and the negative charges are compensated by the presence of two Li^+ cations in the surrounding of the SO_4 tetrahedra which have two NBO atoms. So these two Li^+ cations do not participate to the ionic transport.

In glasses of weak connectivity (B_2O_3-$0.2Li_2O$-yLi_2SO_4), the addition of lithium sulphate creates new BO_4 tetrahedra by the connection of the sulfate ions with the boron-oxygen network, and increases the connectivity of the glass matrix so that it becomes more tridimensional. The small decrease of the Structural Correlation Length from 6 Å to 5.5 Å is an indication that they are probably included in isolated triborate rings without correlations between them. The increase of the number of BO_4 units is probably the principal parameter in the variation of the ionic conductivity with lithium sulphate content.

In the second series of glasses (B_2O_3-$0.6Li_2O$-yLi_2SO_4), with a highest degree of network connectivity, the lithium sulphate is also covalently bonded with the network. It changes the number of four-coordinated boron atoms by direct bonding with the boron-oxygen network or by reaction with the BO_3 with non-bridging oxygen atoms and their transformation into B_{IV} atoms. No changes are observed in the SCL which remain near 5 Å because the BO_4 units are grouped in pairs in diborate groups without correlation between them. In sulphoborate glasses, under these structural modifications, the activation energy is decreased when the BO_4 tetrahedra are grouped by pairs but less than in halogenoborate because of the small correlation between the diborate rings.

8 Conclusion

Fast ionic conductivity in glasses is now a well established fact in an importatnt number of noncrystalline materials and glasses. Compositional changes have been tested in order to obtained very good fast ionic conductors in view of their application in electrochemical devices. However, the experimental determinations of the conduction parameters have revealed an independent behaviour of the prefactor σ_o with the composition. This suggests that the classical interpretation with the simple diffusion model of isolated ions breaks down.

An other interpretation is suggested by the composition-dependence of the activation energy which correlates structural changes with changes in ionic transport in glasses. This latter idea has been discussed in borate glasses which have the interesting peculiarity of possessing three different ranges of structural organisation in their covalent network unlike other glass systems such as silicate or chalcogenide glasses.

The result of the addition of a network modifier into vitreous boron oxide is the conversion of the triangular BO_3 into BO_4 tetrahedra and/or the formation of non-bridging oxygen atoms. These elementary units constitute the first step of the atomic ordering in the glass (Short Range Order). The BO_4 and the BO_3 triangles are negatively charged with different binding energy in regard of the modifier cation. The first one with the lowest binding energy, receives the cations during the ionic transport, while the second one, because its greater binding energy can be considered as a trap for the mobile species.

Superstructural units comprising the basic BO_3 triangles and the BO_4 tetrahedra are incorporated into the network. They describe the second step of the structural organisation (Medium Range Order). The most simplest is the boroxol ring which is a regular ring with three BO_3 and is the main structural unit found in the vitreous B_2O_3. The other superstructural units formed are borate groups containing one (triborate or pentaborate) or two (diborate) BO_4 tetrahedra. It is interesting to note that the diborate groups are the main constituent of the boron oxygen network in glasses with higher conductivity. It contains two BO_4 in neighbouring position forming an internal B_{IV}-O-B_{IV} bridge into the group. In these glasses, the highest values reached by the ionic conductivity can be understood by the appearance of a new type of negative site composed by BO_4 tetrahedra grouped by pairs. In these pairs, because of the important overlapping of the potential wells associated with each BO_4, the energy barrier over which the mobile ions must jump to be free, is lowered.

The studies of the Short and Medium Range Order in the borate glasses have permitted the identification of two kinds of negative sites with different activation energy: isolated BO_4 tetrahedra and pairs of BO_4 tetrahedra. A continuous change from one type to the other can be obtained by structural changes in the network associated with the changes in the composition of the glasses. This is not the only one process which increases the ionic conductivity. The existence of structural correlation into the network (Long Range Order) contributes also to the enhancement of the ionic transport. The formation of clusters containing BO_4 tetrahedra or diborate groups favours the lowering of the activation energy by increasing the electrostatic interaction between the negative species.

An attempt has been made in this paper to develop an approach to ionic conduction process in glasses in terms of glass structure. One of the essential points of this approach is the existence of Short and Medium Range Order. This has permitted the caracterisation of the negative site in the vicinity of which are found the alkali cations. The other one is the existence of Long Range Order from wich it is possible to obtained a description of the pathway were the cations move during the ionic transport. Many fundamental questions about the motion of the mobile ions inside such a network still remains answered. Further experimental and theoretical efforts would be required to raise this challenge.

References

1. C. Julien and G.A. Nazry, Solid State Batteries: Material design and optimisation, (Kluwer Academic Publishers, Hingham,1994).
2. D.P. Button, L.S. Mason, H.L. Tuller and D.R. Hulmann, Solid State Ionics **9&10** (1983) 585.
3. W. Jeitschko, T.A. Bither and M. Bierstedt, Acta. Cryst. **B33** (1977) 2769.
4. J. Zarzycki, Les Verres et l'Etat Vitreux, (Masson) 1982.
5. J.D. Makenzie and W.F. Claussen, J. of Amer. Ceram.Soc. **44** (1961) 79.

6. E.I. Katmitsos and G.D. Chryssikos in European Material Reseach Society Report "Solid State Ionics", edited by M. Balkanski and C. Delmas (European Material Reseach Society, Strasbourg, 1991) 87.
7. M. Kbala, Thèse de l'Université de Bordeaux 1, (1979).
8. M. Irion, M. Couzi, A. Levasseur, J.M. Réau and J.C. Brethous, J. Sol. St. Chem. **31** (1980) 285.
9. D.P. Button, R.P Tandon, H.L. Tuller and D.R. Uhlman, J. Non-Cryst. Solids **42** (1980) 297.
10. W.H. Zachariasen, J. Am. Chem. Soc. **54** (1932) 3841.
11. A.H. Silver and P.J. Bray, J. Chem. Phys. **29** (1958) 984.
12. R.L. Mozzi and B.E. Warren, J. Appl. Crystallogr. **3** (1970) 251.
13. M. Massot, C. Julien and M. Balkanski, Infrared Physics, **29**, (1989) 775.
14. L.A. Kristiansen and J. Krogh Moe, Phys. Chem. Glasses, **9** (1968) 96.
15. P.J. Bray and J.G. O'Keefe, Phys. Chem. Glasses, **4** (1963) 37.
16. P.J. Bray, S.A. Feller, G.E. Jellison Jr. and J.H. Yun, J. Non-Cryst. Solids **38&39** (1980) 93.
17. J. Swenson, L. Börjesson and W.S. Howells, Phys Rev. **B52** (1995) 9310.
18. W.L. Konijnendijk and A.M. Stevels, J. Non-Cryst. Solids **18** (1975) 307.
19. M. Massot, C. Julien, M. Balkanski, A. Krol and W. Nazarewicz, Mater. Sci. Eng. **B3**, 307 (1989).
20. J. Lorösch, M. Couzi, J. Pelous, R. Vacher and A. Levasseur, J. Non-Cryst. Solids **69** (1984) 1.
21. N. Ushida, N Maekawa and T. yokokawa, J. Non-Cryst. Solids, **85**(1986)290.
22. P.J. Bray, E.J. Holupka, J. Zong, R.V. Mulkern and C.J. Brinker, Naturwiss R.**36** (1987) 735.
23. D.P. Button, R. Tandon, C. King, H.L. Tuller and D.R. Uhlmann, J. Non-Crystalline Solids **49** (1982) 129.
24. A.E. Geissberger, F. Bucholtz and P.J. Bray, J. Non-Cryst. Solids **49** (1982) 117.
25. M. Massot, E. Haro, M. Oueslati and M. Balkanski, Mat. Res. Soc. Symp. Proc., **135** (1989) 207.
26. D.P. Button, unpublished, D. Phil. Thesis, Cornell University, Ithaca, New-York, (1978) .
27. D.E. Turcotte, W.E. Risen Jr. and E.I Katmirtsos , Solid State Comm. **51**, (1984) 313.
28. Y. Wang, A. Osaka, Y. Miura, and K. Takahashi. J. Non-Cryst. Solids **95-96** (1987) 571.
29. D. Kline and P.J. Bray, Phy. Chem. Glasses 7 (1966) 41.

30. J.E. Shelby, J. Am. Ceram. Soc. **66** (1983) 225.
31. A. Levasseur, R. Olazcuaga, M. Kbala, M. Zahir, M. Couzi and P. Hagenmuller, Solid State Ionics, **2** (1981) 205.
32. P. Mustarelli, S. Scotti, M. Villa and P.R. Gandhi; Solid State Ionics **39** (1990) 217.
33. M. Massot, E. Haro, M. Oueslati, M. Balkanski, A. Levasseur and M. Menetrier, Mat. Sci. Eng. **B3** (1989) 57.
34. J. Goubeau and H.Z. Keller, Inorg. Chem. **272** (1953) 303.
35. F.L. Galeener, G Lucovsky and J.C. Mikkelsen Jr., Phys. Rev. **B22** (1980) 3983.
36. W.M. Risen, J. Non-Cryst. Solids **76** (1985) 97.
37. J. Krog-Moe, Phys. Chem. Glasses **6** (1966) 46.
38. F.L. Galeener, R.A. Barrio, E. Martinez and R.J.Elliot, Phys. Rev. Lett. **53** (1984) 2429.
39. P.J. Bray, S.E. Feller, G.E Jellison and Y.H. Yun, J. Non-Cryst. Solids **38&39** (1980) 93.
40. A.C. Hannon, D.I. Grimley, R.A. Hulme, A.C. Wright and R.N. Sinclair, J. Non-Cryst. Solids **177** (1994) 299.
41. J.W. Zwanziger, R.E. Youngman and S.L. Tagg, J. Non-Cryst. Solids **192&193** (1995) 157.
42. J. Krogh-Moe, Phys. Chem. Glasses, **3** (1962) 101.
43. T.W. Brill, Philips Research Report, Suppl. **2** (1976).
44. P.J. Bray, J. Non-Cryst. Solids **95-96**(1987) 45.
45. M. Tatsumisago, M. Takahashi, T. Minami, N. Usumeki and N. Iwamoto, Yogyo Kyokai Shi **94** (1986) 464.
46. F.A Fusco, M. Massot, M. Oueslati, E. Haro, H.L. Tuller and M. Balkanski, Mat. Res. Soc. Symp. Proc., **135** (1989) 189.
47. W. Soppe, J. Kleerbezem and H.W. den Hartog, J. Non-Cryst. Solids **93** (1987) 142.
48. M. Irion, M. Couzi, A. Levasseur, J.M. Réau and J.C. Brethous, J. Solid Stat. Chem. **31** (1980) 285.
49. A. Levasseur, B. Cales, J.M. Réau and P. Haggenmuller, Mat. Res. Bull. **13** (1979) 921.
50. A. Levasseur, M. Kbala, J.C. Brethous, J.M. Réau, P. Haggenmuller and M. Couzi, Solid State Comm. **15** (1979) 839.
51. M. Massot, A. Ayyadi, M. Balkanski, "Science and technology of fast ionic conductors", Nato ASI series, edited by M. Balkansi and H.L. Tuller (Plenum Press, New York, 1989) 331.

52. M. Massot, E. Haro, R. Darvich and M. Balkanski, Proceeding of the 11th Conference on Raman Spectroscopy, eds. by R.J.H. Clark and D.A. Long (John Wiley Editors, 1988) 525.
53. J..P. Bronswijk and E. Strijks, J. Non-Cryst. Solids **24** (1977) 145.
54. G.E. Gurr, P.W. Montgomery, C.D. Knutson and B.T. Gorres, Acta. Cryst. **B26** (1970) 906.
55. W.H. Zachariasen, Acta. Cryst. **17** (1964) 749.
56. S.P. Amaral Souto, unpublished, Thèse de l'Université P. et M. Curie, Paris (1994).
57. E.I. Katmitsos,M.A. Karakassides and G.D. Chryssikos, J. Phys. Chem. **90** (1986) 4528.
58. E. Cazanelli and R. Fresh, J. Chem. Phys. **79** (1983) 2615.
59. A. Ayyadi, unpublished, Thèse de Doctorat ès Sciences, Paris (1989).
60. M. Massot, C. Julien, M. Balkanski, A. Krol and W. Nazarewicz, Mat. Sci. Eng. **B3** (1989) 367.
61. S.R. Elliott, J. Phys. IV, Colloque C2, supplément au Journal de Physique **III**, (1962) 79.
62. R. Shuker and R.W.Gammon, "Light Scattering in Solids" ed. by M. Balkanski, (Flammarion Sciences, Paris, 1971)
63. R. Shuker and R.W.Gammon, Phys. Rev. Lett. **25** (1970) 222.
64. V.K. Malinowski, V.N. Novikov, P.P. Parshin, A.P. Sokolov and M.G. Zemlyanov, Europhys. Lett. **11** (1990) 43.
65. S.R. Elliott, Europhys. Lett, **67** (1992) 201
66. A.P. Sokolov, A. Kisliuk, M. Soltwish and D. Quitmann, Phys. Rev. Lett. **69** (1992) 1540.
67. S.R. Elliott, Phys. Rev. Lett, **67** (1991) 711.
68. L. Börjesson, A.K. Hassan, J. Swensson, L.M. Torell, Phys. Rev. Lett. **70** (1993) 1275.
69. L. Börjesson, A.K. Hassan, J. Swensson, L.M. Torell, Phys. Rev. Lett. **70** (1993) 4027.
70. C. Levelut, N. Gaimes, F. Terki, G. Cohen-Solal, J. Pelous and R. Vacher, Phys. Rev. **B51** (1995) 8606.
71. V. K. Malinowski and A.P. Sokolov, Solid State Commun. **57** (1986) 757.
72. U. Strom and P.C. Taylor, Phys. Rev. **B32** (1977) 5512.
73. U. Buchenau, M. Prager, N. Nücker, A.J Dianoux, N. Ahmad and W.A. Phillips, Phys. Rev. **B34** (1986) 5565.
74. N. Ahmad, K.W. Hutt and W.A. Phillips, J. Phys. **C19** (1986) 3765.
75. J.E. Graebner, B. Golding and L.C. Allen, Phys. Rev. **B34** (1986) 5696.

76. E. Akkermans and R.Maynard, Phys.Rev. **B32** (1985) 7850.
77. A.J. Martin and W. Brenig, Phys. Status Solidi **B64** (1974) 163.
78. V.K. Malinowski, V.N. Novikov, P.P. Parshin, A.P. Sokolov and M.G. Zemlyanov Europhys. Lett. **11** (1990) 43.
79. A.P. Sokolov, A. Kisliuk, D. Quitmann and E. Duval, Phys. Rev. **B48** (1993) 7692.
80. T. Achibat, A. Boukenter and E. Duval, J. Chem. Phys **99**, 2046 (1993).
81. R.J. Nemanich, Phys. Rev. **B16** (1977) 1655.
82. A.C. Hannon,R.N. Sinclair and A.C. Wright, Physica **A201** (1993) 375.
83. M. Massot, S. Souto and M. Balkanski, Solid State Ionics **70/71** (1994) 350.
84. M. Massot, S. Souto and M. Balkanski, J. Non-Cryst. Solids **182** (1995) 49.
85. Y.V. Denisov and P. Rylev, JETP Lett. **52** (1990) 412.
86. V.K. Malinovsky, V.N. Novikov and A.P. Sokolov, Phys. Rev. Lett. **A153** (1991) 63.
87. W. Soppe, W. Ebens and H.W. den Hartog, J. Non-Cryst. Solids **105** (1988) 251.
88. L. Börjesson, Dynamic of Disordered Materials, Springer Proceeding in Physics, **37** (Springer Verlag, Berlin, 1989) 316.
89. A. Chehaidar, M.D. Rouhani and A. Zwick, Proceedings of the 6th Int. Conf. on the Structure of Amorphous Materials (Praha, 1994).
90. R.A.Barrio and F.L. Castillo-Alvarado, Phys. Rev. **B46** (1992) 13779.
91. J. Kroeg-Moe Acta Cryst. **15** (1962) 190.
92. J. Kroeg-Moe Acta Cryst. **B24** (1968) 179.
93. K. Otto, Phys. Chem. Glasses, 7 (1966) 29.
94. J.E. Bauerle, J.Phys. Chem. Sol., **30** (1969) 2657.
95. C. Julien and M. Massot, Microionics - Solid State Integrable Batteries, ed. M. Balkanski, (Elseviers Science Publishers, B.V., 1991) 173.
96. H.L. Tuller, D.P. Button and D.R. Uhlmann, J. Non-Cryst. Sol. **40** (1980) 93.
97. H.L. Tuller and D.P. Button, Proc. Int. Conf. on Transport–Structure Relations in Fast Ion and Mixed Conductors, eds. F.W. Poulsen, N.H. Anderson, K. Clausen, S. Skaarup and O.T. Sorensen, (Riso Nat. Lab., Denmark, 1985) 119.
98. F.A. Fusco, H.L. Tuller and D.P. Button, Proc. Symp. on Electro-Ceramics and Solid State ionics, eds. H.L. Tuller and D.M. Smyth, (The Electrochemical Society, Pennington, N.J., 1988) 167.
99. A. Levasseur, J.C. Brethous, J.M. Réau, P. Hagenmuller and M. Couzi, Solid State Ionics **1** (1980) 177.

100. A. Levasseur, J.C. Brethous, J.M. Réau and P. Hagenmuller, Mat. Res. Bull., **14** (1979) 921.
101. A. Levasseur, M. Kbala, J.C. Brethous, J.M. Réau and P. Hagenmuller, Solid State Commun., **32** (1979) 839.
102. I. Darianian, unpublished. Thèse de Doctorat de l'Université Pierre et Marie Curie (1988).
103. Y. Ito, K. Miyauchi, T Oki, J. Non-Cryst. Sol.,57 (1983) 389.
104. E.I Kamitsos, M.A. Karakassides and G.D. Karakassides, Phys. Chem. Glasses, **30** (1989) 19.

Chapter 5

Theory of Atomic Diffusion Across Interfaces

J. Deppe

1 Introduction

Atomic diffusion across material interfaces is a problem of growing interest in many fields of physics. In crystalline or quasi-crystalline materials [1], an understanding of the structure of grain boundaries is essential to understanding interfacial diffusion. These structures have been studied using computer modeling of extended defects [2] guided by scanning tunneling electron microscopy [3]. However, this would be difficult to apply to our case of interest (a solid-state battery) in which one material (the separator) is amorphous. An essential difference between interface diffusion in fast ionic conductors and other materials [4] is that the interface in the former systems almost always inhibits diffusion, whereas diffusion may be enhanced in the latter. This is so because the structure of fast ionic conductors is in some sense optimized for ionic diffusion. Therefore, an interface or boundary layer will tend to reduce the ionic diffusivity. For many other materials, it is diffusion along grain boundaries which may dominate mass transport [5]. One example of this may be found in the use of lithium alloy layers which are used to improve contact between the anode and solid amorphous electrolyte in micro-batteries [6]. In the study of ion transport in micro-batteries there are two such interfaces to be considered: the anode-separator, e.g., Li metal and a lithium doped borate glass; and the separator-cathode, e.g., the lithium doped borate glass and an intercalation material, $InSe$ [7]. This interface diffusion problem has been addressed previously [8] with the aid of a one-dimensional lattice gas model. The lattice sites are specified by l, which takes on integer values. The interface is the layer $l = 0$. The parameters which characterize the model are the hopping rates in the two materials, denoted by α for hopping to the right of the interface ($l > 0$), and by β for hopping to the left ($l < 0$). This type of model is particularly appropriate to the problem of interest because: 1) the ionic diffusion is essentially one-dimensional, and 2) the lack of any information regarding the atomic structure at the interface makes the use of a simple model desirable. The calculation of the frequency dependent resistance for a simple version of this model is given in Ref. [9].

In this presentation, we detail three related approaches to studying the problem of ion transport across material interfaces. In particular, we focus on the interface between fast-ion-conducting glass and an intercalate, e.g., InSe, that arises in the study of solid state batteries. First, we generalize the one-dimensional lattice gas model of particle diffusion across material interfaces to include a finite width interface region, and an energy barrier at the interface[10]. We use overvoltage measurements on $InSe$ and dielectric loss measurements on $B_2O_3 - 0.5Li_2O - 0.15Li_2SO_4$ to determine the hopping rates in the two materials. Numerical and analytical results for the particle density vs. position are presented for representative values of the energy barrier and a range of hopping rates. The presence of the energy barrier is found to be a dominant factor in the problem. Next, the ion transport across the this interface is studied by a structural one-dimensional lattice gas model, which takes explicitly into account hopping rates to first and second neighbors, short range interaction of the diffusing ions, and long-range Coulomb correlations[11]. The hopping to second nearest neighbors is included to avert the problem with one-dimensional models of the mean square displacement varying as $t^{1/2}$ due to the severe effect of site blocking. Numerical examples are given for material parameters of the real interface between doped ternary borate glass and indium selenide. It is shown that the ability of ions to get around each other brings additional structure into the one-dimensional model, while preserving the simplicity of the model. Finally, we present an investigation of two two-dimensional (2D) models of ion diffusion across material interfaces[12]. The geometry of the interface is appropriate to the separator-cathode interface in solid state batteries. The first model is a deterministic lattice gas (LG) model explicitly including the site blocking effect, the second employs an atomistic (stochastic) kinetic Monte-Carlo (KMC) simulation. We include an energy barrier, allow an ion-ion Coulomb interaction, and investigate the effect of a chemical potential, found by solving Laplace's equation with an applied potential in a geometry representing the battery. Numerical results are presented for several values of the energy barrier, and two values of the applied potential both with and without the Coulomb interaction. The results of these two models are similar; quantitative differences are discussed within the context of the models. It is found that charging effects change the nature and relative importance of the effects of the other parameters in the problem. Therefore, a determination of the true Debye screening length, and therefore the number density of charge carriers, seems crucial.

2 One-Dimension Lattice Gas Model of Ion Diffusion across Material Interfaces

2.1 Introduction

We have generalized the model presented in Ref. [8] by including a finite width interface region specified by the two layers $l = 0$ and $l = 1$. The hopping rate between these two layers is α', taken to be the average of α and β. There is no compelling reason for this choice, but it is the simplest for comparison with the results of Blender et al. [8]. It should be noted here that this is a potentially very important generalization because the problem of ion diffusion across the micro-battery is essentially one-dimensional. The ions begin at the leftmost interface, between anode and separator, and diffuse to the rightmost face, between separator and cathode. Therefore, the effects of the relative width of the Van der Waals gap in the intercalation material, surface defects etc., may be incorporated into α'. We also introduce an energy barrier in the interface region. The effect of the barrier is manifested by modifying the hopping rate in the interface region to $\alpha'_+ = \alpha' exp(\epsilon/k_B T)$ for hopping to the left, and $\alpha'_- = \alpha' exp(-\epsilon/k_B T)$ for hopping to the right, where 2ϵ is the energy barrier, T is the temperature and k_B the Boltzmann constant. There is now a positive energy barrier impeding particle flow to the right across the interface.

There are certainly other possibly important effects which could be incorporated into this model of ionic diffusion, among them the Coulomb interaction between the diffusing Li^+ ions. It has been shown [13] that the incorporation of the $Li^+ - Li^+$ Coulomb interaction has a relatively small effect on the ionic conductivity in lithium doped borate glasses. This is essentially because of the extremely small percentage of dopant molecules which ionize to give free Li^+ ions. Some effects which may be important near the interface, including an image charge interaction [14] (the InSe becomes somewhat more metallic as Li is intercalated into it) and Coulomb interactions near large density gradients [15], are also not included. It has been found [16], that the inclusion of Coulomb interactions, making the equations of motion and the induced electric field self-consistent, changes our results quantitatively, but not qualitatively.

With these limitations in mind, we are able to draw two important conclusions from this investigation. The first is that the energy barrier is an extremely important quantity and must be well characterized before any more complete theory of interface transport may be considered. The second regards the interface region, which may be thought of as a first approximation to a graded junction. We find little improvement in the ability of particles to diffuse across the interface with the introduction of a transition region with this simple choice for α'.

The rest of this section is organized as follows. In the next section we define the equations of motion (EOM), specify the initial conditions, and solve the resulting difference equations which result from taking the Laplace transform of the EOM. Following this, we present an analytic solution of the equations of motion in the large time limit. Next we use representative hopping rates for lithium doped borate glasses and $InSe$, and present numerical results for these rates with several values of the energy barrier. We also find very good agreement between the numerical results and the analytical solutions in the large time limit. Finally, we find that the introduction of a transition region, characterized by a hopping rate

$$\alpha' = (\alpha + \beta)/2, \tag{1}$$

has a negligible effect on the ability of particles to diffuse across the interface. Note, this may not be so if α' were chosen differently. The presence of the barrier is found to be a dominant factor in the problem.

2.2 Equations of Motion

The equations of motion for the system defined in the introduction, without blocking (for a discussion of one version of this model with blocking, see Ref. [17]), can be written,

$$\dot{n}_l = \alpha(n_{l+1} + n_{l-1} - 2n_l) \quad l > 1, \tag{2}$$

$$\dot{n}_l = \alpha(n_{l+1} - n_l) + \alpha'_- n_{l-1} - \alpha'_+ n_l \quad l = 1, \tag{3}$$

$$\dot{n}_l = \alpha'_+ n_{l+1} - \alpha'_- n_l + \beta(n_{l-1} - n_l) \quad l = 0, \tag{4}$$

$$\dot{n}_l = \beta(n_{l+1} + n_{l-1} - 2n_l) \quad l < 0, \tag{5}$$

where $\dot{n}_l$ is the time derivative of n_l, l indicates the lattice site, and α'_+ and α'_- are as defined above. The initial conditions are taken to be

$$n_l(t=0) = 0 \quad l \geq 1, \tag{6}$$

$$n_l(t=0) = 1 \quad l < 1. \tag{7}$$

Writing the Laplace transform of $n_l(t)$ as $N_l(s)$, we find after taking the Laplace transform of Eqs. (2)-(5)

$$sN_l = \alpha(N_{l+1} + N_{l-1} - 2N_l) \quad l > 1, \tag{8}$$

$$sN_l = \alpha(N_{l+1} - N_l) + \alpha'_- N_{l-1} - \alpha'_+ N_l \quad l = 1, \tag{9}$$

$$sN_l = \alpha'_+ N_{l+1} - \alpha'_- N_l + \beta(N_{l-1} - N_l) + 1 \quad l = 0, \tag{10}$$

$$sN_l = \beta(N_{l+1} + N_{l-1} - 2N_l) + 1 \quad l < 0. \tag{11}$$

Using a slight generalization of the procedure outlined in Ref.[8], this set of difference equations is solved using the ansatz

$$N_l = N^{(1)} v^{l-1} \quad l \geq 1, \tag{12}$$
$$N_l = (N^{(0)} - 1/s) w^{|l|} + 1/s \quad l < 1, \tag{13}$$

where $N^{(1)}$ and $N^{(0)}$ are two functions of the hopping rates and the energy barrier and are to be determined.

Substituting Eq. (12) into Eq. (8), and Eq. (13) into Eq. (11), we find

$$v^2 - (s/\alpha + 2)v + 1 = 0, \tag{14}$$
$$w^2 - (s/\beta + 2)w + 1 = 0. \tag{15}$$

We note here that, from the form of these equations for v and u, $1/v$ and $1/w$ are also solutions. If Eq. (12) is substituted into Eq. (9) and eq. (13) into Eq. (10), we can determine the constants $N^{(1)}$ and $N^{(0)}$ as follows:

$$N^{(0)} = \frac{(s - \alpha(v-1) + \alpha'_+)(\beta(1-w)/s + 1)}{(s - \beta(w-1) + \alpha'_-)(s - \alpha(v-1) + \alpha'_+) - (\alpha')^2}, \tag{16}$$

$$N^{(1)} = \alpha'_- \frac{N^{(0)}}{s - \alpha(v-1) + \alpha'_+}. \tag{17}$$

2.3 *Large Time, or Small s, Expansion*

We are interested in the long time limit solution of these equations, and therefore make a small s expansion of the relevant expressions. Using the solutions to Eqs. (14) and (15), we can write

$$\alpha(v-1) = s/2 - (s/2(s/2 + 2\alpha))^{1/2} \approx -(s\alpha)^{1/2} + s/2 + O(s^{3/2}) \tag{18}$$
$$\beta(w-1) = s/2 - (s/2(s/2 + 2\beta))^{1/2} \approx -(s\beta)^{1/2} + s/2 + O(s^{3/2}) \tag{19}$$

One may then substitute Eqs. (18) and (19) into Eqs. (12) and (13) respectively and expand the expressions to order $1/s^{1/2}$. Using the fact that the inverse Laplace transform of $1/s$ is 1, and that of $1/s^{1/2}$ is $1/(\pi t)^{1/2}$, defining $r = (\alpha/\beta)^{1/2}$, we find to order $1/t^{1/2}$

$$n_l(t) = \frac{\alpha'_-}{r\alpha'_- + \alpha'_+} \tag{20}$$
$$(1 + \frac{\alpha'_-(r-1)/2 - r\beta - (l-1)(\alpha'_- + \alpha'_+/r)}{(\pi\beta t)^{1/2}(r\alpha'_- + \alpha'_+)}) \quad l \geq 1,$$

$$n_l(t) = \frac{\alpha'_+}{r\alpha'_- + \alpha'_+} [1 + \frac{\alpha'_-}{\alpha'_+}(\frac{\alpha'_+(r-1)/2 + \alpha + |l|r(r\alpha'_- + \alpha'_+)}{(\pi\beta t)^{1/2}(r\alpha'_- + \alpha'_+)})] \quad l < 1. \tag{21}$$

2.4 Hopping Rates and Numerical Solution

We estimate the hopping frequencies, β for a lithium doped borate glass, and α for $InSe$, using dielectric loss measurements and overvoltage measurements respectively. The lithium doped borate glass separator, $B_2O_3 - 0.5Li_2O - 0.15Li_2SO_4$, displays a maximum in the loss peak [18] at $\beta = 1 \times 10^4 s^{-1}$ to $1 \times 10^6 s^{-1}$ for $T = 560K$ to $650K$. The device is assumed to operate at one temperature, $T = 560K$, so that the hopping frequency for the glass is constant at $\beta = 1 \times 10^4 s^{-1}$. For the cathode, Li_xInSe, the diffusion constant [19] has been determined to be $D = 10^{-8} - 10^{-11} cm^2/s$ when the amount of intercalated lithium per formula unit, x, varies from 0.0 to 1.0. Using $D = 0.5L^2/\tau$, where L is the hopping distance ($4\mathring{A}$ from Ref. [20]) and τ the hopping time, we find $\alpha = 1/2\tau = 1 \times 10^4 s^{-1}$ to $1 \times 10^7 s^{-1}$ for $x = 0.0$ to 1.0. As mentioned earlier, the hopping frequency in the interface region, $\alpha' = (\alpha + \beta)/2$.

In addition to the approximate long time solutions presented previously, the equations of motion can be directly integrated using a suitable integration scheme. We have employed a fourth order Runge-Kutta routine and have found the results obtained in this way agree with those from Eqs. (21) and (22) to within 3% near the interface. It will be noticed that equations (2)-(5) represent a system constantly supplied with diffusing particles from the left. This can be most easily seen by examining Eqs. (21) and (22) in the infinite time limit and with no barrier present. In this case, the particle densities to the left and right of the interface are equivalent and independent of l, and attain the value $\beta^{1/2}/(\beta^{1/2} + \alpha^{1/2})$. Note that this is a function of the hopping frequencies, and is not in general 1/2, as one might expect from the initial conditions. In our numerical simulations, impenetrable boundaries are placed a finite distance from the boundary. This difference places limits on the agreement one can expect from the two approaches. Nevertheless, we have found the results obtained from these two methods to be in good agreement.

In Fig. 1 we present results for the case $\alpha = 5\beta$ and 50β respectively, where β was held constant at $1 \times 10^4 s^{-1}$. The energy barrier was taken to be zero. The time step in our simulations was $5 \times 10^{-7} s$, and all results were obtained over 4000 time steps unless otherwise stated. Two comments may be made regarding these figures. The first is that particle flow deep into the right hand region is quicker in the latter case, as one expects physically. This, however, results in a much larger particle gradient at the interface. The second is that the discontinuity in slope follows the current continuity equation,

$$\alpha \frac{\partial n(0^+,t)}{\partial x} = \beta \frac{\partial n(0^-,t)}{\partial x}. \tag{22}$$

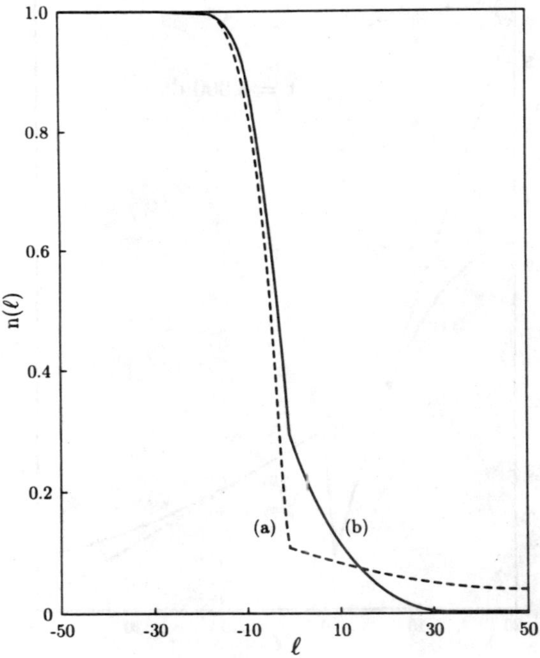

Figure 1: a) Variation of n_l with l for $\alpha = 5\beta$ after 4000 time steps. Results were obtained numerically from Eqs. 1-4 and agree to within 3% with those obtained from Eqs. 21 and 22 near the interface. b) Variation of n_l with l for $\alpha = 50\beta$ after 4000 time steps.

This relation can be verified using the expressions for the particle densities to the left and right of the barrier as given in Eqs. (21) and (22). We finally note that the difference between these results and those obtained with no finite interface region, that is, the model of Blender et al.[8], is negligible for this particular choice of α'. However, it may be that α' is significantly different from both α and β, in which case a model including this possibility becomes essential.

In Figs. 2 and 3 we show a comparison between the numerical results and the analytical results obtained from Eqs. (21) and (22) after 8000 and 16000 time steps. As may be seen, the agreement between these results is very good both well away from the interface as well as in the region closest to the interface. In addition, we find the area of agreement to increase in size with the length of the simulation, as is expected.

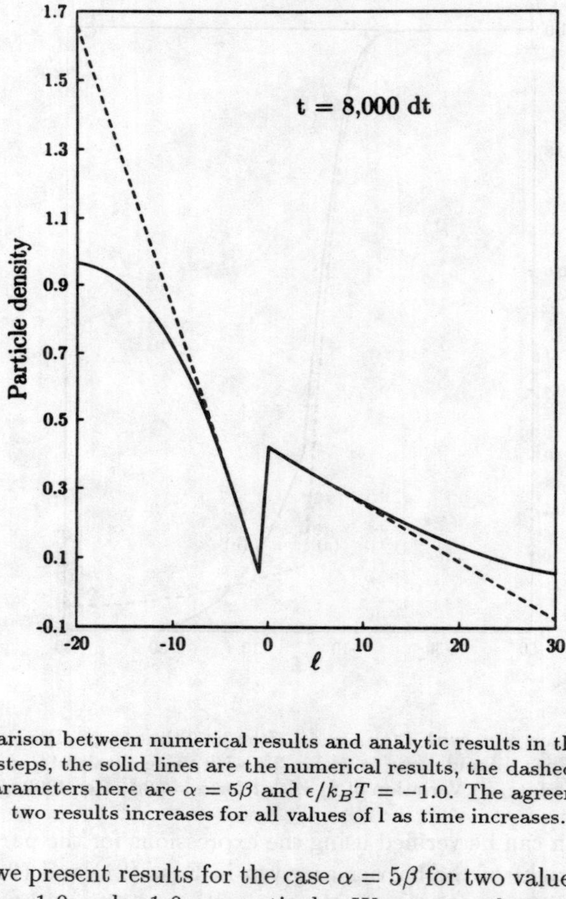

Figure 2: Comparison between numerical results and analytic results in the long time limit after 8000 time steps, the solid lines are the numerical results, the dashed lines the large t solution. The parameters here are $\alpha = 5\beta$ and $\epsilon/k_BT = -1.0$. The agreement between the two results increases for all values of l as time increases.

In Fig. 4 we present results for the case $\alpha = 5\beta$ for two values of the barrier height, $\epsilon/k_BT = 1.0$ and -1.0, respectively. We can see that, as expected, the presence of the positive energy barrier restricts particle flow and could lead to a charge accumulation region at the interface. The negative energy barrier sweeps particles out of the interface region, leaving a depletion region which has been found to grow in magnitude with the size of the barrier.

2.5 Conclusion

We have reported on our investigation of a model of hopping diffusion without site blocking. The inclusion of site blocking will not change Eqs (2) or (5), only the second and third will be modified. In general, the equations of motion with and without site blocking are identical if the hopping rates in the forward and reverse directions are equivalent. We have also investigated this model with the inclusion of the site blocking effect. The results, for the same parameters

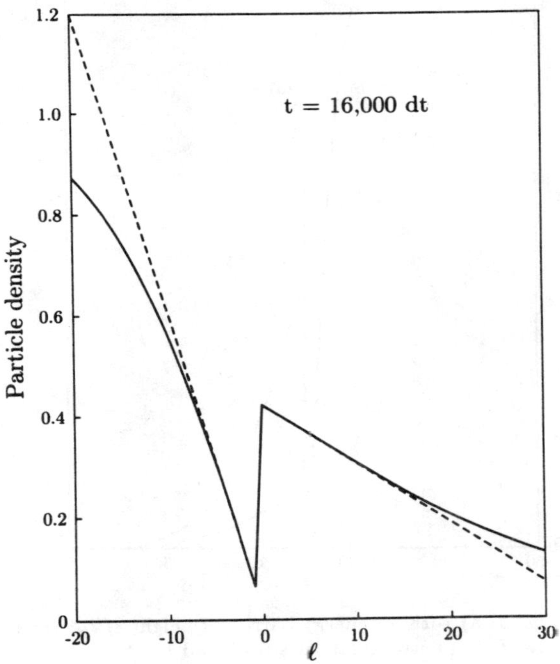

Figure 3: Comparison between numerical results and analytic results in the long time limit after 16000 time steps, the solid lines are the numerical results, the dashed lines the large t solution. The parameters here are $\alpha = 5\beta$ and $\epsilon/k_BT = -1.0$. The agreement between the two results increases for all values of l as time increases.

used to produce Figs. 2 and 3 and are nearly indistinguishable from those shown there. We find a very slight reduction in particle density to the right of the interface, and a small increase (both $< 1\%$) to the left. This small effect may be attributed to the extremely small region where the blocking effect is manifested. As mentioned above, only sites 0 and 1 are affected as they are the only two which have asymmetric hopping rates. The blocking effect was found in to be an important one in Ref.[17]. This, however, is a result of incorporating asymmetric hopping sites over an extended area of the simulation cell. We believe that screening in the cathode, and the small number of free lithium ions in the separator, will limit the spatial extent of the energy barrier to a region very near to the interface.

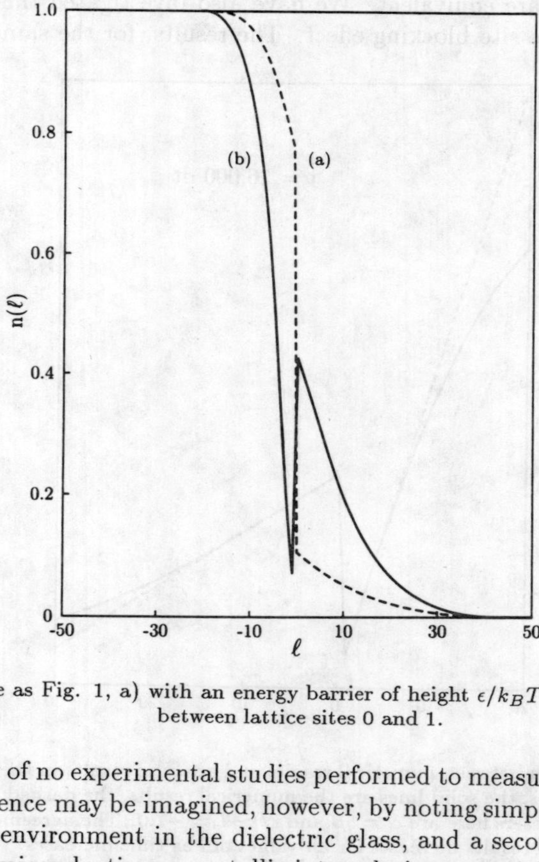

Figure 4: Same as Fig. 1, a) with an energy barrier of height $\epsilon/k_BT = 1.0$ and b) = -1.0 between lattice sites 0 and 1.

We know of no experimental studies performed to measure this energy barrier. Its existence may be imagined, however, by noting simply that the lithium ion is in one environment in the dielectric glass, and a second, very different, one in the semiconducting or metallic intercalation material. It is hoped that XAFS studies currently being carried out in Paris to measure the penetration of lithium into the InSe from the borate glass will show the existence of this barrier.

We have found that a large mismatch in hopping frequencies, as may occur as more and more lithium is intercalated into the *InSe*, results in a much larger particle gradient at the interface. The transition region we have introduced, which may be thought of as a first approximation to a graded junction, resulted in little improvement in the ability of particles to diffuse across the interface. This result, which may be modified with the investigation of more sophisticated models, is not encouraging as far as the introduction of graded junctions in solid state batteries is concerned. In addition, the presence of an energy barrier was found to be a dominant factor in determining the ability of particles to diffuse through the interface region. In particular, a positive barrier, inhibiting particle flow, could lead to a charge accumulation region at the interface, and a negative barrier sweeps charge out of the interface region leading to the formation of a depletion region which grows with the size of the barrier.

3 A Monte-Carlo Model of Lattice Gas Diffusion

3.1 Introduction: An Improved Lattice Gas Model

In this section, an improved lattice gas model is developed and presented. The improvements are as follows: i) the long-range Coulomb interactions are accounted for in such a way as to make the resulting computations more efficient than was possible previously [21], and ii) it takes into account hopping rates up to next-nearest neighbor sites, this makes the local structure of the hopping more elaborate and the model is called 'structural'. As before [17] the short-range hard-core ion interaction is accounted for through the site-blocking effect. Then the model is applied to calculate the ion distribution along the chain, $n_i(t)$, the lattice potential, $\phi_i(t)$, the electric field, and the polarization. They are computed for material parameters of the real interface between doped ternary borate glass and indium selenide, which is used in lithium thin film batteries. Emphasis is placed on the new feature of this model, the ability of ions to circumvent each other.

3.2 Specification of the Structural Lattice-Gas Model

In the framework of the lattice-gas model, the general form of the kinetic equation of hopping particles is given by

$$\frac{dn_p}{dt} = \sum_q [b_{qp} n_q (1 - n_p) - b_{pq} n_p (1 - n_q)], \tag{23}$$

where p and q denote configurations, b_{pq} are hopping rates, and the site-blocking effect is accounted for by the factors $1 - n$. The positive terms in Eq. 23 denote the ways the configuration p is created from other configurations q; the negative terms denote the ways p is changed by hopping. A one-dimensional chain with N sites is considered. Part of this chain represents the glass, the other part represents the intercalate and the interface between them is situated inside the chain. Since the diffusing ions form a one-dimensional charged Coulomb gas, the concomitant potential $\phi_i(t)$ is given by

$$\phi_i(t) = -\lambda \sum_{j=1}^{N-i} j n_{i+j}(t), \tag{24}$$

where i=1,...N, $\lambda = e/(\epsilon_0 \epsilon_r N^2 d)$ is a material factor, and the potential ϕ is given in volts. The expression, Eq. 24, is actually a solution of the Poisson equation with boundary conditions $\phi_N(t) = 0$ and $d\phi_N/dx = 0$, and the meaning of the symbols is as follows: n_i is the probability for the occupation of site i, e is the elemental charge, ϵ_0 and ϵ_r are the absolute and relative permeability, respectively, and d is the average lattice spacing. The derivation of this equation is shown below.

Consider the one-dimensional Poisson equation

$$\frac{d^2\phi(x,t)}{dx^2} = -\frac{e\rho}{\epsilon_0 \epsilon_r} n(x,t), \tag{25}$$

in the interval $0 \geq x \geq$ L, with boundary conditions $\phi = 0$ and $d\phi/dx = 0$ at $x = L$. These boundary conditions are usually assumed when one considers the contribution of the mobile charge, induced by the field effect, in semiconductor accumulation or inversion layers to the surface band bending. They are assumed here due to the analogy with this problem. The notation in Eq. 25 is standard except the density, $\rho = N/L^3$, where $L = Nd$ is the length of the lattice, which is introduced to respect the correct physical dimensions since $n(x,t)$ is a probability of the site occupation when the lattice is discretized in the x direction. Then, Eq. 25 has the form

$$\phi_{i+1} - 2\phi_i + \phi_{i-1} = -\lambda n_i. \tag{26}$$

Consider the successive nodes of the lattice in reverse order. Starting from $i = N$, and keeping in mind that due to the boundary conditions, $\phi_N = \phi_{N+1} = 0$, one obtains from Eq. 26 the following relations:
for node N

$$\phi_{N-1} = -\lambda n_N,$$

for node $N-1$

$$\phi_{N-2} = -\lambda(n_{N-1} + 2n_N),$$

for node $N-2$

$$\phi_{N-3} = -\lambda(n_{N-2} + 2n_{N-1} + 3n_N), \tag{27}$$

and so on, where n_i is the discrete form of the site occupation and is considered here as a known array. Proceeding in this way, one obtains

$$\phi_{N-k} = -\lambda \sum_{j=1}^{N=i} j n_{i+j}(t) \tag{28}$$

for i=1,...,N which is the required expression shown above. This is a closed solution of the Poisson equation, which incorporates the boundary conditions. This solution was tested by comparing ϕ with the output of several standard methods for solving the Poisson equation and the results were identical. However, it is much more convenient to use Eq. 28 both in practical numerical computations and in general theoretical speculations.

3.3 Charge balance

In the present model the fast-ion-conducting glass is considered as a perfect electron separator, i.e., it allows the diffusion of the positive lithium ions from the anode to the cathode of the battery, and inhibits the reverse electron flow. This is an approximation since in practice there is always a reverse electron current, but it may be ignored because, for good materials, its magnitude is several orders lower than the ion current. Thus, when the positive ions diffuse in the fast-ion-conducting glass, an equivalent negative charge is accumulated in the metal anode for the case without external load. The situation is to some extent similar to the field effect in semiconductor surface layers where a mobile charge is induced through a metal electrode. This mobile charge gives rise to a potential barrier, which influences its transport in the glass, and this influence is accounted for here through a self-consistent formulation of the basic equation of motion.

3.4 Equations of Motion

For the self-consistency of the solution it is necessary to incorporate the potential in Eq. 24 in the kinetic equation for the ion dynamics resulting from Eq. 23 when details of the hopping are specified. For hopping to nearest and next-nearest neighbors considered here, the self-consistent system of kinetic equations is given by

$$\begin{aligned}
\frac{dn_i}{dt} &= c_i^- [n_{i-2}(1-n_i)exp[\lambda' \sum_{j=1}^{N-i+2} (2-\delta_{j,1})n_{i+j-2}] \\
&\quad -n_i(1-n_{i-2})/exp[\lambda' \sum_{j=1}^{N-i+2} (2-\delta_{j,1})n_{i+j-2}]] \\
&\quad +b_i^- \Big[n_{i-1}(1-n_i)exp[\lambda' \sum_{j=1}^{N-i+1} n_{i+j-1}] \\
&\quad -n_i(1-n_{i-1})/exp[\lambda' \sum_{j=1}^{N-i+1} n_{i+j-1}]\Big] \\
&\quad +b_i^+ [n_{i+1}(1-n_i)exp[\lambda' \sum_{j=1}^{N-i+1} n_{i+j}] \\
&\quad -n_i(1-n_{i+1})/exp[\lambda' \sum_{j=1}^{N-i+1} n_{i+j}]] \\
&\quad +c_i^+ \Big[n_{i+2}(1-n_i)exp[\lambda' \sum_{j=1}^{N-i} (2-\delta_{j,1})n_{i+j}] \\
&\quad -n_i(1-n_{i+2})/exp[\lambda' \sum_{j=1}^{N-i} (2-\delta_{j,1})n_{i+j}]\Big], \qquad (29)
\end{aligned}$$

where $i = 3, 4, .., N-2$, $\lambda' = \lambda e/2k_BT$, $b_i^{\pm}$ are hopping rates from site i to its first neighbors $i \pm 1$, and $c_i^{\pm}$ are hopping rates to the second neighbors, $i \pm 2$. This notation for the hopping rates results from the balance of the eight contributions to the current at site i, and is more convenient than the general two-index notation in Eq. 23. Note that $b_i^+ = b_{i+1}^-$, $b_i^- = b_{i-1}^+$, $c_i^+ = c_{i+2}^-$, and $c_i^- = c_{i-2}^+$.

The first two and the last two kinetic equations of the system in Eq. 29, i.e. for sites $i = 1, 2$ and $i = N-1, N$ are given by expressions similar to Eq. 29 when the boundary conditions of the problem are specified: reflecting, absorbing, or containing an ion source. This point demonstrates the flexibility of the lattice-gas model to simulate essential peculiarities of ion transport in superionic devices by simple means. Consider the following example: it is well known [7] that, for the efficient performance of a thin film battery, the electrolyte must be a good separator of the ionic current. Although 'separator' generally implies consideration of the electron current too, here the meaning of this term is restricted to 'current in a specified direction' and this is simulated easily with the present model by specifying the following boundary conditions on the left-hand side($i = 1, 2$) of the lattice in Fig. 5, and similarly for sites $N-1$ and N. In Fig. 5, the numbers in the squares denote lattice sites, the double lines with $b_i^{\pm}$ below them represent the contributions from first-neighbor sites, the double lines with $c_i^{\pm}$ above them the contributions from the second neighbors, and the arrows with b_0, c_0 below them are contributions from an ion source like the anode of a microbattery.

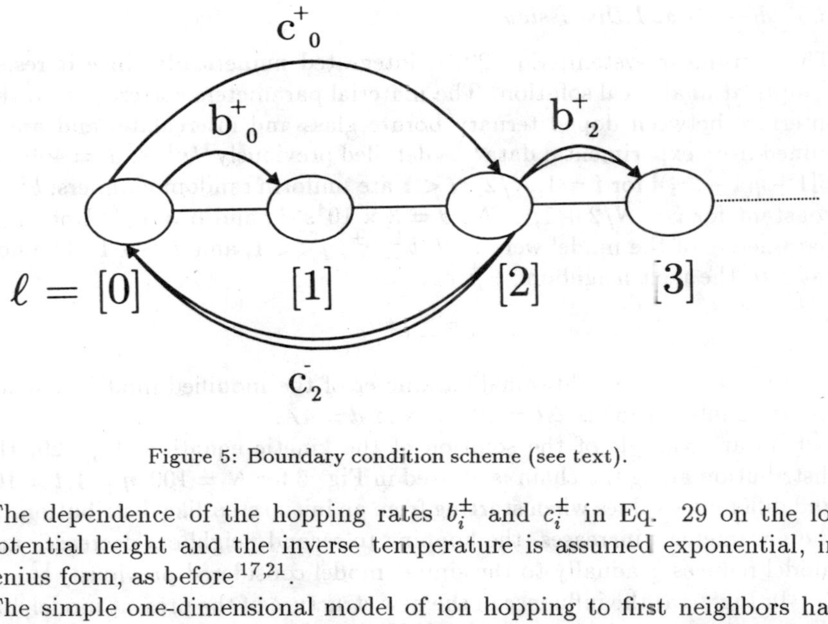

Figure 5: Boundary condition scheme (see text).

The dependence of the hopping rates $b_i^{\pm}$ and $c_i^{\pm}$ in Eq. 29 on the local potential height and the inverse temperature is assumed exponential, in Arrhenius form, as before [17,21].

The simple one-dimensional model of ion hopping to first neighbors has the advantage of being exactly soluble [21,10], but its limitation depends on the fact that it is devoid of much structure. If one lattice site is blocked, for instance, this will have dramatic consequences on the ion transport - it will be interrupted. A straightforward generalization is to consider directly models with higher dimensions, which are more realistic, but more complex. As an intermediate step in this direction the present model with hopping to first and second neighbors is proposed. This introduces some additional structure, which is seen from the fact that if a site is blocked the transport will be modified, but will not stop, and keeps at the same time the simplicity of the treatment characteristic of one-dimensional physics. This approach is inspired by the exemplary investigations of properties of interacting particles in one dimension [22].

Having the solution $n_i(t)$ of the system Eq. 29 it is possible to calculate the lattice potential $\phi_i(t)$ from Eq. 24, its electric field $-d\phi_i(t)/dx$, and the local and total polarizations, given by [23]

$$P(t) = \sum_i p_i(t) = \frac{e}{V} \sum_i x_i n_i(t), \tag{30}$$

where x is the direction along the chain.

3.5 Results and Discussion

The non-linear system, Eq. 29, is integrated numerically since it resists attempts at analytical solution. The material parameters correspond to the real interface between doped ternary borate glass and intercalate, and are determined from experimental data [7] as detailed previously [17,21,10]: $T = 460K; b_i^{\pm} = \beta[1 + \eta(1 - 2r_i)]$ for $i = 1, N/2$, $r_i < 1$ are uniform random numbers; $b_i^{\pm} = \alpha =$ constant for $i = N/2 + 1, ..., N$, $\beta = 3 \times 10^4 s^{-1}$; and $\alpha = \psi\beta$. Until now the parameters of the model were $N, T, b_i^{\pm}, c_i^{\pm}, \eta << 1$, and $\psi >> 1$. The hopping rates to the next neighbors, $C_i^{\pm}$ are

$$c_i^{\pm} = b_i^{\pm}/s, \tag{31}$$

where $s >> 1$ is an additional parameter of the modified model. The step in the time integration is $\Delta t = 10^{-7}s$, and $d = 4\AA$.

As an example of the solution of the kinetic equation, Eq. 29, the ion distribution along the chain is plotted in Fig. 6 for $N = 100, \eta = 4, t = 1000\Delta t$ and different s values when starting from an initial step-like distribution. When the parameter s increases, the hopping to second neighbors decreases and the model reduces gradually to the simple model considered previously [17].

In Fig. 7(a) the influence of the new structure of the present model, namely the hopping to the next-nearest neighbors, on the lattice potential, ϕ_i, given by Eq. 24, is demonstrated for $N = 100, t = 5000\Delta t, \psi = 5, \eta = 0.03$, and two values of the parameter s, which determines the magnitudes of the hopping rates $c_i^{\pm}$, according to Eq. 31. When s is low, $c_i^{\pm}$ is larger, the ions diffuse more efficiently and the corresponding potential is larger. In Fig. 7(b), the corresponding electric field is plotted for the same values of the parameters.

The time development of the local polarization $p_i(t)$, given by Eq. 30, is plotted in Fig. 8 for $N = 100, \psi = 4, \eta = 0.03, s = 2$, and $t = 100\Delta t, 200\Delta t$. This quantity is useful when estimating the current density and the conductivity of the system [23], but this will have more meaning in higher dimensions.

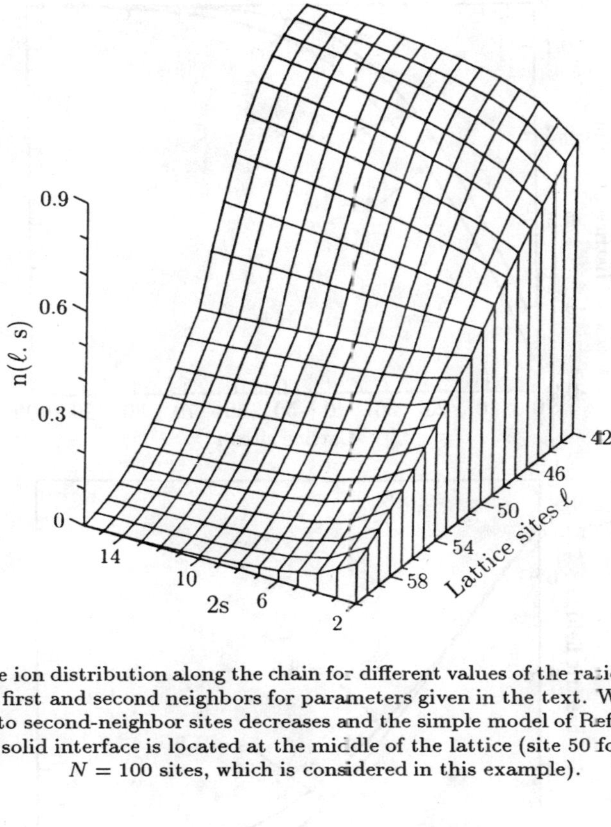

Figure 6: The ion distribution along the chain for different values of the ratio s between hopping rates of first and second neighbors for parameters given in the text. When s increases, the hopping to second-neighbor sites decreases and the simple model of Ref. [17] is recovered. The internal solid interface is located at the middle of the lattice (site 50 for the chain with $N = 100$ sites, which is considered in this example).

Of course, one cannot simulate a thin film battery by a one- dimensional model; however, such models do show certain significant results and might be useful as

4 Two-Dimensional Lattice Gas and Kinetic Monte-Carlo Models

an intermediate step toward the development of more realistic models in higher dimensions, which we currently turn to.

4.1 Introduction

One conclusion of the work presented in the second section of this chapter was that the existence of the finite width interface had little effect on the

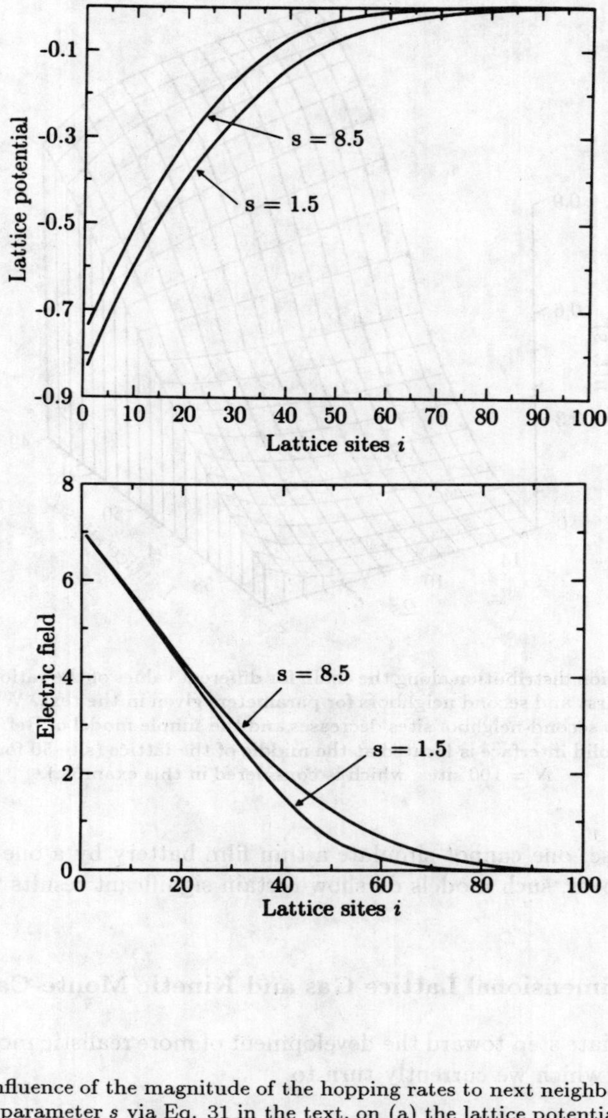

Figure 7: The influence of the magnitude of the hopping rates to next neighbors $c_i^{\pm}$, determined from the parameter s via Eq. 31 in the text, on (a) the lattice potential and (b) the electric field for values of the parameters given in the text. For $_1 < s_2$, $c_i^{\pm}$ is larger and the hopping diffusion is more efficient.

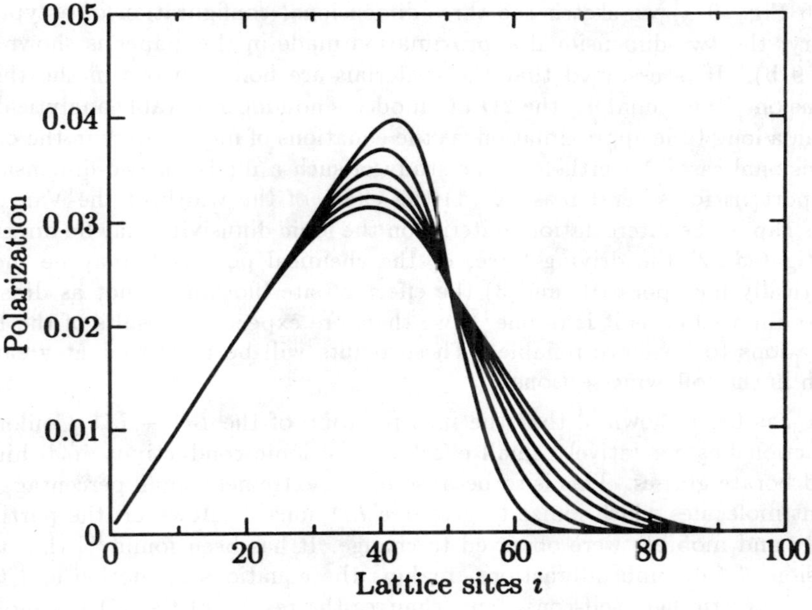

Figure 8: The time dependence of the lattice polarization. The parameters are given in the text and the higher curves correspond to shorter diffusion times.

ability of ions to diffuse into the cathode when the hopping frequency in the interface was taken as the average of the hopping frequencies on either side of the interface. Although the hopping frequency in the interface region may be quite different than this simple assumption yields, this complication is not considered in the present study. In this section, we expand the model to two dimensions and include a spatially varying chemical potential, which gives rise to a driving force, and a Coulomb interaction. The resulting model is solved using the deterministic LG equations of motion, and the stochastic KMC method. Note that it is necessary to employ the KMC method as opposed to the MC method as we wish to compare these results with those of the LG model which have a specific time dependence. In addition, the information we have available is limited at best. In such a situation, KMC simulations are useful in that they allow for a quick determination of qualitative features of the kinetic processes which should dominate the experimental situation. Finally, being a fully atomistic model, it allows us to investigate the importance of the site blocking effect as a function of ion density and applied potential.

In Fig. 9 a) we sketch the three-dimensional configuration of a typical battery; the two-dimensional approximation made in this paper is shown in Fig. 9 b). It is assumed that the materials are homogeneous in the third dimension. Unfortunately, the 2D LG model is no longer solvable analytically, even in a long time approximation, as the equations of motion were in the one-dimensional case. Nevertheless, the study of such a model in two-dimensions is important for several reasons: 1) the effect of the width of the Van der Waals gap in the intercalation material on the ionic diffusivity may be directly investigated; 2) the driving force, or the chemical potential, may be more realistically incorporated; and 3) the effect of site blocking is not as drastic in two-dimensions as it is in one [17], we therefore expect the results of the LG simulations to be more reliable. These points will be addressed at greater length in the following sections.

It has been shown [13] that the incorporation of the $Li^+ - Li^+$ Coulomb interaction has a relatively small effect on the ionic conductivity in lithium doped borate glasses. This is so because of the extremely small percentage of dopant molecules which ionize to give free Li^+ ions [24]. However, the particle density and mobility were observed to change. It has been found [16], that the inclusion of Coulomb interactions, making the equations of motion and the induced electric field self-consistent, changes the results of the 1D LG model quantitatively, but not qualitatively. Nevertheless, we include this interaction here as the potential is possibly very long ranged, depending on the Debye screening length, and the buildup of particle density near the interface can be expected to give rise to important space charge effects. We have also included an energy barrier between sites in the cathode and those in the separator denoted by E_{bar}. We are able to draw two important conclusions from this investigation. The first is that the energy barrier is an important quantity and must be well characterized before any more complete theory of interface transport may be considered. The second regards the effect of the ion-ion interaction near the interface. Any discontinuity in particle density at the interface, due to an energy barrier or a narrow Van der Waals gap for example, may result in either a large increase or decrease in effective diffusivity across the interface. The effect of the Coulomb interaction is to decrease this change in diffusivity at the interface. In essence, this interaction smooths out all kinks in the charge density.

The rest of this section is organized as follows. In the next section we define the LG equations of motion (EOM), specify the initial conditions, describe how site-site energy differences are incorporated into the hopping frequencies and describe the numerical method used to solve the EOM. Next, we give boundary conditions for the chemical potential and present a representative

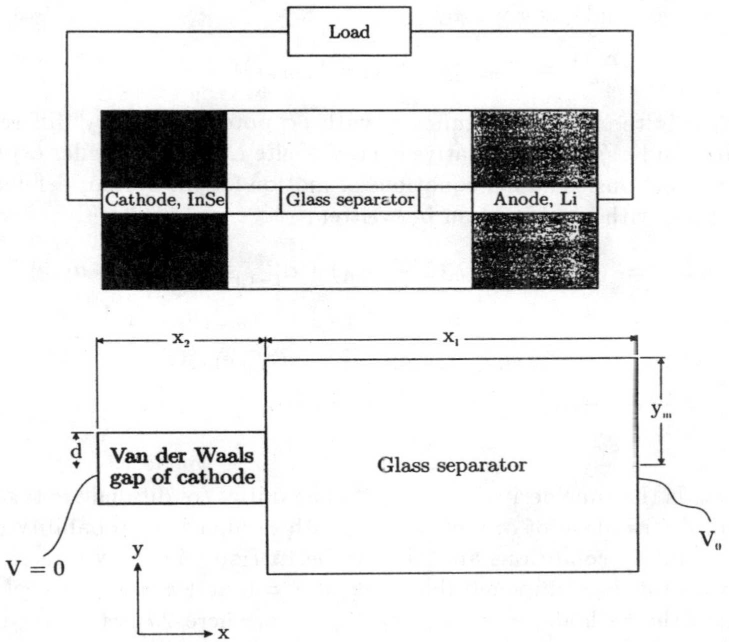

Figure 9: a) Plot of a typical solid state battery showing the relation of anode, separator and cathode; b) our two-dimensional model of the region near the separator-cathode interface.

solution to Laplace's equation for the geometry shown in Fig. 9b). Also the calculation of the charge-charge Coulomb interaction is described, and the boundary conditions defined. We then use representative hopping rates for lithium doped borate glasses and $InSe$, and present numerical results for these rates both with and without an applied potential with several values of the energy barrier. We then show the main effects of the Coulomb interaction, and discuss its significance at such an interface. Following this, the KMC method is very briefly discussed, after which we present results for the same sets of parameters used in the lattice-gas model. Finally we give conclusions.

4.2 Equations of Motion

The quantity of importance in LG models is the hopping frequency, α. In what follows, we will assume that the rate describing hopping from site l, m to

$l, m+1$, for example, is given by

$$\alpha_{l,m}^{\Delta m+} = \alpha_{l,m} exp[e(V_{l,m} - V_{l,m+1})/2k_B T], \tag{32}$$

where $\alpha_{l,m}$ is the hopping frequency with no potential energy difference between sites and $eV_{l,m}$ is the relative energy of site l, m, with similar expressions for other directions. The LG equations of motion for the system defined in the introduction, with blocking, can be written,

$$\begin{aligned} \dot{n}_{l,m} \quad = \quad & \alpha_{l+1,m}^{\Delta l-} n_{l+1,m}(1-n_{l,m}) + \alpha_{l-1,m}^{\Delta l+} n_{l-1,m}(1-n_{l,m}) \\ & +\alpha_{l,m+1}^{\Delta m-} n_{l,m+1}(1-n_{l,m}) + \alpha_{l,m-1}^{\Delta m+} n_{l,m-1}(1-n_{l,m}) \\ & -\alpha_{l,m}^{\Delta l+} n_{l,m}(1-n_{l+1,m}) - \alpha_{l,m}^{\Delta l-} n_{l,m}(1-n_{l-1,m}) \\ & -\alpha_{l,m}^{\Delta m+} n_{l,m}(1-n_{l,m+1}) - \alpha_{l,m}^{\Delta m-} n_{l,m}(1-n_{l,m-1}). \end{aligned} \tag{33}$$

where $\dot{n}_{l,m}$ is the time derivative of $n_{l,m}$. The initial conditions are taken to be a random distribution of occupied sites, with occupation probability denoted p. The boundary conditions are taken to be: periodic in the y direction, with lattice constant $2y_m$, impenetrable walls at $x=0$ and $x=x_m$, except for the opening to the cathode, $x=x_m$, $-d \le y \le d$, where $2d$ is the width of the Van der Waals gap. Particles are assumed "lost" into the cathode when they reach $x = x_m + x_{cath}$.

The equations of motion can be directly integrated using a suitable integration scheme. We have employed a fourth order Runge-Kutta routine. For the 1D problem, in which we can obtain an approximate long time analytic solution, we have found very good agreement between the solutions obtained by these two methods [10].

4.3 Chemical Potential, Hopping Rates and Numerical Solution

In order to determine the chemical potential acting on the ions as they move to the cathode, we must solve Laplace's equation, $\nabla^2 V(x,y) = 0$, along with the appropriate boundary conditions (see Fig. 9b),

$$\begin{aligned} V(0,y) &= V(0), \\ \frac{dV}{dy}(x \le x_m, y = \pm y_m) &= 0, \\ \frac{dV}{dx}(x = x_m, -d > y > d) &= 0, \\ V(x = x_m + x_{cath}, y) &= 0. \end{aligned}$$

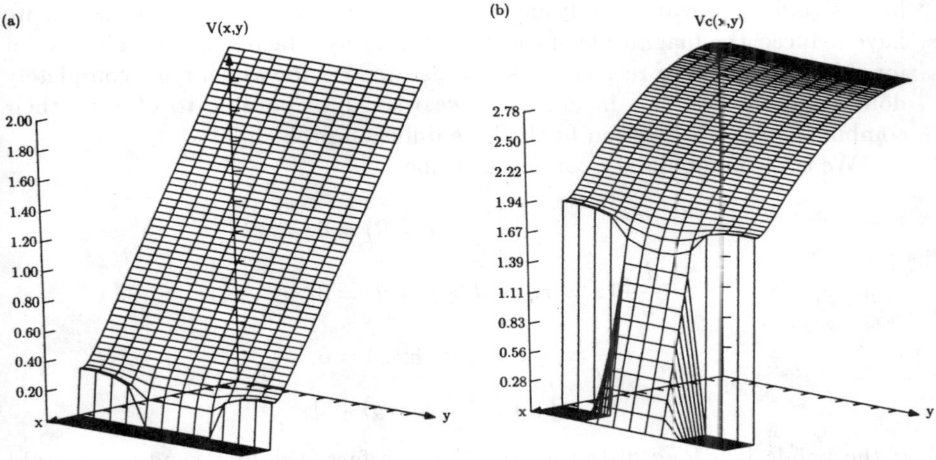

Figure 10: a) Typical plot of the chemical potential, $V(x, y)$. The cathode is located at the mid-foreground of the image. b) The contribution from a 10% distribution of charge, $V_c(x, y)$. The total potential used in our simulations is then $V(x, y) + V_c(x, y)$.

This boundary value problem has been solved using a simple iterative technique [25]. A typical plot of the potential vs. position is shown in Fig. 10a). Note that in our simulations we use the scaled potential $eV(x, y)/2k_BT$.

Clearly a solid state battery will have a linear extent much greater (in relation to the width of the Van der Waals gap) than that shown in Fig. 9b). It may therefore be asked how these boundary conditions, and the results which follow, are to be interpreted. We are primarily interested in the effects of this particular interface on ion diffusion in the battery. We therefore contend that, near the interface, there will exist a chemical potential, possibly an energy barrier, and charge-charge interactions. By imposing the above boundary conditions, we need only ensure that the influence of the boundary be small at the position $x = 0$ in reference to the interface. As the results to follow clearly show, this condition is well satisfied.

When charged particles are present, we have an additional potential given by

$$\nabla^2 V_c(x, y) = -\rho(x, y)/\epsilon, \tag{34}$$

where ϵ is the static dielectric constant, which is taken [26] to be 20, and $\rho(x, y)$ is the two-dimensional charge density. Note that, again, the quantity employed in our simulation is the scaled potential, given by $eV_c/2k_BT$. All of the energies have significant meaning only in relation to each other. For this reason, we have reduced the magnitude of the potential V_c by a factor of five. The reason for this was to have a reasonable set of parameters for which none completely dominated the others. In addition, we wished to be able to observe their combined effects in relation to the bare diffusion problem.

We take the boundary conditions to be:

$$\frac{dV_c}{dx}(x = 0) = 0,$$

$$\frac{dV_c}{dx}(x = x_m, -d > y > d) = 0,$$

$$\frac{dV_c}{dy}(x \leq x_m, y = \pm y_m) = 0,$$

$$V_c(x_m + x_{cath}, y) = 0.$$

If the anode is a long distance from the interface, the first condition should be well satisfied, and we can always add a constant to this potential, thereby justifying the last condition. Note that similar boundary conditions have been used in the past [27]. It is further assumed that the semi-metallic character of the cathode leads to a strong effective screening with a screening length on the order of 20% of the length of the cathode region, which corresponds to 4% of the length of the separator. We neglect the possibility of strong screening in the separator itself. A weak electrolyte theory of ionic conductivity [13] in $B_2O_3 - 0.5Li_2O - 0.15Li_2SO_4$ predicts screening lengths on the order of 30Å, which would correspond approximately to the length of our simulation box. However, in the present case, the particle density may be significantly higher than that predicted for the bulk glass. Our main motivation for neglecting this effect is the relative lack of precise data on particle densities.

The technique used to calculate V_c, described in Ref. [25], is an iterative one. Therefore, the calculation of V_c is relatively fast so long as the charge density does not change drastically from one determination of V_c to the next. This allows for very efficient computation. It has been found that updating V_c each 10 time steps yields good results. The total potential present is simply $V(x, y) + V_c(x, y)$. In Fig. 10b) we show only this contribution to the total potential. We plot here the potential due to a random distribution of charge for which the probability of a site being occupied was taken as $p = 0.1$. This is to be compared to the applied potential shown in Fig. 10a). Clearly, even for relatively low densities, the potential arising from the charge-charge interaction may be a significant factor. As mentioned, the total potential is then the sum of those shown in Figs. 10a) and b).

We use the same hopping frequencies and parameters as were used for the one-dimensional model. As mentioned earlier, we do not consider the effect of a finite width interface region in this work.

In all of the simulations reported on here, the initial particle density in the separator was taken to be 10%. This is probably a rather high density as compared to that in the bulk glass separator [24]. However, as we are close to the cathode, such large concentrations of charge may not be unreasonable. We will see that in such a regime the charge-charge or Coulomb interaction tends to be the dominant effect in the problem. The time step in our simulations was $1 \times 10^{-7}s$, and all results were obtained over 10,000 time steps and were averaged over five sets of initial conditions unless otherwise stated.

In Figs. 11a) and b) we present results for the case with $eV(0)/2k_BT=0$ and $eV(0)/2k_BT = 2$, with $\alpha = 5\beta$ where β was held constant at $1 \times 10^4 s^{-1}$. The energy barrier was taken to be zero and space charge effects were ignored. As expected, one observes a charge accumulation at the interface when the potential is present. The net effect of the potential, however, is to cause a much larger flow of particles into the cathode region and finally out of the simulation area. As shown in Table 1, approximately 50% more particles leave the cathode altogether when the potential is applied. The inclusion of ion-ion interactions can be expected to inhibit the accumulation of charge near the interface; however, this also tends to drive the particles into the cathode as the Coulomb interaction and charges are screened there. Both effects are in fact present, although for these densities, the latter dominates. In Fig. 11c), we see just this effect, where V_c has been included, $V(0) = 0$, and we have used the same parameters as in Fig. 11a). The total amount of charge being transferred is now 13.9% compared to 6.8% for the case without V_c. The Coulomb interaction inhibits particle flow to regions of high particle density, i.e. the separator-cathode interface, but also enhances flow to regions of low effective density, that being the cathode itself. The effect of the applied potential, see Table 1, is found to be somewhat moderated by the Coulomb interaction in that the buildup of charge at the interface is much less than observed in Fig. 11b). The total amount of charge transferred is, again, much larger than without the charge-charge interaction, although the percentage increase due to the applied potential is less.

Similarly, the effects of an energy barrier are also softened. In Figs. 12a) and b) we present results for the case $E_{bar}/2k_BT = -1$ and $E_{bar}/2k_BT = 1$ respectively, where E_{bar} is the energy barrier, with $eV(0)/2k_BT = 2$, $\alpha = 5\beta$ and no ion-ion interaction considered. Again, we take $\beta = 1 \times 10^4 s^{-1}$. The positive barrier sweeps charge out of the interface region, pulling more particles into the cathode, thus leading to a large increase in net diffusivity at the interface. As shown in Table 1, the percentage of particles swept out of

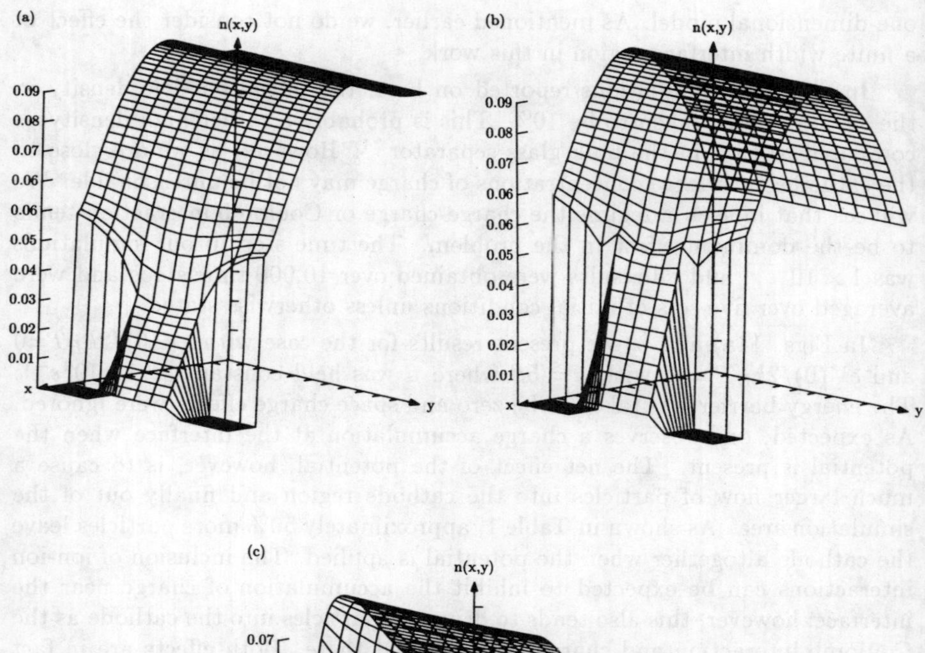

Figure 11: Plot of particle density vs. position after 10,000 time steps of $dt = 1 \times 10^{-7}$ seconds. In all figures to follow, we have $\alpha = 5\beta$ and no energy barrier present at the interface. a) No applied potential, b) $eV(0)/2k_BT = 2$, with no charge-charge interaction and c) $eV(0)/2k_BT = 0$ with the charge-charge interaction, where we take the static dielectric constant to be $\epsilon = 20$.

Table 1: Results of LG simulations for various applied potentials. % change is the percentage of particles which leave the simulation area during simulation time, $V(0)$ is the applied potential and E_{bar} the energy barrier. Results both with and without a charge-charge interaction are given.

$eV(0)/2k_BT$	$E_{bar}/2k_BT$	q-q interax	% change LG	% change KMC
0	0	N	6.8	6.7
0	0	Y	13.9	13.6
2	0	N	10.1	10.0
2	0	Y	18.2	18.7
2	1	N	13.8	13.9
2	1	Y	19.8	19.9
2	-1	N	3.7	3.9
2	-1	Y	12.2	12.3

the simulation area is increased by nearly 50% compared to the case without a barrier. This is comparable to the effect of the long-ranged applied potential. The negative barrier has just the opposite effect, there is a large charge accumulation region near the interface. Note that the negative energy barrier completely dominates and reverses the effects of the driving potential. With both the potential and an energy barrier of $E_{bar}/2k_BT = -1$, only 3.7% of the particles leave the battery, less than that obtained with neither the applied potential nor the energy barrier.

With the Coulomb interaction included, we expect both of these effects to be lessened. The results of our simulations, with the same parameters as used in Figs. 12a) and b), but with a charge interaction present are shown in Table 1 (the plots of the charge density are similar to that shown in Fig. 9c). It is found that the effects of the energy barrier are much less pronounced. For the case of a positive barrier, we observe almost 20% of the particles leaving the simulation area; however, as before, the increase in this which is due to the energy barrier is much less. The effects of a negative barrier, that is a barrier which inhibits particle flow, are almost completely overwhelmed by this interaction. With no charge interaction and a negative barrier, 4% of the particle are swept out of the simulation area in 10,000 time steps, whereas with the interaction, 12% are.

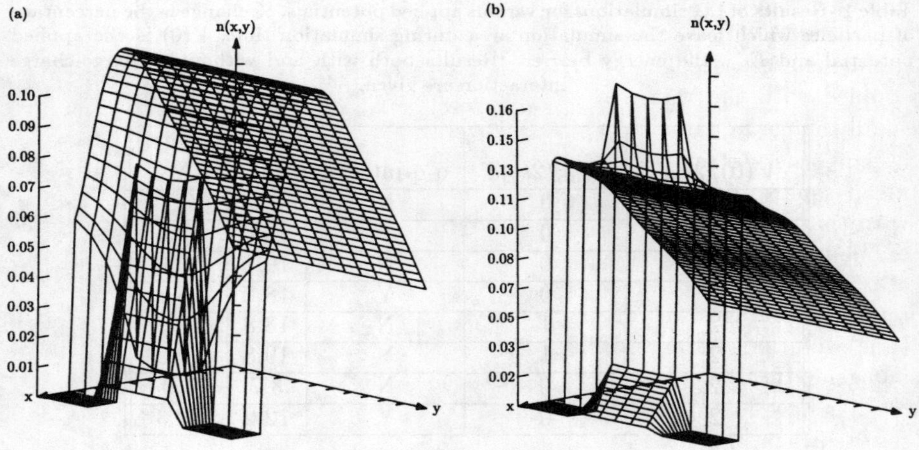

Figure 12: Same parameters as used in Fig. 3, but here we have $eV(0)/2k_BT = 2$ for all of the following and include an energy barrier as follows: a) $E_{bar}/2k_BT = 1$, and b) $E_{bar}/2k_BT = -1$, both with no charge-charge interaction.

4.4 Kinetic Monte-Carlo Method

Detailed descriptions of the KMC method have been available in the literature for some time now; therefore we refer the reader to the appropriate references for more details [28]. In the following, we simply give details of the simulations reported in this section.

At $t = 0$, atoms are deposited onto a 2D lattice having a geometry as shown in Fig. 9b). The fastest process is then determined using the given values of α, β through Eq. (32), including the applied potential and the induced potential through Eq. (34) if applicable. This process then serves to set the time scale for the problem. If the LG equations of motion are to be integrated out to a time $t = t_{max}$, then we must query each atom on the KMC surface which can diffuse $w_{max}t_{max}$ times, in order to account for all processes which can possibly occur within that time period. The number of atoms able to diffuse changes during the simulation time, although neither w_{max} or t_{max} does. This is due to particles either leaving the simulation area, or becoming blocked or unblocked by the walls or by other particles. Therefore, the number of required queries is continuously updated during the simulation. Acceptance of given moves is determined using the standard Metropolis Algorithm. In addition, particle exclusion is enforced, each site may be occupied by, at most, one particle. The boundary conditions are taken the same as used in the lattice-gas model. In the cases in which we consider a charge interaction, the potential due to all mobile charges need only be calculated each time a move is accepted. This, coupled with the small number of iterations needed for convergence after a small change in particle density, allows for efficient computation.

In order to ensure that we were observing an "average" behavior, the mean square displacement of an atom was first calculated, then compared with that from the LG equations of motion. It was found that on the order of one hundred KMC simulations were required to obtain results to within 1% of those of the deterministic lattice gas. Therefore, the results presented here represent averages over one hundred different initial conditions and subsequent time evolutions for each parameter set. A summary of our results is presented alongside the LG ones for comparison in Table 1.

We have found that, as expected, the two models give essentially identical results when sufficient averaging is done. In addition, we have performed simulations with $p = 0.5$, and found similar agreement. In this latter case, with no charge-charge interaction, the percentage of particles leaving the simulation area was found to be nearly identical to that found when $p = 0.1$. With the charge-charge interaction included however, we find a much larger percentage of particles forced out into the screened cathode area. In particular, with $eV(0)/2k_BT = 0$, $E_{bar} = 0$ and without the charge interaction, $\approx 6.6\%$ left, and with this interaction, $\approx 26\%$ did. This is nearly twice the number obtained for the simulation with $p = 0.1$. This is not surprising given that the large density of particles simply results in a large effective Coulomb potential, thus driving more particles into the screened part of the cathode.

4.5 Conclusion

We are able to draw two important conclusions from this part of our investigation. The first is that the energy barrier is an important quantity and must be well characterized before any more complete theory of interface transport may be considered. The second regards the effect of a charge interaction near the interface. Any mismatch in hopping frequency at the interface, due to an energy barrier or a narrow Van der Waals gap for example, may result in either a large increase or decrease in effective diffusivity across the barrier. The effect of the Coulomb interaction is to decrease this change in diffusivity at the interface. In essence, this interaction smooths out all kinks in the charge density. It is also evident that this interaction may well overwhelm the effects of both the chemical potential and the energy barrier. It is therefore crucial to determine the absolute magnitudes of these various quantities.

5 Conclusion

In this chapter we have reviewed the generalization of a one-dimensional lattice gas model. First, we have introduced a finite width interface region where the hopping rate, α', is taken to be the average of α and β, and by introducing an energy barrier in the interface region. We have found that a large mismatch in hopping frequencies, as may occur as more and more lithium is intercalated into the $InSe$, results in in a much larger particle gradient at the interface. The transition region we have introduced, which may be thought of as a first approximation to a graded junction, resulted in little improvement in the ability of particles to diffuse across the interface. This result is not encouraging as far as the introduction of graded junctions in solid state batteries is concerned. In addition, the presence of an energy barrier was found to be a dominant factor in determining the ability of particles to diffuse through the interface region.

Next, an improved lattice gas model was presented. The improvements in this model are as follows: i) the long-range Coulomb interactions are accounted for in such a way as to make the resulting computations more efficient, and ii) it takes into account hopping rates up to next-nearest neighbor sites, this makes the local structure of the hopping more elaborate and elimates the problem that the site blocking effect introduces in one-dimensional problems. The model has been applied to calculate the ion distribution along the chain, $n_i(t)$, the lattice potential, $\phi_i(t)$, the electric field, and the polarization. These were computed for material parameters of the real interface between doped ternary borate glass and indium selenide, which is used in lithium thin film batteries.

Finally, we have investigated particle diffusion across material interfaces in two-dimensions using a deterministic LG model and a stochastic KMC model. We included a spatially varying chemical potential which gives rise to a driving force, a charge interaction, and an energy barrier at the interface between the two materials.

We know of no experimental studies performed to measure the energy barrier introduced in the first and last models discussed in this chapter. It is hoped that XAFS studies currently being carried out in Paris to measure the penetration of lithium into the InSe from the borate glass will show the existence of this barrier.

We are able to draw two important conclusions from these investigations. The first is that the energy barrier is an important quantity and must be well characterized before any more complete theory of interface transport may be considered. The second regards the effect of a charge interaction near the interface. Any mismatch in hopping frequency at the interface, due to an energy barrier or a narrow Van der Waals gap for example, may result in either a large increase or decrease in effective diffusivity across the barrier. The effect of the Coulomb interaction is to decrease this change in diffusivity at the interface. In essence, this interaction smooths out all kinks in the charge density. It is also evident that this interaction may well overwhelm the effects of both the chemical potential and the energy barrier. It is therefore crucial to determine the absolute magnitudes of these various quantities.

Acknowledgments

The author would like to thank Drs. R. F. Wallis and M. Balkanski with whom all of this work was performed. In addition, we would like to dedicate this chapter to the memory of Dr. Ivo S. Nachev, who passed away in August, 1995. The insight, talent, wit, and friendship that Dr. Nachev brought to our collaboration will be sorely missed by us and all who knew him.

1. P. W. Tasker, J. Chem. Faraday Trans. 86 (1990) 1311.
2. M. P. Puls and M. J. Norgett, J. Appl. Phys. 47 (1976) 466.
3. see e.g. K. L. Merkle and D. L. Smith, Ultramicroscopy 22 (1987) 57.
4. A. Atkinson, Sol. St. Ionics, 28-30 (1988) 1377.
5. A. Atkinson, Sol. St. Ionics 12 (1984) 309.
6. M. Hiratani, K. Miyauchi and T. Kudo, Sol. St. Ionics, Diff. and Reactions 28-30 (1988) 1406.
7. M. Balkanski, edt. *Microionics* (North Holland, Amsterdam, 1991).
8. R. Blender, W. Dieterich and H. L. Frisch, Phys. Rev. B 33 (1986) 3538.
9. R. Blender and W. Dieterich Sol. St. Ionics 18&19 (1986) 240.
10. J. Deppe, R. F. Wallis and M. Balkanski, Surf. Sci. 297 (1993) 91.
11. I. Nachev, J. Deppe, R. F. Wallis and M. Balkanski, Mat. Sci. and Eng. B 31 (1995) 275.
12. J. Deppe, R. F. Wallis, I. Nachev, and M. Balkanski, J. Phys. Chem. Solids 55 (1994) 759.
13. M. Balkanski, R. F. Wallis and J. Deppe, Mat. Sci. and Eng. B 3 (1989) 65.
14. P. J. Jennings and R. O. Jones, Adv. Phys. 37 (1988) 341.
15. L. Heyne, in *Mass Transport in Ionic Solids*, eds. F. Beniére and C. A. A. Catlow, NATO ASI Series B, Vol. 97 (Plenum, NY 1983) pp. 425-456.
16. M. Balkanski, I. Nachev, J. Deppe and R. F. Wallis, to appear in Mat. Sci. and Eng. B.
17. I. Nachev, J. Deppe, R. F. Wallis and M. Balkanski, Mat. Sci. and Eng. B 15 (1992) 260.

18. I. Darianian, Thesis, Paris 1988.
19. C. Julien, in *Microionics*, edt. M. Balkanski (North-Holland, Amsterdam, 1992), p. 309.
20. K. Kunc and R. Zeyher, Europhys. Lett. 7 (1988) 611.
21. M. Balkanski, I. Nachev, J. Deppe, and R. F. Wallis, in Solid State Ionics III, Mater. Res. Symp. Proc., 293 (1993) 193.
22. E. H. Lieb and D. C. Mathis (eds.), *Mathematical Physics in One Dimension*, (Academic Press, New York, 1966).
23. W. Brenig, G. Dohler, and P. Wolfle, Z. Phys. 246 (1971) 1.
24. M. Balkanski, R. F. Wallis, I. Darianian and J. Deppe, Mat. Sci. and Eng. B 1 (1988) 15.
25. S. E. Koonin, *Computational Physics* p. 142. Addison-Wesley Publishing Company, Inc., Amsterdam, (1986).
26. P. Dzwonkowski, M. Eddrief, C. Julien and M. Balkanski, Mat. Sci. and Eng. B 8 (1991) 193.
27. A. Mao and R. E. White, Proc. 6th Int. Meeting on Lithium Batteries, p. 84 Munster Germany, (1992).
28. See e.g. H. Metiu, Y. T. Lu and Z. Zhang, Science 255 (1992) 1088; G. H. Gilmer, J. Crystal Growth 42 (1977) 3; and Y. T. Lu and H. Metiu, Surf. Sci. 245 (1991) 150.

Chapter 6

Applications of Superionic Conductors in Microbatteries and Elsewhere

M. Balkanski

1 Introduction

Atomic motion in solids opens to a wide field of applications related to energy transfer, storage and conversion. Most of the applications involve interplay between chemical and electrical energies. Devices based on fast ion transport can be classified into three categories :

- High energy density batteries with an alkali metal as negative electrode : mainly lithium batteries,
- Solid state ionic sensors,
- Electrochromic displays or smart windows.

In all these applications the basic component is the fast ionic conductor. Different types of fast ion conductors have been investigated : crystalline superionic conductors, polymer electrolytes and glass ionic conductors. We shall examine the basic properties of the different categories of materials in view of their applications in solid state devices and systems.

In the particular application of solid state batteries, the fast ion conductor plays the role of a separator, an electronic insulator, between the ion source, the anode, and the cathode which has to be a mixed ionic and electronic conductor. The most efficient cathodic materials, presently, seems to be the intercalation compounds. We shall devote section 3 to review the present status of the art of the intercalation mixed conductors.

The remaining part of this presentation will deal with the applications in terms of solid state batteries and microbatteries, solid state electrochromics, and chemical sensors based on solid state ionics.

2 Fast Ion Conductors

2.1 *Crystalline Materials*

The fast ion conductors are solid compound materials which present an anomalously high ionic conductivity, comparable, in order of magnitude, to that of liquid electrolytes. They also present effective atomic diffusion coefficients of the same order as in liquids or gases. In general, the conductivity increases with temperature according to the Arrhenius law. Systems with conductivity of the order of 1 Ω^{-1} cm^{-1} at room temperature are ideal for practical applications.

The fast ion conductors are often referred to as superionic conductors. In these materials one species of ions is more mobile than the others. The less mobile ions are arrangedin a crystalline skeleton, more or less rigid, denominated the « cage ». The mobile ions can wonder through vacant and intersitial sites of the cage as a consequence of thermal activation.

2.1.1 *Introduction*

When particular phases of certain solids possess an anomalously high ionic conducting (~1 $ohm^{-1}.cm^{-1}$) they become superion or fast ion conductors. In most superionic conductors the mechanism of the ionic motion is that of classical hopping between various neighbouring sites. In some materials this model may break down, particulary when the mobile ion is H^+ when quantum tunnelling may contribute to the conductivity. In other materials the motion may be more liquid like in the sense that the time the ion spends in its equilibrium position is comparable to the time of flight between equilibrium sites.

There are various types of hopping motion. The lowest energy hopping mechanisms are the « vacancy » mechanism and the « interstitial » mechanism. In the vacancy mechanism the ion can only hop to a nearest-neighbour lattice site when that is unoccupied. In the interstitial mechanism the ion hops between unoccupied sites which are not sites of the original lattice but interstitial sites. The interstitial mechanism is essentially equivalent to the vacancy mechanism as the interstitial motion can be considered to be motion via vacancies on a different sublattice. Higher energy hopping mechanisms (and thus less probable) are the « exchange » mechanism where the ion moves by swapping sites with a neighbouring ion, and the « interstitialcy » mechanism where the ion hops onto a site by knocking its occupant onto an interstitial site.

2.1.2 *Different Systems Exhibit High Ionic Conductivity*

a) ß-aluminas [1,2,3]

The ß-aluminas are generally compounds with the formula $M_2O.xAl_2O_3$ where M = Na, K, Rb, NH_4, Tl, Ag, etc. and x is in the range 5 to 11. The principle compound is $Na_2O.11Al_2O_3$ (11 is the ideal stochiometry) which occurs with two phases, the stable ß phase and metastable ß'' phase. The ß'' phase can be stabilised by doping with different ions (Li, Mg, Cd...). ß-alumina has the space group P63/mmc with a hexagonal layered structure. The Na ions occur only in these layers within which the Na motion occurs. It is because of this layered structure that the ß-aluminas have a high (anisotropic) ionic conductivity as well as a high vacancy concentration due to several equivalent Na sites in the layers. The layers are separated by two « spinel blocks » containing the Al and O. The ß''-alumina has a rhombohedral layered structure with space group R3m. Generally an increase or decrease of ionic radius compared with Na of M-ion reduces the conductivity.

b) Silver Ion Conductors [4, 5, 6, 7, 8]

Most silver ion conductors are based on AgI. The conductors are made by partially substituting different cations (e.g. K, Rb, NH_4) for the silver ion and partial substitutions of the anion (e.g. S, SO_4, PO_4, P_2O_7, WO_4. etc.). The best studied silver ion conductor is α-AgI (the ß phase has a much lower conductivity) which has cubic symmetry with space group Im3m containing a bcc iodine sublattice. The central structural feature responsible for the very high conductivity of α-AgI is the nature of the Ag sublattice. The unit cell only requires 2 silver ions, however, thre are 42 available sites of 3 types : 6 octahedral, 12 tetrahedral and 24 trigonal bipyramidal. Single crystal studies suggest the ions can be found preferentially on the tetrahedral sites but the potential barriers between the other types of sites are very low resulting in an appreciable probability of the ion being found in the other sites. The way the silver sites in adjacent unit cell are connected results in almost linear channels with very low potential barriers for the ions to move along further enhancing the conductivity.

Another very important silver ion conductor with a very high temperature conductivity are those of the form MAg_4I_5 (M = Rb, K, NH).

c) Copper Ion Conductors [9, 10]

The principal copper ion conductors are the cuprous halides (CuI, CuCl and CuBr) and to a lesser extent the cuprous chalcogenides (Cu_2S and Cu_2Se). The copper ion conductors are not as well studied as the ß-aluminas and silver ion conductors and generally are only good ion conductors well above room temperature. CuI has three ionic conducting phases, a high temperature cubic phase, an intermediate ß hexagonal phase and a low temperature γ cubic phase. The reason for the ionic conductivity is similar to that of the α-AgI although why CuI is a poorer conductor than AgI is not clear. Only ß-Cu_2S has relatively high conductivity.

d) Oxygen Ion Conductors [11, 12, 13, 14]

The principal group of oxygen ion compounds are those of the form MO_2,

M = Ce, Th, Hf, Zr. These all have a fluorite structure in which the metal ions form a fcc unit cell with four oxygen ions forming a cube within that cell. Generally the ionic conductivity is only reasonably good at high temperatures (~1000°C). Many of the conductors show a significant electronic contribution to the conductivity and the total conductivity is also quite sensitive to the pressure. The ionic conductivity can be increased by doping with divalent (e.g. Ca^{2+}) and trivalent (e.g. Y^{3+}) ions which increases the number of oxygen site vacancies.

Another important class of oxygen ion conductors are those based on Bi_2O_3 which when suitably doped, can have ionic conductivities several times greater than zirconia (ZrO_2).

e) Alkali Ion Conductors [15, 16]

Apart from the alkali ß-aluminas discussed above there are several other types of alkali ion conductors. These are dominated by lithium, the other alkali metals generally producing not very good ionic conductors. There are three principle groups of Li conductors : those with a « tetrahedral » anion of the general form Li_2XO_4 where X = S, Si, P, Ge, Al, Zn, Ti, Ga, etc, a second important material is Li_3N, and there are also the boracities of the general form $Li_4B_7O_{12}X$, X = Cl, Br, S.

An example of the tetrahedral group is Li_2SO_4 of which the high temperature ß phase is the better ionic conductor. In this case the high conductivity arises from the excess of cation sites in the unit cell. There are three equivalent sites which are only occupied by two sulphate ions. Doping Li_2SO_4 with other sulphates generally increases the conductivity.

Lithium nitride is a good ionic conductor. It has a hexagonal layered structure with the layers containing both Li and N which are bridged by the lithium. The lithium movement is within the layers resulting in an anisotropic conductivity.

The boracite conductors show a good isotropic conductivity and that of $Li_4B_7O_{12}Cl$ exceeds the Na-ß-alumina value at high temperatures. In this material the high conductivity arises because one type of lithium site has only 25 % occupancy and there are also easy channels for the Li to move along.

f) Fluorine Ion Conductors [9, 17]

There are several classes of fluorine ion conductors. There are those with a fluorite structure with the general formula MF_2, M = Ca, Sr, Ba, Pb, etc. The mobile ion is the fluorine anion and its high conductivity arises for similar reasons to that of the fluorite structured oxygen ion conductors.

Another class of fluorine ion conductors with a high conductivity have the general formula $M_{1-x}M_xF_{2+x}$ with M = Pb, Ca and M = Y, Bi which have an isotropic conductivity which is generally maximal in the range $0.2 < x < 0.3$. Their conductivity is sensitive to the cationic polarizability which affects the number of cation vacancies.

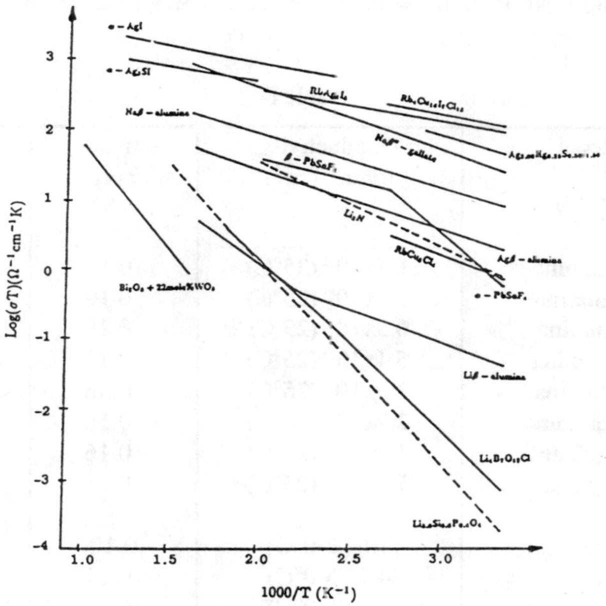

Figure 1: Variation of ionic conductivity for different fast ion conductor versus temperature.

2.1.3 *Values of Ionic Conductivities*

A graphic representation of the variation of the conductivity versus temperature is given in Figure 1. All follow the Arrhenius law

$$\sigma = A \exp(-V/kT)$$

where V is the activation energy, A is the pre-exponential factor, k is Bolzmann's constant and T is the temperature.

A summary of the values of ionic conductivity of typical samples are given in the Table I. Results vary between samples due to structural imperfections and impurities, but this gives an indication of the relative values. The dominant factor is the activation energy and if this is high the conductivity becomes significant only at temperatures well above room temperature. The best conductors at low T are the various iodides of silver but these are not stable enough to be used in practical devices. For these purposes the sodium beta aluminates seem the best available to date.

Table I

Material	Conductivity (ohm^{-1} cm^{-1})	Activation Energy (eV)	References
Naß-alumina	1.4×10^{-2} (25°C)	0.16	1
Liß-alumina	1.3×10^{-4} (25°C)	0.19	1
Kß-alumina	6.5×10^{-5} (25°C)	0.29	1
Agß-alumina	6.4×10^{-3} (25°C)	0.17	1
Tlß-alumina	2.0×10^{-6} (25°C)	0.36	1
Naß''-alumina	2.7×10^{-2} (25°C)	0.18	2
Naß-gallate	3.7×10^{-2} (25°C)	0.16	3
Naß''-gallate	7.1×10^{-2} (25°C)	0.18	3
α-AgI	1.6 (200°C)	0.10	4
α-Ag_2S	4.1 (200°C)	0.11	4
α-Ag_3SI	1.0 (250°C)	0.08	4
ß-Ag_3SI	0.01 (25°C)	0.21	4
Ag_3SBr	2.0×10^{-3} (20°C)	0.28	4
$RbAg_4I_5$	0.30 (30°C)	0.09	4
KAg_4I_5	0.21 (20°C)	0.10	5
α-AgI + (20 mole%) Ag_2SO_4	0.02 (-20°C)	-	5
$Ag_7I_4PO_4$	0.01 (25°C)	0.16	5
$Ag_{26}I_{18}W_4O_{16}$	4.7×10^{-2} (25°C)	0.16	5
$Ag_{2.00}Hg_{0.25}S_{0.50}I_{1.50}$	0.147 (25°C)	0.12	5
$Ag_{1.85}Hg_{0.40}Te_{0.65}I_{1.35}$	0.094 (25°C)	0.15	5
$Ag_{1.80}Hg_{0.45}Se_{0.70}I_{1.30}$	0.10 (25°C)	0.15	5
$Ag_{2.00}Hg_{0.5}Se_{1.0}I_{1.0}$	0.045 (25°C)	0.16	5
α-Ag_2HgI_4	2.0×10^{-3} (60°C)	0.33	6

Material	Conductivity (ohm^{-1} cm^{-1})	Activation Energy (eV)	References
α-CuI	0.09 (450°C)	0.19	7
ß-CuI	0.06 (400°C)	1.1	7
γ-CuI	<10^{-7} (25°C)	1.6	7
α-CuBr	~5.0 (480°C)	-	7
ß-CuBr	~1.0 (450°C)	-	7
γ-CuBr	<10^{-7} (25°C)	-	7
KCu_4I_5	2.0×10^{-6} (25°C)	-	7
KCu_4I_5	0.6 (280°C)	0.22	7
Cu_2S	0.2 (400°C)	0.15	7
ß-Cu_2Se	1.1×10^{-1} (150°C)	0.05	7
CuTeCl	~10^{-6} (25°C)	0.46	7
CuTeBr	~10^{-5} (25°C)	0.41	7
CuTeI	~10^{-5} (25°C)	0.75	7
$Rb_3Cu_7Cl_{10}$	4.0×10^{-3} (25°C)	-	8
$RbCu_3Cl_4$	2.5×10^{-3} (25°C)	0.19	7,8
$Rb_4Cu_{16}I_7Cl_{13}$	0.34 (25°C)	~0.09	8
α-$PbSnF_4$	2.0×10^{-3} (25°C)	0.41	9
ß-$PbSnF_4$	3.5×10^{-2} (100°C)	0.14	9
$RbBiF_4$	3.0×10^{-3} (25°C)	0.37	9
$KBiF_4$	~1.7×10^{-4} (25°C)	0.38	9
α-$TlBiF_4$	~2.0×10^{-5} (25°C)	0.38	9
ß-$TlBiF_4$	3.0×10^{-2} (320°C)	0.30	7
ß-PbF_2	~5.0×10^{-6} (20°C)	0.60	7
KSn_2F_5	5.0×10^{-6} (20°C)	0.55	10
$RbSn_2F_5$	2.0×10^{-5} (20°C)	0.68	10
$TlSn_2F_5$	6.0×10^{-4} (20°C)	0.52	10
LaF_3	2.2×10^{-4} (20°C)	0.43	11
LaF_3	~1.0 (600°C)	0.84	7
$La_{1-x}Ba_xF_{3-x}$ (x=0.071-0.095)	8.3×10^{-3} (25°C)	0.36	11
$Pb_{0.75}Bi_{0.25}F_{2.25}$	3.0×10^{-4} (50°C)	0.39	12
ß-$Pb_{0.96}K_{0.04}F_{1.96}$	~2.0×10^{-3} (25°C)	0.17	14
$Rb_{0.50}Bi_{0.50}F_{2.00}$	1.0×10^{-3} (50°C)	0.37	12
$K_{0.5}Bi_{0.5}F_{2.0}$	7.0×10^{-4} (50°C)	0.38	12

Material	Conductivity (ohm^{-1} cm^{-1})	Activation energy (eV)	References
$K_{0.3}Bi_{0.7}F_{2.4}$	$3.0x10^{-6}$ (50°C)	0.48	12
$K_{0.93}Bi_{1.07}F_{4.14}$	$\sim 2.4x10^{-3}$ (50°C)	0.28	13
$Na_{0.4}Bi_{0.6}F_{2.2}$	$1.5x10^{-4}$ (50°C)	0.46	12
$Na_{0.3}Bi_{0.7}F_{2.4}$	$1.0x10^{-5}$ (50°C)	0.62	12
$La_{0.95}Sr_{0.05}F_{2.95}$	~3 (600°C)	-	7
$Ca_{0.75}Y_{0.25}F_{2.25}$	$1.0x10^{-4}$ (200°C)	-	7
$BiO_{0.09}F_{2.82}$	$2.0x10^{-3}$ (100°C)	0.39	7
CaF_2	$4.0x10^{-2}$ (700°C)	-	7
SrF_2	$\sim 1.0x10^{-2}$ (700°C)	-	7
YF_3	0.5 (1000°C)	-	7
LuF_3	0.15 (840°C)	-	7
CeF_3	$6.0x10^{-4}$ (200°C)	0.20	7
ThO_2 + 7mole % CaO	$5.0x10^{-3}$ (1000°C)	1.20	15
ZrO_2 + 12mole % CaO	$5.5x10^{-2}$ (1000°C)	1.09	7
ZrO_2 + 8mole % Yb_2O_3	$8.8x10^{-2}$ (1000°C)	0.75	7
ThO_2 + 8mole % Yb_2O_3	$4.8x10^{-3}$ (100°C)	1.10	7
ThO_2 + 5mole % CaO	$4.7x10^{-3}$ (1000°C)	1.10	7
CeO_2 + 11mole % Ca_2O_3	$8.0x10^{-2}$ (1000°C)	0.91	7
CeO_2 + 7mole % SrO	0.11 (1000°C)	0.58	7
HfO_2 + 8mole % Y_2O_3	$2.9x10^{-2}$ (1000°C)	1.12	7
HfO_2 + 12mole % CaO	$4.0x10^{-3}$ (1000°C)	1.43	7
La_2O_3 + 15mole % CaO	$2.4x10^{-3}$ (100°C)	0.88	7
Bi_2O_3 + 20mole % Er_2O_3	0.45 (730°C)	-	19
Bi_2O_3 + 22mole % WO_3	$1.0x10^{-2}$ (500°C)	0.59	16
Bi_2O_3 + (22-28)mole % MoO_3	$\sim 9x10^{-3}$ (600°C)	-	17
LiI (pure)	$5.5x10^{-7}$ (30°C)	0.43	7
LiI (2mole % CaF_2)	$1.0x10^{-5}$ (30°C)	0.43	7
LiI (40mole% Al_2O_3)	$1.2x10^{-5}$ (25°C)	0.43	7
LiI (1mole % CaI_2)	$\sim 10^{-7}$ (25°C)	-	7
LiI(25wt % Al_2O_3+11wt% H_2O)	$4.0x10^{-5}$ (25°C)	-	19

Material	Conductivity (ohm^{-1} cm^{-1})	Activation energy (eV)	References
LiCl	5.0×10^{-10} (25°C)	-	19
LiCl(35wt % Al_2O_3 + 19wt % H_2O)	1.0×10^{-4} (25°C)	-	19
LiF	~1.0×10^{-13} (25°C)	-	19
LiF(34wt % Al_2O_3 + 10wt % H_2O)	1.1×10^{-4} (25°C)	-	19
Li_3N	2.5×10^{-3} (27°C)	0.25	20
LiS(Li electrode)	3.0×10^{-7} (100°C)	0.44	21
Li_2O	~2.0×10^{-7} (200°C)	0.94	21
Li_5AlO_4	2.3×10^{-7} (200°C)	1.24	21
Li_5GaO_4	7.5×10^{-7} (200°C)	0.89	21
Li_6ZnO_4	1.2×10^{-6} (200°C)	0.76	21
$Li_{1.8}N_{0.4}Cl_{0.6}$	1.0×10^{-6} (27°C)	0.49	20
Li_2SO_4	0.86 (550°C)	0.36	22
$Li_{1.33}Zn_{0.33}SO_4$	1.0 (550°C)	0.36	22
$LiNaSO_4$	0.93 (550°C)	0.44	22
$LiAgSO_4$	1.17 (550°C)	0.40	22
$Li_{1.1}S_{0.9}W_{0.1}O_4$	1.25 (600°C)	0.51	23
Li_4SiO_4	2.4×10^{-10} (25°C)	-	7
Li_4SiO_4	2.2×10^{-3} (400°C)	0.74	7
$Li_4Si_{0.7}Ti_{0.3}O_4$	4.0×10^{-3} (400°C)	0.62	7
$LiAlSiO_4$ (paralel : c-axis)	1.4×10^{-5} (200°C)	0.78	20
$Li_{3.6}Si_{0.6}P_{0.4}O_4$	6.2×10^{-7} (27°C)	0.57	20
$Li_{14}Zn(GeO_4)_4$	4.0×10^{-3} (227°C)	0.56	20
$LiAlCl_4$	1.2×10^{-6} (25°C)	0.47	18
$Li_4B_7O_{12}Cl$	1.0×10^{-2} (25°C)	0.49	24
$Li_4B_7O_{12}Br$	5.0×10^{-3} (300°C)	0.52	24
$Li_4B_7O_{12}S$	3.0×10^{-4} (300°C)	0.42	24
ß-$LiTa_3O_8$	1.5×10^{-2} (400°C)	0.78	7
$Li_{0.80}Zr_{1.80}Ta_{0.20}(PO_4)_3$	1.5×10^{-3} (200°C)	0.43	7
$LiHf(PO_4)_3$	1.05×10^{-3} (200°C)	0.43	7

Table I - References

1. J.H. Kennedy, in « Solid Electrolytes », ed. by S. Geller, Topics in Applied Physics vol. **21**, Springer-Verlag, Berlin (1977).
2. J.B. Bates, G.M. Brown, T. Kaneda, W.E. Brundage, J.C.Wang, H. Engstron, in « Fast Ion Transport in Solids », ed. by P. Vashishta, J.N. Mundy and G.K. Shenoy, North-Holland, Amsterdam (1979).
3. L.M. Foster, in « Fast Ion Transport in Solids », ed. by P. Vashishta, J.N. Mundy and G.K. Shenoy, North-Holland, Amsterdam (1979).
4. J.S. Kasper, in « Electrolytes, Characterization, Materials, Applications », ed. by P. Haegenmuller and W. van Gool, Academic Press, New-York (1978).
5. T. Takahashi, J. Appl. Electrochem. **3** (1973) 79.
6. K.W. Browall, J.S. Kasper, J. Sol. State Chem. **15** (1975) 64.
7. S. Chandra, in « Superionic Solids, Principles and Applications », North-Holland, Amsterdam (1981).
8. T. Takahashi, in « Fast Ion Transport in Solids », ed. by P. Vashishta, J.N. Mundy and G.K. Shenoy, North-Holland, Amsterdam (1979).
9. P. Haegenmuller, A. Levasseur, C. Lucat, J.M. Réau, G. Villeneuve, in « Fast Ion Transport in Solids », ed. by P. Vashishta, J.N. Mundy and G.K. Shenoy, North-Holland, Amsterdam (1979).
10. S. Vilminot, R. Bachman, H. Schulz, Sol. State Ion. **9/10** (1983) 559.
11. A. Roos, A.F. Aalders, J. Schoonman, A.F-M. Arts, H.W. Wijn, Sol. State Ion. **9/10** (1983) 571.
12. J.M. Réau, J. Grannec, C. Lucat, C. Chartier, S. Matar, J. Portier, P. Haegenmuller, J. of Fluorine Chem. **19** (1982) 363.
13. M.W. Shafer, G.V. Chandrashekhar, Sol. State Ion. **5** (1981) 629.
14. G.W. Mellors, J.R. Akridge, Sol. State Ion. **5** (1981) 605.
15. H.S. Maiti, E.C. Subbarao, J. Electrochem. Soc. **123** (1976) 1713.
16. T. Takahashi, H. Iwahara, J. Appl. Electrochem. **3** (1973) 65.
17. T. Takahashi, T. Esaka, H. Iwahara, J. Appl. Electrochem. **7** (1977) 31.
18. W. Weppner, R.A. Huggins, J. Electrochem. Soc. **124** (1977) 35.
19. H. Rickert, in « Electrochemistry of Solids », Springer-Verlag, Berlin (1982).
20. U. von Alpen, M.F. Bell, in « Fast Ion Transport in Solids », ed. by P. Vashishta, J.N. Mundy and G.K. Shenoy, North-Holland, Amsterdam (1979).
21. R.A. Huggins, Electrochim. Acta **22** (1979) 773.
22. A. Lundén, A. Bengtzelius, R. Kaber, L. Nilson, K. Schoeder, R. Tarneberg, Sol. State Ion. **9/10** (1983) 89.
23. A. Lundén, Sol. State Ion. **28/30** (1988) 163.
24. J.M. Réau, A. Levasseur, G. Magniez, B. Calés, C. Fouassier, P. Haegenmuller, Mat. Res. Bull. **11** (1976) 1087.

2.2 *Polymer Electrolytes*

2.2.1 *Organic Polymer Electrolyte* [18, 19, 20]

Complexes between polyethylene oxide (PEO) and alkali metal salts have shown considerable potential as polymer electrolytes in solid state battery. The basic principle of these compounds resides in the solvating power of a polymer matrix for salts and strong mineral acids in anhydrous media through the direct interaction of the cation with unshared electrons pairs carried by an heteroatom : - O - for ether, - S - for sulfide, - N - for amine, - P - phosphor (21). In Table II, formulae of the most common solvating polymers are given.

Table II : Formulae of the most common solvating polymers.

Linear poly ether oxides	
$[-\ CH_2\text{-}CH_2O\ -]_n$	poly (ethylene oxide)
$[-\ CH\text{-}CH_2\text{-}O\ -]_n$ CH_3	poly (propylene oxide)
$[-\ CH_2\text{-}CH_2\text{-}NH\ -]_n$	polyethylene imine
$[(-\ CH_2)_n\text{-}S\ -]_{2<n<6}$	poly (alkylene sulfide)
$[-\ CH_2\text{-}CH\ -]_n$ N O	poly vinlyl pyrrolidone
$[(CH_3O\text{-}C_2H_4\text{-}OC_2H_4O)_2PN]$	poly (bis methoxy ethoxy) ethoxy phosphazene

The properties of the polymer with the salt addedarise from several factors : the polarity of the heterotoms, the distance between them for the solvation of alkali metal ions and strong mineral acides, the flexibility of the polymer chain, the cohesive energy of the macromolecular network and the lattice energy of the salt.

Linear polymer chains like PEO (poly ethylene oxide), PPC (poly propylene oxide), PEI (poly ethylene imine) have a great ability to crystallise at room temperature. The polymer electrolytes show good ionic conductivity when the polymer phase has a low T_g and is amorphous or in a very disordered state. The sequential motions of the polymer chains must be of high amplitude and high frequency, since they control the ionic motion related to the conductivity. In order to improve conductivity performances at room temperature, the occurrence of crystalline phases can be suppressed by using crosslinked or branched polymers.

In the case of PEO, for example, improvements of conductivity are achieved by the formation of cross linked network with branched 2-methoxy ethyl poly (ethylene glycol) methacrylate.

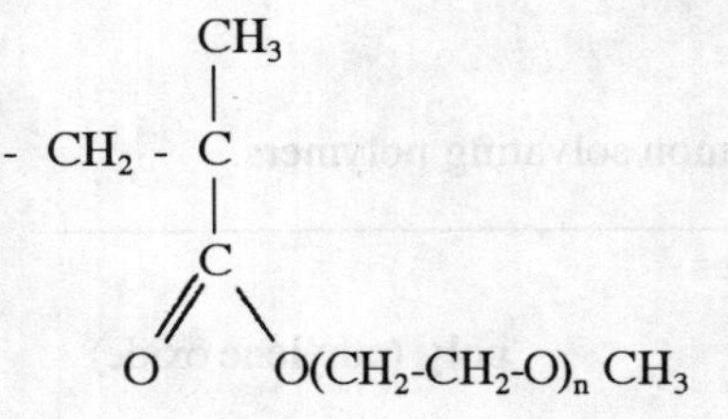

with $n < 22$

2.2.2 *Organic-Inorganic Polymers*

Introduction

The synthesis with the sol-gel process of silicate compounds allows preparation at low cost of compounds of great purity, great homogeneity, and having specific properties. The sol-gel process is also an useful tool in order to obtain related combination structures of organic and inorganic components giving the new family of ORMOSIL (ORganically MOdified SILicates). Dissolution of an ionic compound (salt, acid, base) into this solid host material leads to a family of new solid electrolytes. The silica backbone provides mechanical strength, an amorphous state and transparency in the visible spectrum. The organic groups offer solvating properties with respect to the guest ionic compounds. Materials such $SiO_{3/2}$-R (MX) are synthetised where R, an organic chain, is able to dissolve the ionic compound MX.

Aminosil

Aminosil is an example of an ORMOSIL ionic conductor obtained by hydrolysis and condensation of a trimethoxy amino propyl silane, in the presence of a strong mineral acid (HX) and water. Aminosils are obtained as monolithic transparent films, hard and non-porous, which are stable under atmosphere conditions. They exhibits a glass transition around -30°C and are thermally stable in air up to 180°C. The conductivities, measured by impedance spectroscopy, are of the order of 10^{-5} $\Omega^{-1}cm^{-1}$ at room temperature.

This example of aminosil as a protonic conductor in anhydrous media demonstrates the potentialities of ORMOSILS as solid ionic conductors. Numerous possibilities exist for combining typical inorganic and organic structures by the sol-gel process : the use of - O - Si - C = bonds as crosslinking points is a important factor in combining the inorganic and organic networks. It can be extended to other metal alkoxydes (Ti, Al, Sn) to give composite materials at the atomic scale : glassy polymers at the border of organic and inorganic chemistry. The wide variability of organic chemistry can be applied : if the influence of the determining parameters on different properties is known, the tailoring of materials with very special properties becomes possible. It leads to the synthesis of large variety of compounds like new materials, for contact lenses scratch resistant or functional coatings, and may be extended to the synthesis of new glassy polymers for solid state electrochemistry.

The good performances of polymer electrolytes like PEO-$LiClO_4$ are already demonstrated in the realization of lithium polymer electrolytes batteries, and in electrochromic displays. With the new ORMOSIL ionic conductors, important developments are to be expected in the area of solid state electrochemistry building upon the present applications of neutral ORMOSIL.

2.3 Lithium Glasses

Glassy solid electrolytes containing lithium have been studied extensively and are found to have some advantages over crystalline conductors, such as isotropic properties, flexibility of size and shape, absence of grain boundaries, and a large electroactivity domain. Another important feature of this type of solid electrolyte is the possibility of continuously changing the composition over a wide concentration range.

The ionic character of the conductivity in oxide glasses has been established by Warburg in 1884. In a general way binary glasses are made by network forming oxides (SiO_2, B_2O_3, P_2O_5, GeO_2) giving rise to a random corner linked tetrahedral arrangements broken in some places by the network modifier alkali oxides. The structure has large « holes » which facilitate ionic migration. These « holes » are often surrounded by non brindging oxygens which trap alkali ions

The conductivity of a glass for temperatures lower than the glass transition shows a behaviour typical of a thermally activated process following the Arrhenius law as discussed in section 1. From transference number determination, it can be shown that the current carriers are exclusively the alkali ions. The ionic conductivity of alkali-containing oxide glasses increases and the activation energy decreases as the alkali oxide concentration is increased. However, the conductivity is not a linear function of the alkali metal ion concentration. This had led to the concept of the weak electrolyte model.

The possibility of increasing the alkali metal ion concentration to enhance the conductivity is limited by the devitrification phenomenon. This problem can be overcome in two ways :

- by using fast quenching technics. The vitreous domain can be enlarged in the high lithium content,
- by increasing the concentration of alkali ions by the addition of an alkali salt.

In these ways quite large concentrations of alkali salt can be incorporated and the conductivity can be increased substantially by about two orders of magnitude for lithium glasses.

2.3.1 *Glass Systems* [22, 23, 24, 25]

a) Oxide Glasses

More than 100 different compositions in the M_2O-B_2O_3-SiO_2 (M = Li, Na, K) systems are investigated. The value of $10^{-1}\Omega^{-1}cm^{-1}$ at 350°C obtained for one of the compositions is still of the highest conductivity reported so far for Li^+ conducting glass.

Alkali Borates

Glasses are obtained in the system B_2O_3-Li_2O-Li_nX (n =1, x = F, Cl, Br, I; N = 2, X = SO_4, MoO_4, WO_4 ; n = 3, X = PO_4). The best of them show a maximum conductivity of $10^{-6}\Omega^{-1}cm^{-1}$ at 25°C reaching 10^{-2} at 300°C with an activation energy of 0.46 eV.

Other glass systems have been studied, like for example Li_2O-LiF-Li_2SO_3-Li_2SO_4-B_2O_3 but the ionic conductivity is less than for the previous ones.

Structural investigations have shown that borate glasses are formed by a covalent framework B_2O_3-Li_2O in which the lithium salt is diluted without notable interaction.

Alkali Silicate Glasses

Glasses in various alkali silicate systems show a sharp decreasein resistivity and activation energy with increasing alkali content. For example electrical conductivity derived from complex impedance analysis in the Li_2O-SiO_2-Li_2SO_4 system show slightly higher values ($10^{-5}\Omega^{-1}cm^{-1}$ at 100°C). Glasses have been reported in the $LiNbO_3$-SiO_2 system and in many other systems, but in no case does the conductivity go beyond this value.

Alkali Phosphate Glasses

High values of ionic conductivity in the unusual glass forming system M_2O-MF-$Al(PO_3)_3$ (M = Li, Na, K, Cs) which does not contain any conventional glass former have been reported. Other glasses have been obtained in the system $LiPO_3$-LiX (X = Cl, Br, I). Many other systems were investigated, but for the best oxide glasses, whatever the system the conductivity is not larger than $10^{-6}\Omega^{-1}cm^{-1}$ at 25°C with an activation energy E = 0.5 eV. In an attempt to improve the ionic conductivity of lithium glasses, several possibilities concerning either the addition of lithium salt of the change of support glass have been considered. Recently new families of glasses have been studied.

b) Sulphide Glasses

Glasses from the systems P_2S_5-Li_2S-LiI, GeS_2-Li_2S-LiI, SiS_2-Li_2S-LiI and B_2S_3-Li_2S-LiI have ionic conductivities of $10^{-3}\Omega^{-1}cm^{-1}$ at 25°C, some 1000 times that of corresponding oxide glasses and the activation energy is about 0.3 eV. The width of their electrochemical stability domain is generally around 4V vs lithium. Many other sulphide glasses, like recently SiS_2-P_2S_5-Li_2S-LiI have been investigated but in no case is the conductivity higher than $10^{-3}\Omega^{-1}cm^{-1}$ at 25°C. These glasses are generally very sensitive to moisture and difficult to handle.

3 Materials for Insertion Cathodes

3.1 *Introduction*

All ionic devices rely on ion transport and electron transfer-exchange. The insertion cathode is the place where the electron exchange occurs. In order for this process to be efficient it is necessary to assure an easy access of the ions and a good electron conductivity. The materials which constitute the insertion cathodes have to be, therefore, electron and ion conductors. Good electron conductors are metals and semiconductors and in order to assure good ion conductivity such materials have specific structures, in which the ion mobility is high.

Crystalline host materials can be classified by the dimensionality of the passage into which the guest ion is inserted as shown in Figure 2. The simplest configurations provide drain like tunnels which present essentially one-dimensional migration paths for guest ions. Examples of one-dimensional structure are $NbSe_3$, MoS_3, TiS_3, $(CH)_x$.

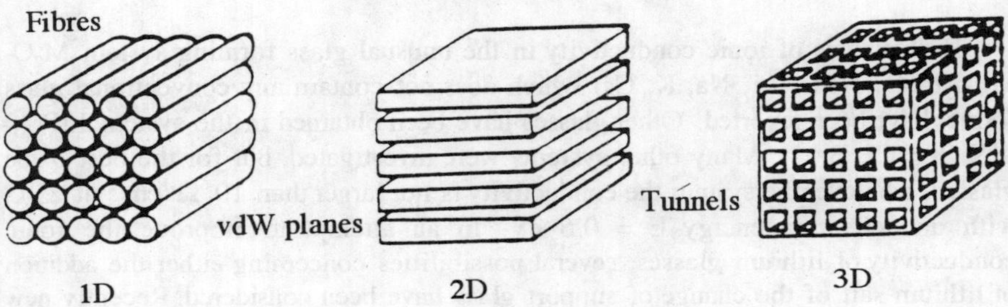

Figure 2: Types of insertion compounds with different structural dimensionality.

The most extensively investigated insertion materials are two dimensional layered compounds such as MX, MX_2, M_2X_3 compounds with M = In, Ga or V, Ti, Mo, Hf... and X = S, Se and also MPX_3 with M = Fe, Ni, Co and X = S, Se, Te. These compounds form alternating structures of covalently bounded layers and Van der Waals gaps. These layer materials are however three-dimensional solids with varying degree of anisotropy in their physical properties. One can establish a general hiearchy of the degree of anisotropy, on the basis of force constants determined from phonon studies, for some of these materials. Such a series would be graphite, MoS_2, $MoSe_2$, HfS_2, $HfSe_2$, $SnSe_2$, CdI_2 in decreasing order of anisotropy.

The weak interlayer forces of these materials offer the possibility of introducing foreign atoms or molecules between the layers. In this process of intercalation, the intercalant species are inserted in the Van der Waals gap, which is generally accompanied by a charge transfer between the intercalant species and the host lattice. It is believed that the tendency for this charge transfer is the driving force for the intercalation reaction. The intercalation process is the basis on which certain layered compounds are used as the cathode of solid state batteries and its practical applications are likely to increase, particularly in microelectronics. At the same time intercalation is now being recognized as an important practical way of controlling oxygen content and introducing cation doping to oxide superconductors.

The simplest layered materials are graphite and BN in which each layer consits of a simple sheet of atoms. Generally layered compounds have layers each consisting of three atomic planes : one metal atom sheet sandwiched between two non-metal atom sheets. These three atomic sheets are strongly bonded by either a covalent or ionic bond, but the layers are only weakly bound with each other by Van der Waals interactions. Such structures are formed by compounds of transition metal dichalcogenides such as ZrS_2, NbS_2 and MoS_2. Numerous other variants with more complex structures such as those of the first row transition metal phosphorous trichalcogenides MPX_3 with M = Fe, Ni, Co and X = S, Se and group III chalcogenides GaSe, InSe, In_2Se_3. Some of the recently discovered oxide superconductors may also be regarded as of this class of materials. Some insertion compounds have intersecting paths that form a three dimensional array and are referred to as framework compounds. Examples for such insertion compounds are MnO_2,WO_3, V_6O_{13}, MoO_2, Mo_6S_8 which have been actively studied in recent years.

3.2 Properties of the Host Materials [26, 27]

3.2.1 Structures of Insertion Materials

a) Graphite

Pristine graphite is a natural material consists of honeycomb layers of carbon atoms strongly linked by covalent bonds with only weak interlayer forces as evidenced by the difference between the in-plane and out of plane carbon-carbon distances shown in Figure 3. These layers form stacks of hexagonal symmetry with

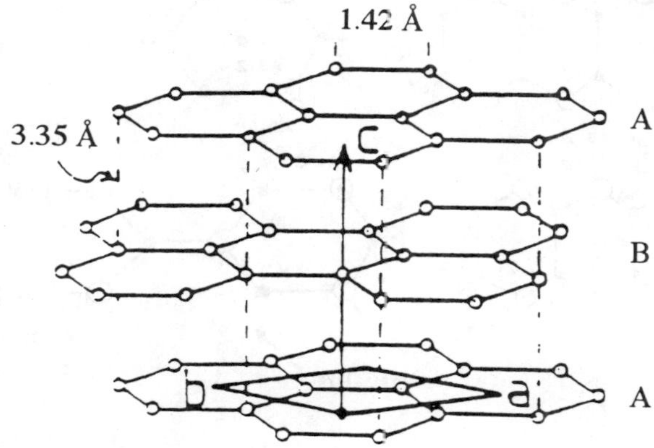

Figure 3: Structure of hexagonal graphite showing the ABAB stacking honeycomb carbon layers (from ref. 27).

a sequence ABAB. The honey comb carbon layers (28) have a space group $P6_{3/m}$ m_c and $a_G = 0.246$ nm, $c_G = 0.670$ nm.

b) Layered III-VI Semiconductors

The III-VI family includes essentially only combinations of type MX and M_2X_3 with M = In, Ga and X = S, Se.

Indium Selenide

Indium monoselenide is a layered semiconductor in which the layers are formed of two InSe formula having the Se atoms on the outside and the In atom on the inside of the layer as shown in Figure 4a and 4b which represent the hexagonal structure of the top view of the upper half of the layer.

According to the method of crystal growth, different polytypes can be obtained. The unit cell of the ß-structure is of hexagonal type and extends over two layers which are rotated by 60° with respect to each other. The crystal space group is C^4_{6v}. The unit cell of the ε-polytype structure is also of hexagonal type and also extends over two layers but this time these layers are translated in parallel to the xy plane with an amplitude of a/2. The ε-polytype space group is D^1_{3h} group. These two polytypes comprise eight atoms in the unit cell.

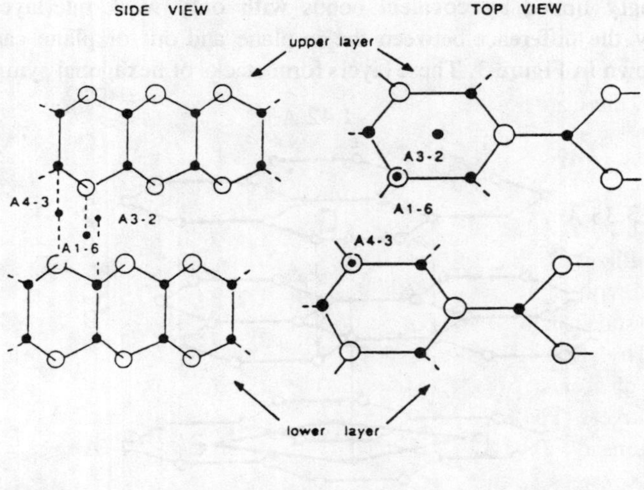

Figure 4: Hexagonal structure of InSe side view (a) and top view (b) with corresponding possible insertion sites.

In Figure 4 is shown the structure of γ -InSe.

The γ -polytype unit cell is rhombohedric and contains four atoms distributed in four adjacent layers. Its space group is C^5_{3v}.The unit cell is rhombohedral with a hexagonal symmetry in the plane of the layers. The layers are stacked in the sequence CBA, ABC separated by the Van der Waals gap.

When InSe crystals are grown by the Bridgman method the polytype γ -3R is obtained with the parameters : a = 0.4002 nm and c = 2.4946 nm. The space group of this structure is R3 m.

III_2-VI_3 Compounds

A typical example of this group of compounds in In_2Se_3 which exists in three different crystalline varieties α, ß and γ with the corresponding phase transition temperatures of 473 K and 923 K for the α-ß and ß-γ transitions respectively. The α-phase of In_2Se_3 stable below 473 K is a hexagonal layered compound with space group $P6_3$ mc (C^4_{6v}). The alternating selenium and indium layers along the c-axis have a repeat unit of the form Se-In-Se-In-Se and two such units form the unit cell with a hexagonal packing of the form BbCcBCcBbC... where B, C denote selenium and b, c, indium positions. The bonding inside the layers is strongly covalent, while the interlayer interactions (Se-Se) is of the Van der Waals type. The unit cell parameters at room temperature are a = 0.4025 nm and c = 1.9235 nm.

The γ -phase, stable above 923 K, is also a hexagonal structure with space group $P6_1$ and with unit cell parameters a = 0.711 nm, c = 1.934 nm. It can be considered as a distorted wurtzite-like structure in which the In atoms are either tetrahedrally or pentagonally coordinated. In this modification Se and In layers alternate and there are no neighboring Se-Se layers.

c) Layered V-VI Compounds

Bismuth selenide, Bi_2Se_3, is a layered narrow-gap semiconductor, E_g = 0.29 eV, which crystallizes in the rhombohedral lattice with D_{3d}^5 (R3m) space group and is isostructural with Bi_2Te_3, and Sb_2Te_3.

The crystal structure of Bi_2Se_3 consists of layers perpendicular to the trigonal c-axis formed by alternating planar arrangements of identical atoms. A layer contains five planes of atoms in the sequence : Se-Bi-Se-Bi-Se. The unit cell is formed by three such layers. The primitive unit cell, containing one formula unit Bi_2Se_3 has the dimensions a_o = 0.414 nm, b_o = 0.984 nm and c_o = 2.864 nm.

d) Transition Metal Dichalcogenides[29]

Almost all transition metals from groups IV, V, VI, VII and VIII of the periodic table form layered structure dichalcogenides. However, only the disulfides and diselenides of groups IV, V and VI have been studied in any detail. Layer metal dichalcogenides, MX_2 are formed with X = S, Se and M = transition metal of following groups.

Group	IV	V	VI
	Ti	V	Cr
	Zr	Nb	Mo
	Hf	Ta	W

In addition to the hierarchy derived from the degree of anisotropy, there is also a varying degree of ionicity in the bonding between metal atoms and chalcogen atoms. In general the sulfides are the most ionic with the sequence being S > Se > Te. Similarly there is a sequence for the transition metal ions where the order of increasing ionic character, for the group IV for example, is Ti < Zr < Hf.

The layer of transition metal dichalcogenide crystals MX_2 is made of a sheet of metal atoms M, sandwiched between two sheets of chalcogen atoms, X. Each sheet consists of atoms in a hexagonally close-packed network. In designating the polytypes one first indicates the number of sandwiches required to obtain a unit cell perpendicular to the plane of the layers and then the overall symmetry of the structure : trigonal T, hexagonal H or rhombohedral R. Lower case subscripts are used to distinguish polytypes otherwise similarly labelled e.g. $2H_a$ and $2H_b$. Thus the simplest polytype with octahedral coordination, labelled 1T, has a repeat perpendicular to the layer, of the sandwich, and trigonal symmetry and the simplest trigonal prismatic coordination polytypes have two sandwich repeats, and are designated $2H_a$ and $2H_b$. The $2H_a$ structure is adopted by metallic group V materials such as NbS_2 and stacks the metal atoms directly above each other along the c-axes, whilst in the $2H_b$ structure, adopted by semiconducting group VI materials such as MoS_2, the metal atoms are staggered.

e) Transition Metal Phosphorus Trisulfides MPS_3[30]

The MPS_3 compounds can be considered either as metal phosphorus trichalcogenides or as metal triphosphates.

In the first case, the phases are taken as double chalcogenides of metal and phosphorus and using the formula $M_{2/3}$ $(P_2)_{1/3}$ S_2 they can be considered as layered disulfides in which a third of the cation sites are substituted by phosphorus atom pairs P_2. In the second case, the phase is viewed more as a salt constituted from M^{2+} cations and $(P_2S_6)^{4-}$ anions. MPS_3 crystallizes in the monoclinic symmetry, space group C2/m. The structure is constructed from an essentially cubic close packed array of sulfurs with every other layer of octahedral sites completely filled by the transition metal ions and the phosphorus atoms as P_2 pairs orderedin a 2:1 ratio.

f) Chevrel-Phase Compounds

The class of molybdenum chalcogenides known variously as ternary molybdenum chalcogenides or as Chevrel phases are the compounds $A_xMo_6X_8$. A specific feature of these phases is a Mo_6X_8 cluster with X (Se or S) atoms approximating a cube and Mo atoms being slightly above the centers of the cube faces. Each cube is rotated approximately 25° about its body diagonal (3 axis). This arrangement of the Mo_6X_8 units leaves large cavities between them. The manner in which the cavities are filled by the A (ternary) atoms depends mainly on the 3 axis, but small ternary atoms occupy tetrahedral sites displaced away from the 3 axis.

3.2.2 Electronic Properties

a) Electronic Structure of Graphite[31]

Graphite has four valence electrons per carbon atom : three of them (2s, $2p_x$, $2p_y$) form covalent in-plane s bonds, the fourth electron, in the $2p_z$ state, gives rise to the conduction and valence π bands. At the some symmetry points of the two dimensional Brillouin zone (BZ), related by time reversal operation, the valence and conduction π bands are degenerate. These bands are of particular interest since in 2D graphite the Fermi level lies at the degeneracy points.

b) Electronic Structure of III-VI Layered Compounds

In this group of compounds GaS, GaSe and InSe have a common crystallographic structure : they are all hexagonal layer semiconductors. The theoretical study of the electronic properties of GaS, GaSe and InSe has been developed some time ago on the basis of some drastic simplifications. Early tight binding calculations were based on a two dimensional approximation, and the layer-layer interaction was introduced later using the empirical pseudopotential method in order to improve the agreement with the optical studies. The relative ionicity of the cation-anion bond is in the order InSe, GaS and GaSe. GaSe is the least ionic whereas the In-Se bond is the most ionic.

For these three compounds the unit cell spreads over two layers. Consequently each electronic state in one layer has a replica in the adjacent layer. These states are said to be « equivalent electronic states ». The splitting of the bands which corresponds to such equivalent states is due to the interaction between the layers. Some electronic states are very sensitive to this interaction and these are called « three-dimensional states ». In particular the two uppermost valence bands which are strongly split at Γ correspond to a combination of s and p_z states over the atoms in the unit cell. As the p_z orbitals of Se spread far in between the layers, it is not surprising that the corresponding bands are sensitive to the interlayer interaction. On the contrary, the « two-dimension bands » are those which are only weakly dispersed by the interaction.

The value for the direct gap in this three compounds follows the rule that the direct fundamental gap is a decreasing function of the dimension of the unit cell.

Table III: Direct gap energy and unit cell dimension of III-VI compounds.

Compound	InSe	GaSe	GaS
V ($Å^3$) 240.49	194.64	172.52	
E_g (eV) 1.4	2.2	3.1	

For γ-InSe for example, the tight binding scheme and first principles calculations account well for the magnitude of the direct energy gap, in spite of the fact that tight binding calculations put it at the center of the Brillouin zone, at the Γ point, whereas first principles calculations give a direct band gap at the Z point. Generalizations of the tight binding scheme to include the effect of intercalated lithium atoms, with the help of experimental optical absorption data, yield the result that the presence of lithium introduces an impurity band in the energy gap between conduction and valence bands. Furthermore, this band is quite flat, which means that the overlap of the Li electron wave functions with the In and Se atoms is small.

c) Electronic Structure of Transition Metal Dichalcogenides

The electronic structure deduced from the optical properties and band structure calculations for the group IV, V and VI transition metal dichalcogenides can be schematized in a density of states representation [29] shown in Figure 5.

The group IV metal, such as Zr, adopts the 1T structure with octahedral coordination of the metal by the chalcogens. Zr d electrons are used in bonding to S so that a d° configuration corresponds to Zr^{4+} and S^{2-} for a fully ionic model. The crystal field splitting of the d orbitals gives the actual form of the conduction bands. The valence band are formed by sulfur p orbitals. The p-d energy gap of these semiconductors is large for the more ionic compounds ZrS_2, $ZrSe_2$, HfS_2, and $HfSe_2$, small for TiS_2 and there is a small band overlap for $TiSe_2$ to give semimetallic properties.

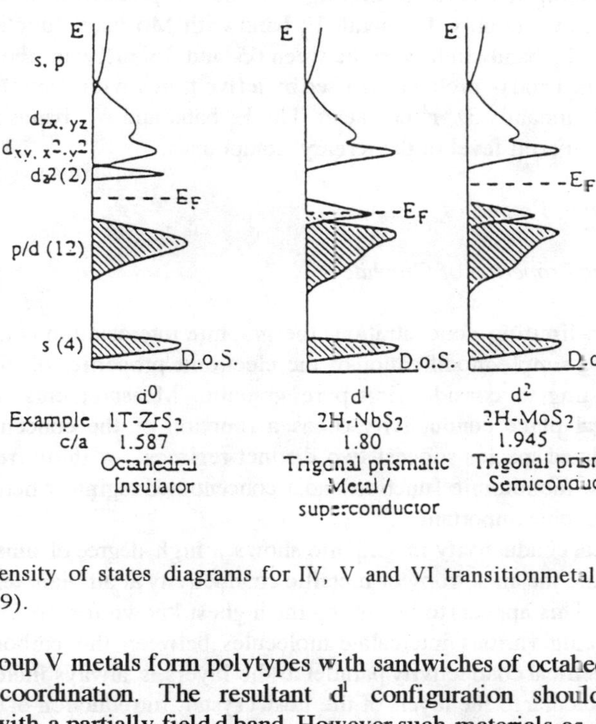

Figure 5: Density of states diagrams for IV, V and VI transitionmetal dichalcogenides (from ref. 29).

The group V metals form polytypes with sandwiches of octahedral and trigonal-prismatic coordination. The resultant d^1 configuration should give metallic properties with a partially-field d band. However such materials as 1T-TaS_2 and 1T-$TaSe_2$ are particularly susceptible to Fermi-surace driven structural distortions, and even at room temperature only weak metallic properties are observed. In the trigonal prism coordination, the 2H structures show rather better metallic properties and some materials are found to be superconductors.

The group VI metals usually adopt trigonal prismatic coordination structures, eg MoS_2, WS_2. These materials are semiconductors.

d) Electronic Structure of Chevrel-Phase Compounds

Using muffin-tin orbitals and the atomic-sphere approximation, the band structures of Chevrel-phase molybdenum chalcogenides, $A_xMo_6X_8$ have been reported (32) and level schemes are computed for a range of Mo and X potentials, for three Mo_6X_{14} clusters appropriate for the crystal structures of Mo_6X_8 and Pb $Mo_6S_{7.5}$ respectively. The cluster levels give the positions of the Mo 4d-like bands, while the widths and dispersion are estimated analytically in the tight-binding approximation taking the covalent mixing with the X p-states into account. The Fermi level falls in a doubly degenerate E_g band with Mo wave functions of x^2-y^2 character and the E_g bandwidths vary between 65 and 35 mRy in the compounds considered. The E_g band is probably crossed by a five times wider, singly degenerate A_{1g} band of predominantly $3z^2$-r^2 character. The E_g band and A_{1g} bands are the only ones crossing the Fermi level in the ternary compounds.

3.2.3 *Transport Properties*

a) Electronic Properties of Graphite

Up to some limiting concentrations the graphite intercalation compounds can be considered as a simple modification of the electronic properties of graphite, it is therefore interesting to consider first pure graphite. Measurements of the room temperature basal plane conductivity σ_a as a function of the concentration x of different intercalated species suggest two distinct regimes : a dilute regime where $\sigma_a(x)$ is a simple monotomic function and a concentrated regime where changes in band structure become important.

The electrical conductivity in graphite shows a high degree of anisotropy. For the parent graphite the anisotropy of metallic conductivity at ambiant temperature is of the order 10^4. This appears to be among the highest known for any crystal.

On introducing various intercalate molecules between the carbon hexagonal network, the electrical conductivity parallel to the layers is always increased. In the direction perpendicular to the layers of the host crystal, introduction of donors such as alkali metal atoms usually increases the electrical conductivity likewise, often to an extent which reduces the anisotropy ratio of conductivity σ_a/σ_e below that of the parent graphite. By contrast, introduction of acceptor molecules in graphite nearly always decreases the electrical conductivity σ_e, in the direction of c-axis, as though the layers, when separated by this insertion, becomes more nearly insulator than the parent graphite.

b) Electrical Properties of n-Type InSe

In layered semiconductors like InSe the crystal anisotropy significantly affects the carrier scattering and consequently the electrical conductivity. The weakness of the binding forces gives rise to various kinds of planar defects : stacking faults or interlayer precipitates of impurities parallel to the layers, or affect only very weakly it if at all the carrier mobility $\mu_{\perp}$ in the layer or perpendicular to the c-axis, but strongly reduce the mobility $\mu_{//}$ parallel to the c-axis. At temperatures below 20 K, in n-type InSe, the electron behaves like a typical 2D system.

c) Transport Properties of Transition Metal Dichalcogenides

TMD Semiconductors

The small gap semiconductors, $ZrSe_2$ and TiS_2, can only be grown with a small excess of metal present and this results in partial filling of the d conduction band and « extrinsic » semiconductor properties. Crystals of $ZrSe_2$ show electron concentrations of ~ 5×10^{19} cm^{-3} and TiS_2 shows a range from 1×10^{20} up to 3×10^{21} cm^{-3} for crystals with an excess titanium of ~ 5 %.

TMD Semi-Metals

$TiSe_2$, $TiTe_2$ and $HfTe_2$ show semimetallic properties with overlap between the top of the chalcogen p valence band and the bottom of the transition metal d conduction band. The transport properties of these materials are determined by both the electrons and holes present, and a more complicated behaviour in the temperature dependence of the Hall coefficient is observed, than in the extrinsic semiconductors.

3.3 Intercalation

3.3.1 Graphite

The most widely investigated and the best known intercalation material is graphite. Graphite intercalation compounds (GICs) can be formed by the insertion of atomic or molecular layers of many different chemical species between layers in the graphite host material. These compounds are of particular physical interest because of their relatively high degree of structural ordering, notably the staging phenomenon characterized by intercalate layers that are periodically arranged in a matrix of graphite layers. GICs are thus classified by a stage index n denoting the number of graphite layers between adjacent intercalate layers.

Intercalation provides a means for controlled variation of many physical properties over wide ranges. Because the free carrier concentration of the graphite host is very low, of about 10^{-4} free carriers/atom at room temperature, intercalation with different chemical species and concentrations permits wide variation of the free carrier concentration and thus of the chemical, thermal and magnetic properties of the host material. The electrical conductivity is the most interesting of these properties, with room temperature values exceeding that of copper. The large increase of conductivity in graphite results from a charge transfer from the intercalate layers, where the carriers have a low mobility, to the graphite layers where the mobility is high.

The GICs can form donor and acceptor compounds. The most common of the donor compounds are formed with the alkali metals K, Rb, Cs and Li, while acceptor compounds are those formed with Br_2 or halogen mixtures. The intercalation generally causes crystal dilatation along the c-axis. Many of these compounds are unstable in air so they usually require encapsulation.

Use of Graphite for Battery Cathode Material

The use of graphite intercalation compounds for battery materials and electrodes has been discussed and specific application have been attempted. GICs are attractive for cathode materials because of their potential for low weight relative to energy density. Graphite fluorides have been studied for use with a lithium anode and a non-aqueous electrolyte. The most serious technical problems associated with batteries based on these electrodes concerns the metastability of the electrolyte near the anode or cathode surface (where graphite could act as catalyst). On the other hand the high conductivity of intercalated graphite is an attractive material property for this application.

3.3.2 *Intercalation in InSe and In_2Se_3*

Layered III-VI compounds such as InSe and In_2Se_3 also provide interesting host materials for intercalation. For example, lithium or silver can be intercalated into these semiconductors by a spontaneous reaction in n-butyl lithium or via an electrochemical process. These systems are of interest as a Li solid-state battery. The Li ions in InSe are intercalated on tetrahedral sites.

Li Intercalated Indium Selenide

Lithium intercalation in the van der Waals gap of InSe has been widely studied. Shifts of the excitonic peaks observed in absorption as well as in Raman scattering indicate a modification of the electronic structure, with a significant evolution of the band gap with increasing lithium concentration. The first exciton E_0 at 1.352 eV shifts towards higher energies while the second exciton E'_1 at 2.52 eV shifts towards lower energies with increasing Li dose. In the photoluminescence spectrum, in the lower energy region of the first set of emission bands, we observe the appearance of new bands upon lithium intercalation.

3.3.3 *Transition Metal Dichalcogenides*

a) Intercalation Procedure and Structures Obtained

The alkali metals Li, Na, K, Rb, Cs and also metals such as Ag and Cu readily enter between the layers of these materials to form intercalate complexes. For high concentration of alkali metals and low temperatures ordered structures can be obtained. For the process of intercalation, the lithium versus TiS_2 electrochemical cell has been particularly carefully studied (33) since it has possible applications for high-energy-density secondary batteries. The open cell voltage is above 2.4 V at lithium molar concentration x = 0.1 and falls only to 1.8 V as x approaches 1. The fall is in part due to the increase in the Fermi energy in the Ti d band ; that it falls by only 0.4 V is a consequence of the high value of the density of states in this d band. As a result of the intercalation there are changes in the lattice parameters, with an expansion in the c direction. To a first approximation this expansion can be attributed to an increase in the van der Waals gap.

The occupation of well defined inter-sandwich sites, together with the relatively high diffusion coefficients usually found for alkali metal intercalation complexes gives the possibility of forming ordered or disordered layers of intercalants, and of achieving order-disorder transitions as a function of composition, temperature and pressure. Observation of ordered intercalant superlattices in lithium complexes has been difficult to achieve. High-resolution electrochemical titration of Li_xTiS_2 has revealed (33) peaks in the inverse derivative of the cell EMF at concentration x = 1/9, 1/4, 5/7, 4/5 and 6/7 and associated these with a degree of superlattice ordering at these concentration. Diffuse X-ray scattering experiments have shown superlattices at x = 1/4 and x = 1/3.

b) Electronic Properties

Alkali metals donate most or all of the outer s valence electron to the host material d band so that a semiconductor like MoS_2 turns into a good metal [34] after intercalation with Na.

Electrical and other transport measurements show that the intercalate complex becomes metallic and a superconductor [35]. The Hall coefficient for Li_xTiS_2 is almost independent of temperature and the electron concentration, n, estimated from the Hall coefficient R_H ($n = 1/R_H e$) follows closely the value of x confirming that charge transfer is close to complete. In all cases good metallic behaviour is observed with a resistivity ratio ρ (300 K) / ρ (4.2 K) of between 6 and 12.

For TiS_2 the value of 2.3 for the slope of the curves relating the log of the resistivity versus temperature has been modelled by a combination of longitudinal acoustic phonon scattering both within and between very small electron pockets at the L points of the Brillouin zone boundary. As the size of the electron pockets is increased with increasing x in Li_xTiS_2 the temperature dependence of the resistivity above ~100 K falls towards the linear dependence expected for a simple metal, since the distinction in size of the phonon wave-vector between intra- and interpocket scattering is rapidly lost as the size of the electron pockets becomes comparable to the dimensions of the Brillouin zone.

4 Solid State Batteries

4.1 Introduction

A solid state battery is an energy converter transforming chemical into electrical energy by means of internal electron exchange. The electron transfer is mediated by mobile ions released from an ion source, the anode, and neutralised in the electron exchanger, the cathode. The positive ion is transmitted through a dielectric which is a good electronic insulator, the separator. The ideal solid state battery should be based on one unique material in which three regions corresponding to the ion source, the separator and the electron exchanger are separated by internal homojunctions. Unfortunately today it is not yet possible to realise this in practice. The classical electrochemical batteries are the exact opposite of such a system. Here the anode and cathode are two different metals separated by an electrolyte solution containing two ions with opposite electrical charges. The contacts are heterogeneous interfaces metal/liquid on which chemical reactions occur accompanied by corrosion and gas formation. Such systems have been continuously improved during the last one and half centuries but they are definitely inadequate for applications in solid state electronics.

The concept of a wholly solid state battery is not new. This idea has arisen repeatedly in connection with the search for very small size batteries. A strictly solid state integrable microbattery would be a battery operating as a solid state device where the positive electrical charge is transported by small ions diffusing through a medium which is a perfect electronic insulator and the negative charges are transported by electrons injected from the external circuit.

Presently existing solid state batteries are formed by three components as shown in Figure 6: *the ion source* is a metal, most frequently lithium ; *the separator* is a superionic conductor and *the electron exchanger* is an insertion compound.

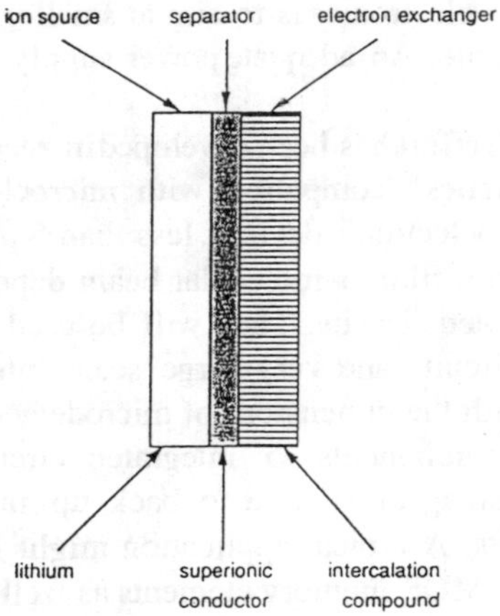

Figure 6: Schematic representation of the structure of a solid state battery.

Different fast ion conductors are used as separators - solid electrolytes - depending on the applications and the type of electrochemical cells. According to their mode of functioning one can distinguish two types of electrochemical cells :

- *The primary battery* is a non-rechargeable system, used only during one discharge ;
- *The secondary battery* is an electrically rechargeable system, used in high power density, high discharge rate applications.

Two main uses are envisaged for solid state batteries : in electrical vehicles and in microelectronics. The size of the batteries for this two uses are quite different. In the first case, the battery is large, uses Na-ß-alumina as separator and operates at relatively high temperature, 350°C. In the second case, for sensors, CMOS memories, credit cards and other microelectronic applications one uses microbatteries with a solid fast ion conductor separator which is amorphous, a glass or a polymer.

Microbatteries generally have energetic densities of the order of 200 Wh/kg. The specific energy by which a battery is usually characterized, is the ratio of the theoretical energy of the generator to the overall mass of the cell and is expressed in watt hour per kilogram. The capacity of microbatteries is in the range of 200 μAh.

4.2 Microbatteries[36, 37, 38]

The present tendency in electronics is to aim at smaller and smaller devices and systems which is also portable. An adequate power supply for such systems is the solid state microbattery.

A considerable research effort has been developed in recent years to demonstrate the feasibility of microbatteries[36] compatible with microelectronic technology and batteries of the size of microelectronic devices, less than 5 μm thick, are now being developed and built using thin film or molecular beam deposition technologies that are compatible with integrated circuits. They will be used in future microdevices, microsensors, integrated circuits and very large scale integration. An integrable microbattery compatible with the dimensions of microdevices will act as a source of power for many single components in integrated circuits and devices. Such microbatteries can be advantageously used to back up random access memories (RAMs) and CMOS circuits. A typical application might be, for example, an on-chip power source driving CMOS memory elements as well as backing-up RAMs.

With the tendency of microelectronics towards further decrease in the size of active elements the distribution of electric power becomes a major limitation for further integration. Three-dimensional integrations will be limited by cross talking problems if the electric power has to be distributed from on external source. This can be avoided by the use of integrated solid state microbatteries to provide local power. New solid state devices based on superconductors with the bias provided by an integrated microbattery function can also be envisaged within the near future.

The design of new microelectronic architectures which have the power generated on the spot where needed, rather than having it distributed from a remote power source avoids electrical conduction links, inevitable power losses and cross talking that are major drown back for higher level integration.

Small batteries have been in use for a long time. The lithium anode primary cells [4], $Li/LiI/I_2$, are used for implantable pacemakers operating at 37°C. These cells have an open circuit voltage V_{oc}, of 2.8 V and an energy density of 100-200 Wh/kg. Such lithium cells have extended battery lives of up to 10 years for 120-250 mAh capacities. More than 4 million cardiac pacemakers have been powered by lithium/iodine batteries since 1972.

Another type of small battery - the polymer electrolyte battery - holds high promise. Experimental studies on small laboratory cells suggest that such batteries could provide specific energy densities of up to 150 Wh/kg with on open circuit voltage of 2.3 V. The potential applications of polymer electrolyte batteries spans the entire range of battery products, from smart credit cards to stand-by power in telecommunications, including Polaroid instant film packs and batteries built into printed-circuit boards of laptop personal computers. Their combination of high energy density and mechanical rigidity make them particularly attractive for applications in space.

The main advantage of polymer electrolytes lies in their mechanical properties : they can deform under stress and can be configured into almost any shape. Deformable polymer electrolytes offer the promise of maintaining good electrode interfacial contact during cell operation. The limitation is that their performances are reduced if they are used in the form of very thin films less than 1 μm. This is because of polarization due to the fact that polymers are weak electrolytes and contain two types of ions : cations and anions which, at very short distances form a double layer and generate an electrical field. There are attempts at present to produce polymer electrolytes with only one type of mobile ion while the other is fixed by being attached to the molecular frame. The ionic conductivity of this type of polymer electrolyte is, so far, still relatively low.

4.3 Electrical Currents and Energy

The electrical charge flow is mediated by ion transport inside the battery and electron transport outside. Positive ions are emitted from the ion source, anode, into the separator. Simultaneously the anode supplies electrons to the external circuits.

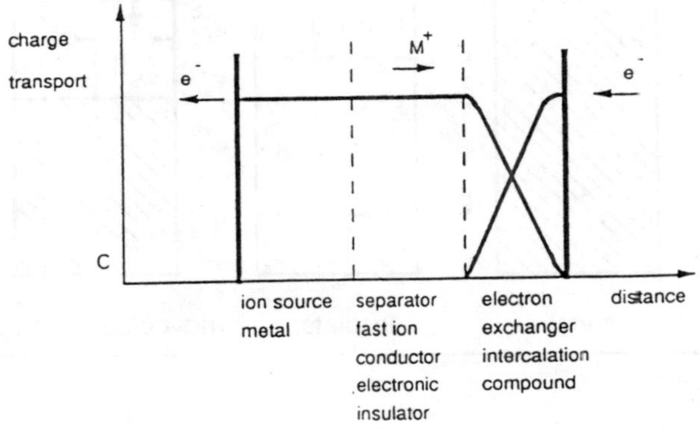

Figure 7: Schematic representation of atom flow (positively charged) across the battery and electron flow in the external circuit.

The separator is permeable only to the positive ions. Thus a current of positive charges is flowing from the anode to the cathode as shown in Figure 7.

Electrons and ions available in the ion source are simultaneously extracted from the two opposite surfaces of the metal. Ions are injected in the battery and electrons into the external circuit. After crossing the separator positive ions are inserted into the ion exchanger. An electron exchange occurs making available free electrons in the conduction band in order to assure overall electrical neutrality. Through this electron exchanger electrochemical energy is made available for work by the electrons in the external circuit. The electron exchanger is thus a mixed, electron and ion conductor, accepting electrons from the external circuit and ions from the bulk of the battery. By the atom motion inside the battery, electrical energy is made available for an external circuit.

4.4 Model of Functioning of the Battery

The solid state battery is formed by three components. At equilibrium, when isolated from each other each one of the components is characterized by an energy band diagram as shown in Figure 8.

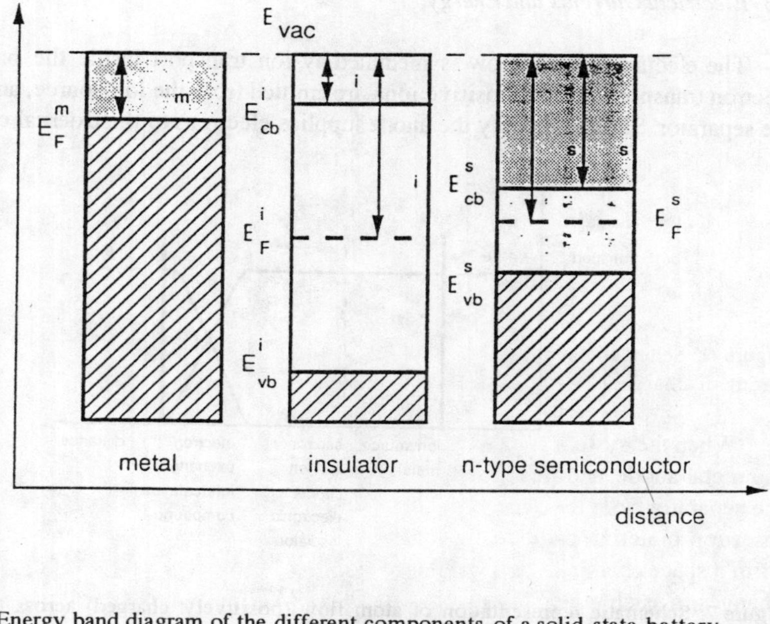

Figure 8: Energy band diagram of the different components of a solid state battery.

Each component is characterized by its work function, which is the energy necessary to extract one electron from the material measured with reference to the energy of the vacuum. ϕ_m is the work function of the metal, ϕ_i is the work function of the insulator and ϕ_s is the work function of the semiconductor. E_{cb}^i and E_{cb}^s are respectively the bottoms of the insulator and semiconductor conduction bands. E_{vb}^i and E_{vb}^s are the tops of the valence bands, and χ_i and χ_s are the electron affinities of the insulator and the semiconductor respectively.

When the three components are brought into contact and are connected to an external load, the energy diagram changes drastically and we now have a dynamical regime. The Fermi level no longer the same across the system, as in equilibrium state, but instead a quasi Fermi level is defined for each component with regard to the charge distribution and the current flow. The energy band diagram for the system during discharge is shown in Figure 9.

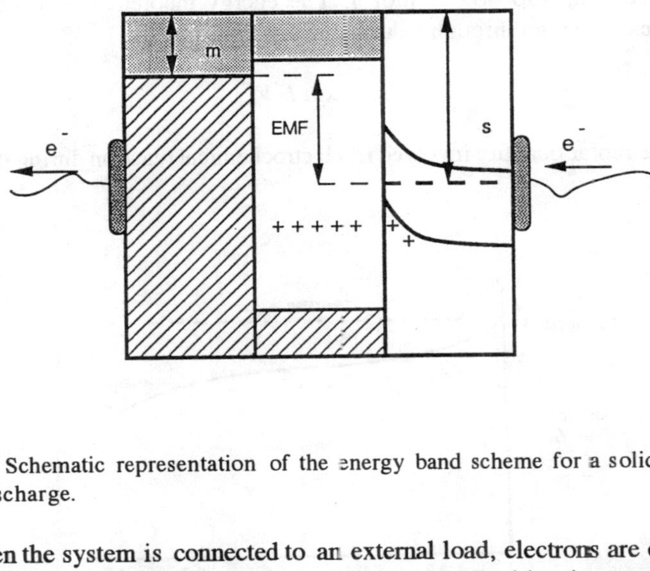

Figure 9: Schematic representation of the energy band scheme for a solid state battery during discharge.

When the system is connected to an external load, electrons are extracted from the metal anode, and flow into the external circuits. Positive ions are injected into the separator from the anode and diffuse toward the cathode. Once transferred into the insertion material, cathode, the positive ions are distributed near the interface and form a space charge layer. The quasi Fermi level now depends on the distribution of changes in each material. A very thin layer of negative charges is formed at the metal - insulator interface to compensate for the positive charge distribution throughout the insulator. A space charge layer is formed at the insulator-semiconductor interface to account for the ion injection into the intercalation semiconductor compound.

The electromotive force (EMF), created across the system, is due to the potential difference between the metal and the intercalation material which are characterized by the work function ϕ_m and ϕ_s

$$EMF = \phi_m - \phi_s = E_{F(m)} - E_{F(s)} = V_{oc}$$

where V_{oc} is the open circuit voltage.

The total energy available for the discharge in a battery is

$$- \Delta G = n\, F\, V_{oc}$$

where ΔG is the Gibbs free energy, n is the number of transferred electrons and F is the Faraday constant (96 487 C mol^{-1}). The energy theoretically available in a electrochemical generator through a discharge process is

$$E_{th} = x\, n\, F\, V_{oc}$$

where x is the molar quantity involved in electrochemical reaction during discharge.

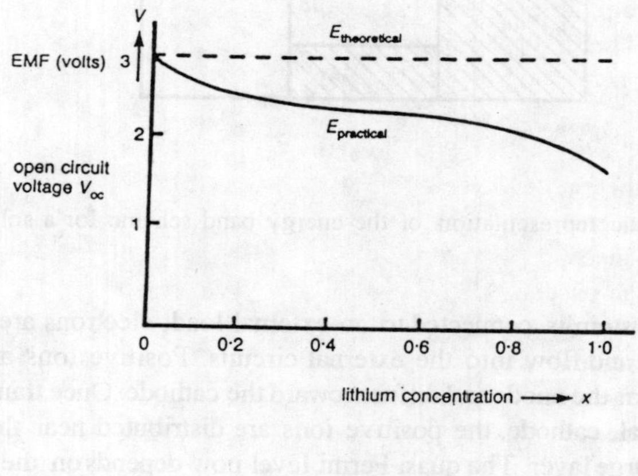

Figure 10: Discharge diagram of a Li/TiS$_2$ battery.

A comparison between the theoretical estimate and a practical discharge curve for the system Li/TiS_2 is given in Figure 10.

A few typical values of the energy density for different battery systems are given in the following Table.

System	Voltage V_{oc} (volts)	Energy density Wh/kg
Li/TiS_2	2.5	564
Li/FeS_2	1.6	1159
$Li/NiPS_3$	2.2	450
Li/InSe	2.8	375

4.5 Solid State Primary Lithium Batteries

Commercialy available solid state lithium batteries are based on LiI electrolyte.

4.5.1 Lithium-Iodine Cell

This has a solid lithium anode and a polyphase cathode of poly - 2-vinyl-pyridine (P2VP). The structure of the cell is [39] :

$$Li \, / \, LiI \, / \, I_2 (P_2VP)$$

It has a specific theoretical energy of 1.9 Wh/cm^3 with an open circuit voltage of $V_{oc} = 2.8$ V and an energy density of 100-200 Wh/kg. The separator LiI is formed and grows as the discharge proceeds. Such batteries are used as power sources for implantable cardiac pacemakers operating at 37°C. Lithium-iodine batteries have extended system lives of 10 years for 120-150 mAh capacities.

Batteries of medium capacities, i.e. up to around 1 Ah, can be used for random access memory power supplies in electronics. Li/Br_2 batteries have also been built. The greater electronegativity of bromine gives a voltage of the order of 3.5 V and energy densities as high as 1.25 Wh/cm^3 [40].

4.5.2 *Lithium, Lithium Iodine Cells with Different Cathodes*

Cells built for low rate operations in applications requiring long life under low drain or open circuit conditions, have the structure [41] :

$$Li \ / \ LiI\text{-}A_2O_3 \ / \ PbI_2$$

In this structure different cathodes have been tested such as PbI_2 + Pb, PbI_2 + PbS + Pb or TiS_2+ S, or As_2S_3. The separator is a dispersion of LiI and LiOH with alumina. The open circuit voltage of this battery is V_{oc} = 1.9 V and the energy density is 75-150 Wh/kg. A three-cell battery delivers 6 V and offers a capacity of 140 mAh for pacemaker power sources.

Another cell envisaged for use as a cardiacpulse generator is the system [42] :

$$Li \ / \ LiI \ (SiO_2, \ H_2O) \ / \ Me_4NI_5 + C$$

The cathode is a mixture of carbone and tetramethyl-ammonium penta iodide (Me_5Ni_5). The open circuit voltage of this cell is V_{oc} = 2.75 V and energy density 0.4 Wh/cm^3.

4.6 *Solid State Secondary Batteries with Insertion Cathodes*

Table IV : Lithium rechargeable batteries (cylindrical cells) developed or commercialized in the world.

Country	Company	Performances (Wh/dm^3)	Cycle	Couple	Market[a]
France	SAFT	175	50	LiV_2O_5	M
Japan	Sony	220	200	$LiMnO_2$	C
	Sanyo	-	-	$LiMnO_2$	C
	Matsushita	220	40	LiV_2O_5	R&D
				$LiMnO_2$	C
Canada	Moli-Energy	120	400	$LiMoS_2$	C
USA	Eveready	180	100	$LiTiS_2$	C+M
	Honeywell	175	50	LiV_2O_5	M
	AT&T	200	350	$LiNbSe_3$	S
	Grace	230	100	$LiMoS_2$	-

[a]M = Military application, C = Commercial, S = Space.

Insertion of M^+ ions may occur by way of three dimensional diffusion in the lattice of a close-packed spinel material, in the van der Waals gap of layered materials, as well as in one dimensional channels. The choice of lithium insertion materials is very large. The most attractive family seems to be the transition metal oxide compounds.

Some examples of commercialy available secondary batteries are presented in Table IV.

4.6.1 *The Li-V_2O_5 Battery*

An interesting realization is the coin shaped lithium V_2O_5 battery. The cathode is a mixture of vanadium pentoxide and 5 % carbon black. During discharge the average voltage is about 2 V. The battery has a relatively flat discharge curve and an electrical capacity of 36 mAh at a discharge rate of 1 mA.

Li-MnO_2 cells have also been fabricated and are commercially available in the form of flat-type batteries.

Interesting results are obtained with the LiI-A_2O_3 polyphase electrolyte. The type of solid state battery built with this fast ionic separator, operating at 300°C is :

$$Li\text{-}Si \; / \; LiI \; (Al_2O_3) \; / \; TiS_2, Sb_2S_3$$

It has a practical energy density of the order of 200 to 580 Wh/kg.

4.6.2 *Silver-Cells*

Some crystalline fast ion conductors such as $RbAg_4I_5$ exhibit unusually high ion conductivity at room temperature. Cells with $RbAg_4I_5$ as separator have been tested. A good example is the structure :

$$Ag \; / \; RbAg_4I_5 \; / \; RbI_3$$

with an open circuit voltage, V_{oc} = 0.66 V and a practical energy density of 4.4 Wh/kg. Silver fast ion conducting glasses are also used in Silver-cells such as :

$$Ag \; / \; AgI\text{-}Ag_2O\text{-}B_2O_3 \; / \; NbS_2$$

The open circuit voltage of this cell is V_{oc} = 0.39 V and discharge can be operated with a current density of 3.3 μAcm^{-2}.

4.6.3 *Rocking-Chair Batteries*

The use of lithium metal as ion source has a certain number of disadvantages among which are:

- The possibility of dendritic formation during charging. The lithium dendrites reacts with the electrolyte and become isolated from the bulk lithium metal which leads to lower cycle life time;
- The unsafe operating characteristics due to be the high reactivity of lithium;
- The high cost of technology due to the need for the assembly equipment to be located in dry room, the relatively expensive separators, and the need for excess Li metal.

To alleviate this problems it has been proposed to use as anode an other lithium insertion compound analogous to that of the cathode.

This new generation of rechargeable lithium cells is called *the rocking chair battery*. During charge, the Li^+ ions are disintercalated from the Li-bearing cathode, and intercalated into a carbon anode. Thus the amount of cathode and anode material are directly correlated. Rocking-chair batteries are made using two different intercalation compounds, in which the chemical potential of the intercalant differs by several electron-volt, for the electrodes. The output voltage of such a cell is determined by the difference between the electrochemical potential of Li in the two insertion compounds. For the anode, the chemical potential of the intercalated Li should be close to that of Li metal, like it is in Li_xC_6.

A lithium ion rechargeable battery based on the couple $Li_xC_6/LiMn_2O_4$ gives 0.4 Ah cycled between 2 and 4.6 V.

A composite polymer electrolyte is used in the cell:

$$C \,/\, PAN\text{-}EC\text{-}PC\text{-}LiN\,(SO_4CF_3)_2 \,/\, LiNiO_2$$

This cell has a discharge voltage of about 3 V and a specific energy of 337 Wh/kg.

4.7 *Thin Film Solid State Microbatteries*

A certain member of solid state microbatteries are presently demonstrated in a prototype stage. One of the earliest reported[43] thin film microbatteries has Ag as ion source. The electrochemical cell is :

$$Ag \,/\, AgI \,/\, Pt \qquad \text{with} \qquad V_{oc} = 0.55\ V$$

Ag is also used as the anode in the cell:

$$Ag\ /\ NH_4Ag_4I_5\ /\ (C + RbAg_4I_5 + KI_3) \qquad with \qquad V_{oc} = 0.6\ V$$

Lithium-electrode thin film batteries show more promise. One of the first proposed[45] solid state electrochemical cells was :

$$Li\ /\ LiI\ /\ AgI \qquad with \qquad V_{oc} = 2.1\ V$$

and short circuit currents greater than 100 μAcm^{-2} at 25°C.

More recently[46] secondary thin-film lithium cells have been built with TiS_2 cathodes :

$$Li\ /Li_{3.6}Si_{0.6}P_{0.4}O_4\ /\ TiS_2 \qquad V_{oc} = 2.45\ V$$

The lithium is deposited by evaporation, the solid electrolyte is a thin film obtained by RF pulverisation and the cathode, TiS_2 is obtained by chemical vapor deposition. More than 2000 cycles have been operated at current densities between 6 μA/cm^2 and 16 μA/cm^2.

Li batteries with TiS_2 cathodes have also been built with other fast ion conductors in cells such as :

$$Li\ /\ P_2S_5\text{-}Li_2S\text{-}LiI\ /\ TiS_2 \qquad V_{oc} = 2.5\ V$$

The conductivity is 2x10^{-3} (Ωcm)$^{-1}$ at 25°C and the energy density[47]: 150-200 Wh/kg.

Designed with 50 mAh capacity such batteries deliver continuous currents up to 0.1 mA/cm^2 and allow pulsed currents up to 10 mA/cm^2. The application envisaged is as power sources, for electronic computers.

Microbatteries have also been built[48] using RF sputtering techniques to form the cell:

$$Li\ /\ B_2O_3\text{-}0.8Li_2O\text{-}0.8Li_2SO_4\ /\ TiS_xO_y \qquad V_{oc} = 2.6\ V$$

which cell can be operated with a current density of 65 μAcm^{-2}.

Other lithium anode batteries are :

$$Li\ /\ Li_2O\text{-}B_2O_3\ /\ V_2O_5\text{-}P_2O_5 \qquad V_{oc} = 3.6\ V$$

These cells are obtained by successive evaporation of thin films. The energy density of the positive electrode material, measured by coulometric titration[49], is 310 Wh/kg.

Another type of Li microbatteries has the structure :

$$Li\ /\ B_2O_3\text{-}0.7Li_2O\text{-}0.5Li_2SO_4\ /\ InSe \qquad V_{oc} = 2.8\ V$$

The thin films are obtained by molecular beam deposition or conventional flash evaporation. Multiple discharge-charge cycles can be carried out at a current density[50] of 50 μA/cm^2. The specific energy of this type of microbattery is of the order of 300 Wh/kg.

Thin film rechargeable lithium batteries with several different kinds of cathodes with open circuit voltages ranging from 2.4 to 4.4 V have been recently fabricated using LIPON/$Li_{2.9}PO_{3.3}N_{0.46}$ as separator.

A more detailed description of the fabrication and functioning of different types of microbatteries can be found in a recent review[51]. Multiple structures can be obtained by successive repetition of the sequence of thin layers forming a microbattery, very much in the way semiconductor superlattices are obtained. Such superstructures can be used where high current densities or high voltage are needed by interconnecting them in series or in parallel.

Further investigations on the electrolyte and cathode materials are necessary in order to improve the performances and to make the technology attractive in practice. The essential parameters to be improved are the ionic conductivity of the electrolyte and the diffusion coefficient of the Li^+ ion in the insertion compound. In order to become able to do so, further fundamental knowledge is necessary on the transport mechanism and the intercalation process. The new challenge is therefore in developing basic research, in materials science, on the electrolyte and cathode materials.

4.7.1 *Polymer Batteries*

Solid polymer electrolytes derived from complexes between poly-ethylene-oxide (PEO) and alkali metal salts MX can be used as separators in solid state batteries. Polymer electrolytes are formed from a polymer matrix with a strong dissolving power for salts and strong mineral oxides in anhydrous media through direct interaction of cation and unshared electron pairs carried by an heteroatom: -O- for ether ; -S- for sulphide ; -N- for amine ; -P- for phosphors. The PEO-MX complexes may be considered as plastic electrolytes. Most of the polymer batteries are of the form Li/PEO-LiX/insertion cathode, while commonly used positive cathodes are V_6O_{13}, TiS_2, MnO_2, Cr_3O_8 or LiV_3O_8.

An example of a battery using a polymer separator and a cathode formed of composite which consists of small particles of insertion compound bound together with polymer electrolyte and carbon is the structure[52] :

$$Li\ /(PEO)_9LiCF_3SO_3/\ V_6O_{13} + PEO + C$$

The open circuit voltage of this cell is $V_{oc} = 2.8$ V. During discharge, at current densities of around 1 $mAcm^{-2}$, the voltage stabilises at 2V. The practical energy density is of the order of 200 Wh/kg, the power density reaches 0.1 W/g.

Another system investigated for a small number of cycles[53] is:

$$Li\ /\ PEO\text{-}LiClO_4\ /\ TiS_2$$

The initial voltage of this cell is $V_{oc} = 3$V and the theoretical energy density is 565 Wh/kg.

A novel class of solid state organic positive electrodes has been introduced. These new electrodes, termed Solid Redox Polymerization Electrodes (SRPE) are based on polysulfide polymers. The redox mechanism of the positive electrode is a redox dimerization/scission reaction which liberates free electrons. The cells have the following structure:

Li or Na / polymeric electrolyte / SRPE

The theoretical energy density[54] for lithium based batteries is 990 Wh/kg and the open circuit voltage is $V_{oc} = 3.0$ V while for sodium based batteries the theoretical energy density is 750 Wh/kg and the open circuit voltage is $V_{oc} = 2.7$ V, at a discharge current density of 0.125 $mAcm^{-2}$.

Rechargeable lithium batteries with high energy density have been developed recently using dimercaptan-polyaniline composite electrodes[55]. The battery is charged at 3.25 to 4.75 V and discharged at 0,1 $mAcm^{-1}$. A capacity of 185 Ah/kg cathode is obtained at a charging voltage of 4.75 V. The gravimetric energy density of the composite cathode is greater than 600 Wh/kg.

4.8 High Energy Batteries

4.8.1 Na-S Batteries

Na-S secondary batteries are remarkable because both energy and power density are high, raw materials are abundant and cheap and lifetime is inherently unlimited as a consequence of the fact that the electrochemical processes are basically reversible.

The sodium-sulphur batteries are also known as the ß-batteries, because they use ß-alumina as a fast ion conductor for Na^+ ions. The cell operation implies emission of Na^+ ions from the anode which then diffuse though the wall of ß-alumina, acting as separator to reach the cathode where reduction of sulphur occurs. The ß-battery is formed with molten Na as the negative electrode, a solid Na^+-ion conducting separator and molten sulphur-sodium polysulfide mixture as the positive electrode. The operating temperature is 300-350°C and the cells are sealed to prevent reactions with the wall and the air.

The reduction of sulphur occurs in two steps. The first reaction gives an open-circuit voltage of 2.08 V whereas the second reaction gives 2.08-1.76 V. Current characteristics are an energy density of 200 Wh/kg and a capacity of 15 Ah. ß-batteries are used for electric vehicle traction and load levelling applications.

Development of high-quality, fine-grained materials with addition of ZrO_2 has helped to improve the batteries performance. Also, special sulphur electrode designs are now been used. The ß-alumina can be replaced by Na^+-conducting borate glass and other fast ion conducting materials such as Nasicon ($Na_{1-x}Si_xZr_2P_{3-x}O_{12}$ with $1.8 < x < 2.4$).

4.8.2 Sodium Chloride Batteries

The major difference from Na-S battery is the positive electrode which consists of an insoluble metal chloride in molten $NaAlCl_4$. The cell configuration is :

$$Na\,/\,\beta''\text{-}Al_2O_3\,/\,NaAlCl_4.MCl_2$$

The operation temperature is 250°C. Energy densities are higher than 130 Wh/kg. Large electrical vehicle batteries of about 25 kWh have been built and tested.

5 Solid State Electrochemical Sensors[55]

One of the most frequent and best developed applications of fast ionic conductor is their use as gas sensors. The ion flow across the solid state fast ion conductor creates a measurable potential difference proportional to the intake of ionic species.

The all-solid-state gas sensors play an important role in the fields of environmental protection and process control. Solid oxygen ion conductors are employed in the exhaust of automobiles and industrial combustion processes. The fuel/oxygen ratio is controlled in this way to reduce pollution and make efficient use of the fuel. In addition, the sensor is employed in-situ during the reduction of ores and the carbonization of metals for hardening. These successful applications have initiated the present interest in the development of sensors for other gases. Since there is a lack of suitable solid ionic conductors for the transfer of other gases than oxygen, especially of complex gaseous molecules, new principles are examined which provide indirect equilibration with the mobile ions. This equilibration may be achieved by the fast ion conductor or by employing auxiliary phases as thin films. Much more sophisticated information may be obtained by both the measurement and/or control of the voltage and the electrical current. This allows high sensitivity and selectivity.

Galvanic cells transduce chemical energy into electrical energy and vice versa. Specifically, electrochemical gas sensors are based on the reaction of the detectable gas with the electroactive component. This may be, in the simplest case, the expansion of the gas from a compartment of high partial pressure into another one of low partial pressure. In more advanced devices the reaction is much more complex and involves one or several auxiliary phases. The reactions may occur virtually in thermodynamic equilibrium, may deviate considerably from this equilibrium for kinetic reasons or may be forced to take place by externally applied currents and voltages.

5.1 Principles of Solid State Electrochemical Gas Sensors

The presently available gas sensors are only capable of detecting gases that may be directly transferred as ions through the solid electrolyte[56, 57] (« type I » solid state electrochemical gas sensors). These devices may be described in a conventional way as galvanic concentration cells as illustrated in Figure 11. The tendency of the ions to diffuse from the high to the low pressure side is compensated by an electrical field which is formed as a result of the charge displacement. The diffusion is proportional to the gradient of the concentration whereas the magnitude in the electrical field is proportional to the absolute value

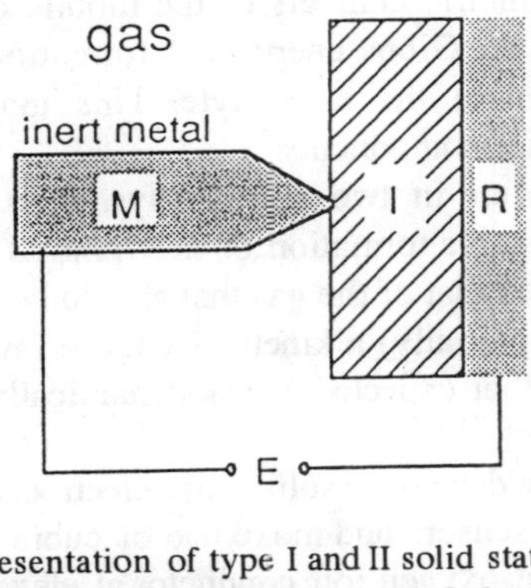

Figure 11: Schematic representation of type I and II solid state ionic gas sensors. The gas equilibrates with the mobile or another component of the solid ionic conductor. This results in a tendency of diffusion of the ions across the fast ion conductor which is compensated by an electrical field as measurable with an inert metal (from ref. 55).

of the concentration of mobile ions. This results in a logarithmic relationship between the cell voltage E and the concentration or partial pressure p_i of the gas :

$$E = -\frac{kT}{z_i\, q}\ln\frac{p_i^{gas}}{p_i^{ref}} \tag{1}$$

where k, T, z_i and q are : the Boltzmann's constant, the absolute temperature, the charge number of the mobile ions of type i and the elementary charge, respectively. The cell voltage is measured against a reference partial pressure of i, p_i^{ref} , at the opposite side of the electrolyte.

The measurable partial pressure is determined by the chemical stability of the electrolyte with respect to the gaseous component. In the case of three or more components, the stability depends also sensitively on the ratio of the stoichiometries of the immobile components[58]. The partial pressure p_i of a N-component electrolyte in equilibrium with N-1 adjacent phases in the phase diagram is given by:

$$p_i = \exp\left[\frac{1}{kTd}\sum_{j=1}^{N}(-1)^{j+1}\, d_{ji}\,\Delta G_{f\,j}^{0}\right] \tag{2}$$

where $\Delta G^{o}_{f,j}$, d and d_{ji} are the standard Gibbs energy of formation of the compound j, the determinant formed by the stoichiometric numbers of the N components of the N compounds and the minor of d formed by elimination of the j-th row and the i-th column of the stoichiometric numbers of the mobile component, respectively. Accordingly, p_i varies with the Gibbs energies of formation of the N-1 compounds which are in equilibrium with the electrolyte. This may be controlled by the composition of the immobile components.

Other gases may interfere in two major ways. First, the gas may cause a chemical reaction with the stable formation of new phases if the partial pressure is sufficiently high compared to that of the gas that should be detected. Second, the gas may cause (exclusively or partially) a kinetically more favourable, but less stable galvanic cell reaction than that expected thermodynamically. This effect lowers the emf.

Practically all presently developed solid state electrochemical sensors are based on the principle of *type I* sensors and make use of cubic stabilized zirconia. This material only becomes a fast oxygen ion conductor at elevated temperatures, but is extraordinarily stable over a large oxygen partial pressure range from >1 atm down to the pressure of the reduction to the metal[59]. This relation between the required temperature and the stability and the inverse situation are quite common for compounds with « simple » crystallographic structures (i.e., small elementary cells) because large binding energies prevent successful atomic jumps and vice-versa. Applications include the control of the air/fuel ratio in combustion processes (e.g., the Lambda probe in automobiles or in heaters), carbonization processes for metal hardening and the control of the oxygen activity during the reduction of ores[57]. In spite of the high stability, an interference is observed with gases containing sulphur or fluorine at low oxygen partial pressures[60].

For the operation of the sensor it is not important that the gas equilibrates with the mobile component of the electrolyte. The gas may equilibrate with one of the immobile components (*type II* electrochemical gas sensor) since the chemical potentials of the components of the electrolyte are related to each other by Duhem-Margules' equation

$$S\,dT - V\,dp + \sum_j n_j\, d\mu_{j^*} = 0 \tag{3}$$

in which S, V, n_j and μ_{j*} are the entropy, volume, concentration of the component j and chemical potential of neutral species j, respectively. In a binary or quasibinary compound $M_{y+\delta}X$, the stoichimetry corresponding to an increased activity of X may either be achieved by uptake of X or a loss of M, depending on the type of mobile ions (Figure 12). The result is the same because of the well defined relationship between the chemical potentials.

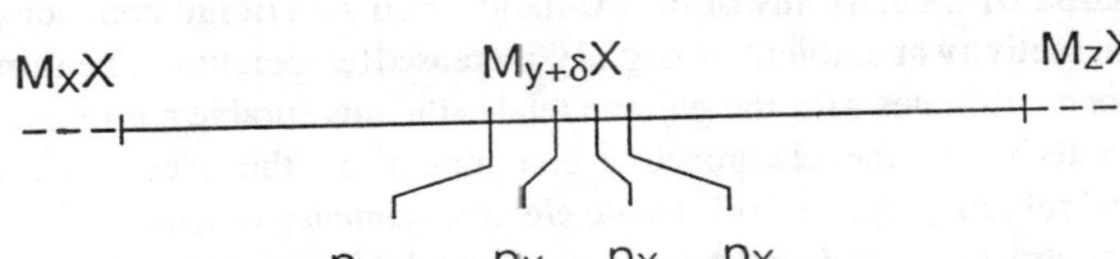

Figure 12: Phase diagram of the binary system M-X with the ionic conductor $M_{y+\delta}X$. The stoichiometry corresponding to the partial gas pressure of X may be either controlled by a variation of the concentration of X (type I) or M (type II gas sensor) (from ref. 55).

A fast ion conducting material of three (or in general, N) independent components requires equilibration with two (N-1) gases. The relationship between the partial gas pressures and the emf is obtained by solving Duhem-Gibbs' equation for the standard Gibbs energy of formation of the electrolyte from the pure elements ($a_{j*} = 1$)

$$\Delta G_f^0 = kT \sum_j \left[n_j \ln a_{j^*} \right] \tag{4}$$

for the measurable activity of the electroactive component i and by feeding this expression into Nernst's law for a reference electrode of defined activity $a_{i^*}^{ref}$ of the mobile component:

$$E = \frac{kT}{n_i z_i q} \left[\frac{\Delta G_f^0}{kT} - \sum_{j \neq i} n_j \ln a_{j^*} - n_i \ln a_{i^*}^{ref} \right] \tag{5}$$

In the case of a multinary compound in which not all activities are completely defined by equilibration with the gas phase, the relationship depends on the stoichiometry of the components which has been selected by the preparation.

The concept of type II electrochemical gas sensors has been mostly applied so far for the measurements of SO_2 (SO_3) partial pressures by employing various cation conducting alkali metal and silver sulphates[61-64]. The ternary compounds equilibrate with both SO_2 (or SO_3) and O_2. The voltage depends accordingly Eq. (5) on both partial pressures. Carbonates are presently being studied for possible application in CO_2 sensors[61], and AgI has been suggested for iodine vapor sensors[66].

In spite of this extension, the number of detectable gaseous species is still very small and limited to high temperatures. These restrictions are overcome by employing thin films of solid mixed ionic and electronic conductors as electrodes in combination with nearly any of the available solid superionic conductors with high ionic conductivity at ambient or slightly increased temperature. The thin film mixed conductor equilibrates with the gas and relates the measurable partial gas pressure (s) to the activity of the electroactive component of the electrolyte by Duhem-Margules' relation (*type III* solid state electrochemical gas sensor)[67-69].

In contrast to *type I* and *type II* sensors which make use of an inert metal to probe the concentration of electrons in the electrolyte, *type III* sensors employ solid ionic conductors to measure the chemical potential of an auxiliary neutral component of a mixed conductor which equilibrates with the gas (Figure 13). The applicable solid fast ion conductors are mostly conductors for alkali metal-, silver- and copper-ions and show ionic conductivities which at room temperature are already as high as that of cubic stabilized zirconia at 1000°C.

The gas sensitive electrode and the fast ion conductor together are composed of at least three components. If the gas equilibrates with only one of the components, the variation of the activities of the other components is dependent on their stoichiometric numbers and is in this way dependent on small variations in the preparation. In order to avoid these poorly defined conditions, compositions along

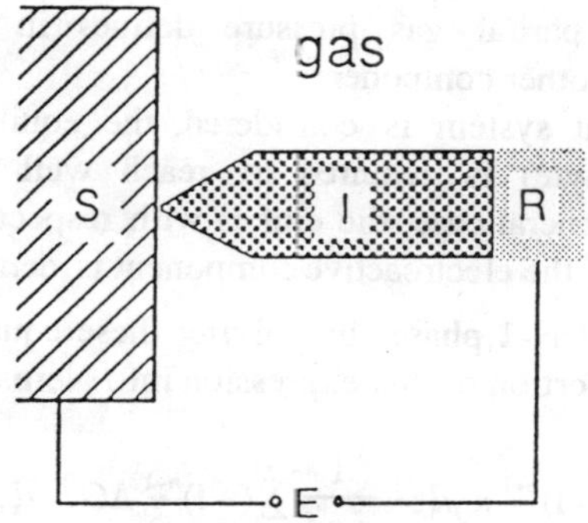

Figure 13: Schematic representation of a type III sensor, the gas equilibrates with one of the components of a mixed ionic and electronic conductor (S). The activity of another component is measured by a solid ionic conductor (I) (from ref. 55).

the border line of the single phase field are employed. This is achieved by the equilibration of two adjacent compounds within the ternary phase diagram (Figure 14). The electrolyte is included in the equilibration process and may serve as one of

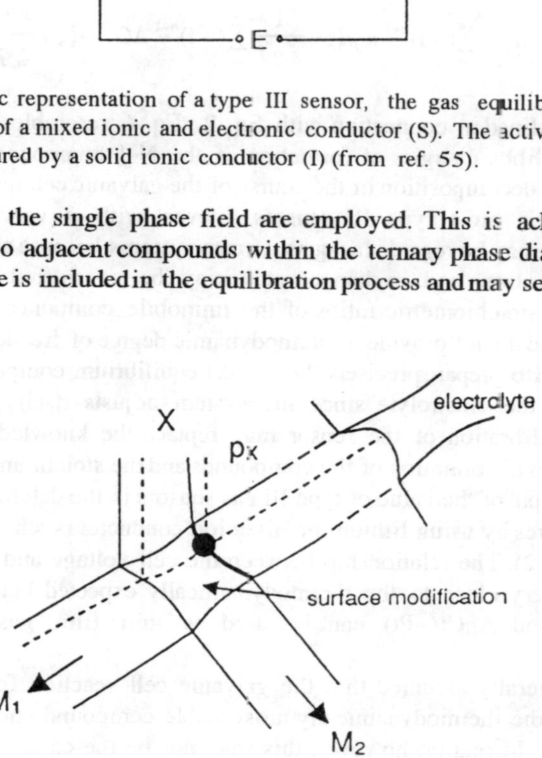

Figure 14: Phase diagram of the equilibrium between the gas sensitive thin film and the ionic conductor. A variation of the partial pressure of X change the composition of the thin surface film along the phase boundary with the 2-phase regime (type III gas sensor) (from ref. 55).

the compounds. The partial gas pressure defines in this way the precise stoichiometries of both other components.

If an N-component system is considered, the equilibration of N-1 phases (including the electrolyte) is required to reach well defined thermodynamic relationships. In this general case, the emf E with respect to a reference electrode with the activity a_i^{ref} of the electroactive component is derived from Duhem-Gibbs' equations Eq. (4) for all N-1 phases by solving these equations for the activity of the component i and insertion of this expression into Nernst's law

$$E = \frac{kT}{z_i\, q\, d} \ln p_j \sum_{n=1}^{N-1} (-1)^{n+1}\, n_{ij} d_{in} - \frac{1}{z_i q d} \sum_{n=1}^{N-1} (-1)^{n+1}\, \Delta G^0_{f,n}\, d_{in} + \frac{kT}{z_i\, q} \ln a^*_{i,\,ref} \tag{6}$$

d and j_i are defined in connection with Eq. 2. Eq. 6 resembles Nernst's law, but includes the Gibbs energies of formation of the N-1 compounds because of their formation and decomposition in the course of the galvanic cell reaction.

In practice, most type III sensors - even with a very large number of components - may be regardedas quasi-ternary systems because all components with very slow motion and exchange rates may be considered thermodynamically inactive. The stoichiometric ratios of the immobile components are fixed by the preparation and do not provide a thermodynamic degree of freedom. Furthermore, it is not requiredto prepare precisely the correct equilibrium composition between the electrode and the electrolyte since the system adjusts itself to the equilibrium condition. Calibration of the sensor may replace the knowledge of the standard Gibbs energies of formation of the compounds and the stoichiometric numbers.

An example of the value of type III gas sensors is the determination of chlorine partial pressures by using lithium or silver ion conductors such as $LiAlCl_4$ or Ag-ß-alumina (70-72). The relationship between the cell voltage and the partial chlorine pressure is very close to the thermodynamically expected behaviour. LiCl(+Pt), $AlCl_3$(+Pt) and AgCl(+Pt) can be used as thin film gas sensitive surface modifications.

It is generally assumed that the galvanic cell reaction follows the path of formation of the thermodynamically most stable compound under the given partial gas pressures. In reality, however, this may not be the case. The deviations from thermodynamic equilibrium may be, on the one hand side, favourable for the selective detection of gases. Since kinetic processes are often very specific for gases, this effect may be employed to overcome problems of cross-sensitivity of different gases and to provide a possibility to detect several different gases by a single galvanic cell. On the other hand side, kinetic deviations may give rise to malfunctions of the sensors.

In any case, the Gibbs energy ΔG_r of the actual galvanic cell reaction for the passage of n elementary charges q is converted into electrical energy :

$$\Delta G_r = nqE \tag{7}$$

at constant total pressure p and temperature T. ΔG_r involves, however, only the redox process that induces the electronic charge carriers from the electronic leads of the galvanic cell. All other reactions that occur at a later stage, e.g., to convert the product eventually into thermodynamically more stable compounds, are not included in the reaction[62]. The chemical energy of those subsequent reactions is converted into heat.

5.2 *Practical Examples of Solid State Electrochemical Gas Sensors*

5.2.1 *Oxygen Sensors*

The use of calcia and, more recently, yttria-stabilized zirconia (YSZ) as a semi-permeable oxygen membrane in an oxygen sensing device has been utilized for a long time to monitor oxygen pressure in experimental equipments. The use of galvanic cells based on zirconia to monitor oxygen in boiler flues, and hence to control and optimize the combustion process has been commercially exploited for nearly two decades.

YSZ and other refractory materials are used to measure oxygen pressure. One type of sensor realized in thin and thick film techniques is shown in Figure 15.

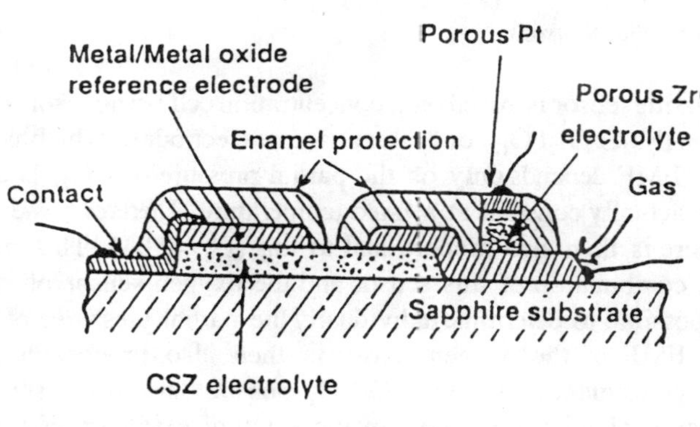

Figure 15: Oxygen micro-sensor using ZrO_2 : CaO thin film (from ref. 73).

It integrates on a small dimension (10x2.5x3 mm^3) substrate, a set of four functions: gas sampling, electrochemical pump, catalysis and oxygen concentration cell. Under low oxygen pressure (below 10^{-5} atom), or high cathodic polarization (over 200 mV), or with particularly thick electrodes, or at high temperatures (above 800°C), the reaction rate at the Pt, O_2(g)/YSZ interface tends to be controlled by the diffusion of oxygen molecules in the gas phase or in the electrode pores. Under high, over 1 V, cathodic overpotentials, the partial ion-blocking effect becomes noticeable and the electronic conduction appears. Further increase in overpotential up to 2 V causes the electrolysis of zirconia. Such devices, called the Lambda Sond, are now installed in many automobiles. Other applications of such devices are oxygen sensors in steelmaking.

5.2.2 *Hydrogen Sensors*

The same concept may be extended in order to make hydrogen sensors. In this case the solid fast ion conductor is a protonic conductor or cationic conductor able to exchange protonic species with the external medium. The reference electrode is a metallic hybrid.

In the Pd/H_x/Na-ß-ß''-Al_2O_3/Pt structure, the electrochemical cell is subjected to various N_2-H_2 gas compositions, from 100 ppm to 10 % H_2, at room temperature. The EMF response, measured with a high impedance voltmeter, shows a Nernstein behaviour and the response times of one to three minutes at room temperature are consistent with their use as hydrogen gas sensors[74]. A planar sensor Pd/H_x/HUP/Pt(H_2-N_2) has also been made, using the protonic conductor (HUP) hydrogen-uranyl-phosphate[75].

5.2.3 *Carbon Dioxide Sensors*

The carbon dioxide sensor is based on a concentration cell using a solid fast ion conductor : Nasicon ($Na_3Zr_2Si_2PO_{12}$) or ß-alumina, one electrode of which is covered with Na_2CO_3. The EMF depends only on the partial pressure of CO_2, because the activity of Na_2O is actually constant at a fixed temperature. Therefore, the EMF in CO-CO_2 atmosphere is then also proportional to log ($P^2_{CO2}P^{-1}_{CO}$). This result has suggested that the combination of this sensor and the oxygen sensor of stabilized zirconia makes it possible to determine individually the partial pressures of CO and CO_2, because the EMF of the oxygen sensor is then also proportional to log ($P^2_{CO2}P^{-1}_{CO}$). The performance of the CO-CO_2 sensor has been successfully demonstrated, by the evolved gas analysis, in the study of oxidation of ceramics[76]. The existence of H_2 with high concentration does not affect the EMF's. This fact proves the applicability of the two-sensor system for the monitoring and the controlling of the reducing atmosphere in industrial processes.

5.2.4 Moisture Sensors

Moisture sensors use the variation of fast ionic conductivity in composite ionic conductors. An example of composite ionic conductor is LiI/Al_2O_3 where, for LiI well-dispersed in very finely-divided Al_2O_3, the Li^+ conductivity may be enhanced by several orders of magnitude over that measured in LiI itself. Composite ionic conductors are used in the monitoring of moisture vapour pressure. LiI is very hygroscopic and LiI/Al_2O_3 is used as moisture sensor. An enhancement of lithium-ion conductivity over that measured in the simple salt is also found for LiF/Al_2O_3. Although LiF is only very sparingly soluble in water, the measured ionic conductivity of LiF/Al_2O_3 around room temperature and above is strongly dependent upon the ambient moisture vapour pressure and can be used as a sensitive humidity monitor.

6 Solid State Electrochromics[77]

Electrochromic light transmission modulators[78-82] are of large practical interest for energy conservation, temperature control in heated or air conditioned rooms and for the control of the light in buildings, automobiles etc. Furthermore, this technology is of great importance for electrochromic displays and mirrors which work in reflection with the counter electrode being invisible behind a light reflecting background. The only purpose of the counter electrode in this case is to be a reversible source of the active component and may be the same material. In the case of electrochromic windows, the counter electrode is within the optical path and has to be optically passive or to be changed in an optically complimentary way to the primary electrochromic layer. A schematic representation of an electrochromic structure is given in Figure 16.

The coloration efficiency η at a given wavelength λ is given by the difference of those of the two electrodes (cathode c and anode a) :

$$\eta_{tot} = \eta_c - \eta_a \qquad (1)$$

In order to reach a large η_{tot}, η_c and η_a should be quite different if the sign is the same or should have different signs (by definition η_c is positive for the cathodic electrochromism of the reduction and negative for the anodic electrochromism of the oxidation).

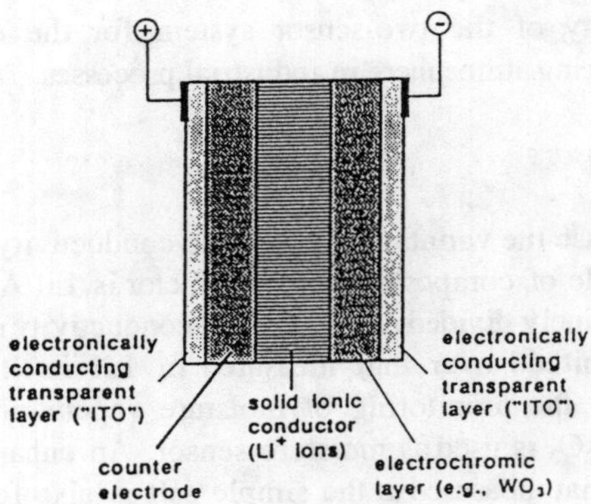

Figure 16 : Schematic arrangement of an electrochromic light transmission modulator based on a solid lithium ion conductor (from ref. 77).

The coloration efficiency multiplied by the charge Q per area (As/cm^2) corresponds to the change Δ of the optical density OD and is proportional to the product of the absorption coefficient α(cm^{-1}) and the thickness d (cm):

$$\Delta\ (\lambda) = \eta Q = \alpha d = \log\ (1/T) \tag{2}$$

where T is the transmission of the electrochromic layer.

Optically transparent electrodes are usually made of Sn-doped In_2O_3 (« ITO »). This material shows a high absorption in the infrared and supports in this way the active electrochromic layers. A suitable and widely used electrochromic active material[83] is WO_3 with the electrochemical insertion of H^+, Li^+, Na^+ or Ag^+. In most cases WO_3 is an amorphous thin layer. The stoichiometry which is required for complete coloration is about M_xWO_3 ($x \geq 0.3$) when M = H, Li, Na, Ag. Major problems occur when hydrogen is used because of the hydrogen gas development and the loss of water since all practically useful proton conductors contain structural water. The resistance of the electrolyte is increased and finally the current is interrupted but at least there is a need to control the humidity.

Besides WO_3 several metal oxides with edge- and corner-linked octahedra (MO_6) show the reversible intercalation of H, Li, Na and Ag [84, 85]. The coloration efficiency is, however, the largest for WO_3 with a value of 50 cm^2/As (depending on the preparation conditions). The oxides TiO_2, Nb_2O_5 and Ta_2O_5 show coloration efficiencies[81, 86] lower than 10 cm^2/As. These materials also show cathodic coloration, but are more useful candidates as anodes because of the low η-value. An important limitation to their applicability is the low diffusion rate of lithium and the other metals. As a reference electrode, the electronic lead In_2O_3 has also been considered. However, only low chemical diffusion rates of the order of 10^{-17} -10^{-13} cm^2/s have been observed[87].

Amorphous IrO_2 colorates anodically with hydrogen with an efficiency[11] of η = -15 cm^2/As. Li is built in only very slowly and apparently irreversibly. Crystalline V_2O_5 is the only presently reported intercalation compound for non-aqueous systems on the basis of lithium electrolytes for electrochromic light modulation[88]. This material shows fast chemical diffusion of lithium and also colorates anodically. The formation of $Li_xB_2O_5$ with x = 0.1 and 0.45 has been reported where the color changes during the reduction from yellow to blue. Together with WO_3 as a cathode, a change is observed from dark green-blue in the colorated state to a weak blue in the decolorated state.

As solid ionic conductors, so far predominantly glasses, especially $LiAlF_6$, $LiNbO_3$ and $LiTaO_3$, have been considered. For the selection of the electrochromic electrodes one has to make sure that the electrochemical reaction is reversible and no degradation takes place over a long period of time. Also, photochemical reactions have to be excluded. As a result, silver compounds have to be nearly completely eliminated (in spite of fast Ag^+ ion conduction in a variety of silver salts).

In order to achieve a high coloration efficiency, cathodic coloration at the cathode and anodic coloration at the anode is preferable. In the one case coloration is achieved by excess electrons and in the other case by defect electrons. An additional important aspect is the chemical stability between the electrodes and the electrolyte in every state of discharge of both electrodes. In view of long term operation, thermodynamic stability is to be preferred over kinetic stability. Future work must be concerned with developing suitable materials in this regard.

The response time of an electrochromic device depends on the chemical diffusion coefficients of the two solid electrode materials. The assumption of typical solid state diffusion rates indicates that the electrochromic display is not suitable for moving pictures. There are, however, many compounds reported which show effective diffusion coefficients of the same order of magnitude as those typical for liquids (~10^{-5} cm^2/s). The mobility of the ions is enhanced by internal electrical fields and the response time of a layer of 1 μm in thickness is in this case of the order of 10^{-3} s. Also, the problem of large currents may be overcome by backing-up the charge locally by electrochemical energy storage. Only the signal to close and open the electrical circuits has to be transferred.

7 Conclusion

Fast ion conductors enter in the structure of different type of devices such as solid state batteries, electrochemical sensors and electrochromics. Optimization of the performances of all this different devices do not depend essentially on the ion mobility but rather on the structure of the materials. For practical reasons one is inclined to use thin films which often means amorphous or disordered materials. Polymer electrolytes became fashionable at one stage. Their advantage is to be easily shaped in different forms which is important in applications. Their drawback at the present stage is the narrow range of the temperature in which this materials have convenient ionic conductivity and also the fact that many of them dissociate into two opposite charges which prevents their use in the form of very thin layers.

Glassy materials are probably the best suited in many applications but they still face problems of stability and ionic mobility. Further efforts in materials science are necessary to promote the progress of the applications, of ionic materials, some of which are of considerable industrial potential.

References

1. J.H. Kennedy in « Physics of Solid Electrolytes », ed. S. Geller (Springer-Verlag, New-York, 1977).
2. R. Galli and F. Garbassi, Chim. Ind. (Milan) **55** (1973) 522.
3. J.T. Krummer, Prog. Solid State Chem. **7** (1972) 141.
4. H. Wiedersich and S. Geller in « The Chemistry of Extended Defects in Non-Metallic Solids », eds. L. Eyring and M. O'Keefe (North-Holland, Amsterdam, 1971).
5. B.B. Owens in « Advanced in Electrochemistry and Electrochemical Engineering », vol. **8**, ed. C.W. Tobias (John Wiley, New-York, 1971).
6. T. Takahashi in « Fast Ion Transport in Solids », ed. W. van Gool (North-Holland, Amsterdam, 1973).
7. K. Funke, Prog. Solid State Chem. **11** (1976) 345.
8. K. Shaki, Phys. Stat. Solidi (a) **41** (1977) 11.
9. S. Chandra in « Superionic Solids, Principles and Applications » (North-Holland, Amsterdam, 1981).
10. T. Matoui and J.B. Wagner in « Solid Electrolytes », eds P. Hagenmuller and W. van Gool (Academic Press, New-York, 1978).
11. T.H. Etsell and S.N. Flengas, Chem. Rev. **70** (1970) 339.
12. R.M. Dell, and A. Hooper, in « Solid Electrolytes », eds P. Hagenmuller and W. van Gool, Academic Press, New-York (1978).

13. B.C.H. Steele in « MPT International Review of Science », ed. L.E.J. Roberts, vol. 10, Ser. **1** (Butterworths, London, 1972).
14. B.C.H. Steele and G.J. Dudley in « MPT International Review of Science », ed. L.E.J. Roberts, vol. 10, Ser. **2** (Butterworths, London, 1975).
15. R.A. Huggis, Electrochim. Acta **22** (1977) 773.
16. J. Schoonman and P.H. Bottelberghs in « Solid Electrolytes, eds P. Hagenmuller and W. van Gool, Academic Press, New-York (1978).
17. J.M. Réau and J. Portier in « Solid Electrolytes, eds P. Hagenmuller and W. van Gool, Academic Press, New-York (1978).
18. D.E. Fenton, J.M. Parker and P.V. Wright, Polymer **14** (1993) 585.
19. P.V. Wright, Brit. Polymer J. **7** (1975) 319.
20. M.B. Armand, J.M. Chabagno and M.J. Duclot, 2nd International Conf. on Solid Electrolytes (St-Andrews, Scotland, 1978).
21. M.B. Armand, Am. Rev. Mat. Sci. **16** (1986) 245.
22. P.J. Bray and J.G. O'Keefe, Phys. Chem. Glasses **4** (1963) 37.
23. J. Krogh-Moe, Ark. Kemi. **12** (1958) 475.
24. W.J. Konijneudijk and J.M. Steveis, J. Non-Cryst. Solids **18** (1975) 307.
25. M. Balkanski in « Science and Technology of Fast Ion Conductors », eds H.L. Tuller and M. Balkanski.
26. M. Balkanski, Physica Scripta **39** (1991) 9-20.
27. « Intercalation in layered materials », ed. M.S. Dresselhaus, NATO ASI Series B, vol. 148 (Plenum, New-York, 1986).
28. R. Moret, in ref. 2 p. 185.
29. W.Y. Liang, in ref. 2 p. 31.
30. R. Brec, in ref. 2 p. 93.
31. C. Rigaux, in ref. 2 p. 235.
32. O.K. Andersen, W. Klose and H. Nohl, Phys. Rev. B **17** (1978) 1209.
33. A.H. Thompson, Phys. Rev. Lett. **40** (1978) 1511.
34. J.A. Acrivos, W.Y. Liang, J.A. Wilson and A.D. Yoffe, J. Phys. C : Solid State Phys. **4** (1971) 418.
35. R.B. Somoano and A. Rembanm, Phys. Rev. Lett. **27** (1971) 402.
36. Solid State Integrable Microbatteries, ed. M. Balkanski (North-Holland, 1990).
37. M. Balkanski « Solid State Micropower » (Physics World, Nov. 1990) p. 29
38. J.R. Akridge and M. Balkanski, eds Solid State Microbatteries, NATO ASI Series B, vol 217 (Plenum, New-York, 1990).
39. C.R. Schlaikjer, C.C. Liang, J. of Electrochem. Soc. **118** (1971) 1441.
40. P. Bro, R. Homes, N. Marincic, H. Taylor, J. Power Sources **5** (1975) 703.

41. E.W. Poulsen, 6th Riso International Symposium on Metallurgy and Materials Science, 1985.
42. K.H. Park, M.H. Miles, D.E. Bliss, D. Stilwall, R.A. Hollins, R.A. Rhein, J. Electrochem. Soc. **135** (1988) 2901.
43. P. Vouros and J. Masters, J. Electrochem. Soc. **116** (1969) 880.
44. S. Chandra, R.C. Agrawal and R.K. Panday, Natl. Acad. Sci. Lett. **1** (1978) 112.
45. C.C. Liang, J. Epstein and G.H. Boyle, J. Electrochem. Soc. **116** (1969) 1452.
46. K. Kanehori, K. Matsumoto, K. Miyauchi, T. Kudo, Solid State Ionics **9-10** (1983) 1445.
47. J.R. Akridge, H. Vourlis, Solid State Ionics **18-19** (1986) 2082.
48. G. Meunier, R. Dormoy and A. Levasseur, Materials Sci. Eng. B **3** (1989) 13.
49. L. Jourdaine, Thesis, Grenoble (1985).
50. M. Balkanski and C. Julien, EEC Progress Report (1986).
51. C. Julien, Mater. Sci. Eng. B **6** (1990) 9.
52. A. Douridah, F. Dalard, D. Beroo and M. Armand, Solid State Ionics **18-19** (1986) 287.
53. W.J. Macklin and R.J. Neat, Solid State Ionics **53-56** (1986) 12.
54. M. Lin, S.J. Visco, L.C. de Joughe, J. Electrochem. Soc. **137** (1990) 367C.
55. N. Oyama, T. Tatsuma, T. Sato and S. Sotomura, Nature **373** (1995) 598.
56. W. Weppner in « Solid State Ionics », European Materials Research Society Network Reports, 1991, eds M. Balkanski and C. Delmas.
57. H. Dietz, W. Haecker, H. Jahnke, Adv. Electrochem. Electrochem. Eng. **10** (1977) 1.
58. K.S. Goto in « Solid State Electrochemistry and its Applications to Sensors and Electronic Devices » (Elsevier, Amsterdam, 1988).
59. W. Weppner in « Electrochemical Detection Techniques in the Applied Biosciences », ed. G.A. Junter (Ellis Horwood Ltd., London, 1988), p. 142.
60. W. Weppner, J. Electroanal. Chem. **84** (1977) 339.
61. W. Weppner, Goldschmidt Inform. **2/83** (1983) 16.
62. W.L. Worrel, Q.G. Liu, Anal. Chem. Symp. Ser. **17** (1983) 332.
63. W.L. Worrel, Q.G. Liu, J. Electroanal. Chem. Interfacial Electrochem. **168** (1984) 355.
64. K. Saito, T. Maruyama, Nippon Kinzoku Gakkai Kaiho **23** (1984) 30.
65. V. Demuysere, C.W. Bale, Solid State Ionics **9-10** (1983) 1285.
66. M. Gauthier, A. Belanger, D. Fauteux, C. Bale, R. Cote, Anal. Chem. Symp. Ser. **17** (1983) 353.
67. K. Nagashima, S. Suzuki, Anal. Chim. Acta **162** (1984) 153.

68. W. Weppner, German Patent n° 2826172 (28 June 1979) ; US Patent 4.352.068 (28 Sept. 1982).
69. W. Weppner, Sensors and Actuators **12** (1987) 107, Proc. 2nd Int. Meet. Chem. Sensors, ed. J.L. Aucouturier et al. (Bordeaux, 1986) p. 59.
70. W. Weppner in « Science and Technology of Fast Ion Conductors », eds H.L. Tuller and M. Balkanski (Plenum Publ. Comp. New-York, 1989).
71. G. Hötzel, W. Weppner in « Transport-Structure Relations in Fast Ion and Mixed Conductors », ed. by F.W. Poulsen, N. Hessel Andersen, K. Clausen, S. Skaarup and O. Toft Sorensen (Riso National Laboratory, Roskilde DK 1985), p. 139.
72. G. Hötzel, W. Weppner, Solid State Ionics **18-19** (1986) 1223.
73. G. Hötzel, W. Weppner, Sensors and Actuators **12** (1987) 449.
74. M. Croset, P. Schnell, G. Velasco and J. Siejka, J. Appl. Phys. **48** (1977) 775.
75. P. Schnell, G. Velasco and P. Colomban, Solid State Ionics **24** (1981) 281.
76. A.T. Howe, S.H. Sheffield, P.E. Childs and M.G. Shilton, Thin Solid Films **67** (1980) 365.
77. T. Maruyama, X.Y. Ye and Y. Saito, Solid State Ionics **24** (1987) 281.
78. W. Weppner in « Solid State Ionics », European Materials Research Society Network Reports, 1991, eds M. Balkanski and C. Delmas.
79. S.K. Deb, Phil. Mag. **27** (1973) 801.
80. R.B. Goldner and R.D. Rauh, Sol. Energy Mater. **11** (1984) 177.
81. T. Oi, Ann. Rev. Mater. Sci. **16** (1986) 185.
82. S.F. Cognan, T.D. Plante, M.A. Parker and R.D. Rauh, Solar Energy Mater. **14** (1986) 185.
83. D.W. Murphy and P.A. Christian, Science **205** (1979) 651.
84. P.G. Dickens and G.J. Reynolds, Solid State Ionics **5** (1981) 331.
85. S.F. Cogan, E.J. Anderson, T.D. Plante and R.D. Rauh, Proc. SPIE **562** (1985) 23.
86. R.B. Goldner, G. Foley, E.L. Goldner, P. Norton, and K.K. Wong, Appl. Opt. **24** (1985) 2283 ; M. Marezio, Acta Cryst. **20** (1966) 723 ; S.J. Golden and B.C.H. Steele, Solid State Ionics **28-30** (1988) 1733.
87. R.D. Rauh and S.F. Cogan, Solid State Ionics **28-30** (1988) 1707.
88. T. Oi and K. Miyauchi, Mater. Res. Bull. **16** (1981) 1281.
89. W. Weppner in « Transport-Structure Relations in Fast Ion and Mixed Conductors », ed. by F.W. Poulsen, N. Hessel Andersen, K. Clausen, S. Skaarup and O. Toft Sorensen (Riso National Laboratory, Roskilde DK, 1985), p. 139.

Index